THE LINGUAL APPARATUS OF THE AFRICAN GREY PARROT, *PSITTACUS ERITHACUS* LINNÉ (AVES: PSITTACIDAE): DESCRIPTION AND THEORETICAL MECHANICAL ANALYSIS

ORNITHOLOGICAL MONOGRAPHS

This series, published by the American Ornithologists' Union, has been established for major papers too long for inclusion in the Union's journal, *The Auk*. Publication has been made possible through the generosity of the late Mrs. Carll Tucker and the Marcia Brady Tucker Foundation, Inc.

Correspondence concerning manuscripts for publication in the series should be addressed to the Editor, Dr. David W. Johnston, Department of Biology, George Mason University, Fairfax, VA 22030.

Copies of *Ornithological Monographs* may be ordered from the Assistant to the Treasurer of the AOU, Frank R. Moore, Department of Biology, University of Southern Mississippi, Southern Station Box 5018, Hattiesburg, Mississippi 39406. (See price list on back and inside back covers.)

Ornithological Monographs, No. 39, xi + 233 pp.

Editor of AOU Monographs, Mercedes S. Foster and David W. Johnston

Special Reviewers for this issue, Richard L. Zusi, Division of Birds, National Museum of Natural History, Washington, D.C.; Kathleen K. Smith, Department of Anatomy, Duke University Medical Center, Durham, N.C.

Author, Dominique G. Homberger, Department of Zoology and Physiology, Louisiana State University, Baton Rouge, Louisiana 70803

First received, 31 January 1983; accepted, 25 July 1983; final revision completed, 12 September 1984

Issued June 12, 1986

Price $25.00 prepaid ($20.00 to AOU members).

Library of Congress Catalogue Card Number 86-71332

Printed by the Allen Press, Inc., Lawrence, Kansas 66044

ISBN: 0-943610-49-4

THE LINGUAL APPARATUS OF THE AFRICAN GREY PARROT, *PSITTACUS ERITHACUS* LINNÉ (AVES: PSITTACIDAE): DESCRIPTION AND THEORETICAL MECHANICAL ANALYSIS

BY

DOMINIQUE G. HOMBERGER
Department of Zoology and Physiology
Louisiana State University
Baton Rouge, Louisiana 70803

ORNITHOLOGICAL MONOGRAPHS NO. 39

PUBLISHED BY
THE AMERICAN ORNITHOLOGISTS' UNION
WASHINGTON, D.C.
1986

TABLE OF CONTENTS

LIST OF FIGURES

LIST OF TABLES

LIST OF APPENDICES

INTRODUCTION

Parrots form one of the largest and best defined avian orders. The Psittaciformes are characterized by a number of features, such as a strongly curved upper mandible that fits over the shorter lower mandible and is connected to the skull through a movable joint, a fleshy tongue, a zygodactyl, usually prehensile foot, a short tarsometatarsus that is entirely covered by reticulate scales, a usually colorful plumage, white eggshells, a specific seed-husking behavior, a complex social behavior, and, in captivity, a tendency to mimic sounds. These similarities among the psittacine species notwithstanding, the order displays a remarkable diversity in many other respects. Parrots have a pantropical distribution occupying Australia, Pacifica, Asia, Africa, and the Neotropics. Within each region, parrots live in a great variety of habitats (e.g., deserts, savannas, rainforests, alpine and littoral regions), and a variety of microhabitats has been exploited by them within the particular habitats (e.g., the ground, tall trees, shrubs, rock cliffs). As a consequence, parrots have undergone remarkable adaptive radiations in various aspects such as size, locomotion, social behavior, and reproduction. Probably the most outstanding adaptive radiation, however, can be observed in the feeding apparatus (Güntert and Ziswiler 1972; Homberger and Ziswiler 1972; Homberger 1978, 1980a, b, 1981) and the digestive tract (Güntert and Ziswiler 1972; Güntert 1981).

Of all the organs associated with feeding in parrots, the tongue has been receiving the greatest attention by scientists for more than four centuries. It is relatively easily observed in living birds and plays a central role in feeding and drinking (Homberger 1980a). For a long time, the tongue was also assumed to take part in the vocalization process (e.g., Kutorga 1832; Denker 1907; but see Nottebohm 1976). Furthermore, the psittacine tongue is unique among those of birds in that it is very "fleshy" and contains a large number of intrinsic muscles. In this aspect and in its ability to modify the relief of the fleshy lingual surface, the psittacine tongue has evolved in convergence to the mammalian tongue. Unlike the mammalian tongue, however, the psittacine tongue is supported by a "bony" skeleton, the hyoid, like all avian tongues.

Interest in the anatomy of the psittacine lingual apparatus is reflected in the large body of information published on the subject (Kutorga 1832; Duvernoy 1835; Thuet 1838; Nitzsch in Giebel 1862; Shufeldt 1886; Mivart 1895, 1896; Mudge 1903; Chaine 1904, 1905; Kallius 1906; Denker 1907; Lubosch 1933; Steinbacher 1935, 1951; Kasai 1957; Dubale and Rawal 1965; Evans 1969; Feder 1969; Rawal 1970; Güntert and Ziswiler 1972; Homberger and Ziswiler 1972; Burton 1974c; Homberger 1978, 1980a and references therein, 1980b. See the bibliographies of Kutorga 1832 and of Shufeldt 1886 for pre-1832 publications). Nevertheless, a complete redescription of the psittacine lingual apparatus to serve as the basis for future comparative studies is needed at this point for several reasons. First, previous studies have more or less arbitrarily divided the complex lingual apparatus into different parts (e.g., skeleton, muscles, epithelium, salivary glands), described and analyzed these parts more or less independently from one another and left interactions and interdependencies among the particular parts unexplored. Secondly, those studies that included the mechanical functioning of the apparatus were restricted to the muscle-bone system, and their functional interpretations were based on intuitive considerations. Functional analyses, such as these, for the most part proved to be inaccurate or even incorrect because the mechanical functioning of the entire lingual apparatus involves the interaction of a multitude of tissues besides bone and muscle. Thirdly, the internal anatomy of the extremely complex tip of the psittacine tongue has been inadequately described. Thus, the anatomy and function of that portion of the tongue which interacts in a most specialized way with the environment (i.e., food, water) remains unexplored. Finally, the majority of the studies on the psittacine lingual apparatus were presented without proper reference to previous work. Although this lack of checking and rechecking of anatomical data is "typical" for avian anatomical studies (Homberger 1980c; Vanden Berge 1982), it is especially noticeable in the numerous studies on the psittacine lingual apparatus.

The deficiencies mentioned above notwithstanding, some of the earlier papers on the skeletomuscular system of the psittacine lingual apparatus have occasionally been used as bases for comparative studies leading to conclusions about the phylogeny and systematics of parrots and other avian orders. The lack of success of these attempts (e.g., Mudge 1903) has led avian evolutionary biologists to conclude that the lingual skeletomuscular system of parrots and, consequently, of birds in general, does not provide useful taxonomic characters.

In contrast to this opinion, however, the surface morphology of the feeding apparatus and specifically of the tongue has been a rich source of characters useful for systematic and phylogenetic studies (Homberger 1979a, 1980a, b, 1985; see also Forshaw 1978). Since the surface morphology of the lingual apparatus exhibits a great diversity among parrots, and since it is possible to explain these structural differences among taxa of higher categories as having resulted from adaptive and other evolutionary processes (Homberger 1980a, b, 1983), it seems reasonable to expect that a proper study of the internal structures of the psittacine lingual apparatus would also be useful for systematic and evolutionary studies. The precondition for such studies, however, is a thorough anatomical description that can serve as a basis for future comparative work.

The lingual apparatus of the African Grey Parrot (*Psittacus erithacus* L.) was chosen for the basic anatomical description of the psittacine tongue for several reasons. First, as a member of the Psittacinae (the "true" parrots), *Psittacus* is a representative of the largest, most diversified and least specialized subfamily of the Psittacidae (Homberger 1980a). Secondly, the adaptations of its feeding apparatus are those of a generalist in the sense that *Psittacus* feeds on fruit as well as on seeds and nuts. Thirdly, the lingual apparatus of *Psittacus* is relatively large, thus facilitating dissections in the microanatomical range. Fourthly, enough anatomical material from this species was available for an adequate study.

Complete description of a morphological system demands a functional analysis because every structure has not only a form but also at least one function (Bock and von Wahlert 1965). With the need for a functional analysis of the lingual apparatus of *Psittacus* arose also a need for the development of a method to analyze the mechanical functioning of complex systems. One of the most widely used approaches in recent years has relied strongly on results derived from the application of electro-physiological techniques (i.e., electromyography). This approach is of limited value for a complex apparatus, such as the lingual apparatus of *Psittacus,* in which the skeletal elements consist not only of hard tissue but also of hydraulic structures, and in which the actively moving elements are represented not only by muscles but also by inflatable vascular tissue. An appropriate analytical method should culminate in the construction of a model for the operation of an integrated complex apparatus. To assure the testability of the model, the method should lead to the formulation of predictions that can be tested against independent, possibly experimental observations. In addition, the method should permit a check on the accuracy and completeness of the anatomical description. To satisfy the latter demand, the analytical method would have to be based on the anatomical description and on theoretical-mechanical and physiological considerations (see also Bühler 1980). For complex morphological systems such an analytical method has not yet been designed, although a number of mechanical and mathematical approaches have been suggested for the analysis of well-defined, simple systems (e.g., Dempster 1961; Bock 1974; Winkler and Bock 1976; Bock and Winkler 1978; Crowninshield and Brand 1981).

The lingual apparatus of *Psittacus* is very well suited as an object on which to base the development and formulation of a theoretical method that would meet the criteria mentioned above. First, the range of possible movements of the psittacine lingual apparatus has already been determined (Homberger 1980a). Secondly, the lingual apparatus of *Psittacus* exhibits a remarkable, yet manageable structural complexity. Thirdly, it has a potential for a great variety of movements. This variety, however, is tempered by external and internal structural constraints; for example, the hyoid apparatus can move only within the limits set by the mandible, and the movements of different parts of the hyoid apparatus are often coupled through ligaments and multiple-joint muscles.

The approach to functional analysis formulated and presented here provides a theoretical, mechanical model for the functioning of the psittacine lingual apparatus. This model makes certain predictions (e.g., about the synchronization of certain muscular actions) that can be tested independently (e.g., by electromyography). By testing the predictions of the model, the independent observations test the functional analysis of the lingual apparatus as well as the validity of the method used for the construction of the model. This clearly divides the procedures of hypothesis formulation and testing into two steps undertaken in sequence.

In summary, the objectives of the present functional-anatomical study of the lingual apparatus of *Psittacus erithacus* are: (1) to provide a basic anatomical description of the psittacine lingual apparatus that amends and completes earlier descriptions, and that can serve as a reference for future comparative studies on the avian feeding apparatus; (2) to develop a theoretical method for the formulation of testable hypotheses explaining the functioning of complex morphological systems; (3) to present such a hypothesis in the form of a theoretical, mechanical model explaining the functioning of the lingual apparatus of *Psittacus*.

MATERIAL AND METHODS

The anatomical descriptions are based on the dissection of adult specimens of *Psittacus erithacus* that had died in captivity; thus, the locality of origin of the particular specimens is unknown. Sexes of these specimens are also unknown because most of the specimens consisted only of heads. The specimens had been stored in a deep freezer, subsequently fixed in 4% formalin and stored in 2% phenoxetol or 70% propyl alcohol.

Illustrated or cited specimens are identified by a catalog number with one of the following abbreviations: AMNH (American Museum of Natural History), LSUMZ (Museum of Natural Science at Louisiana State University), and Z (Zoological Museum of the University of Zürich).

Special attention has been given to the description of individual variations. Nine specimens (AMNH 7203, LSUMZ 1287, Z1–Z7) were dissected for the description of the soft parts; five specimens (AMNH 1968, AMNH 2354, AMNH 3448, AMNH 9029, LSUMZ 94491) were compared for the description of the skeletal parts.

Dissections were performed under a Wild M5-Stereomicroscope and Wild M8-Stereomicroscope with a camera lucida (drawing tube) attachment. The preparations were stained with an iodine solution (Bock and Shear 1972) to increase the contrast between musculature and other tissues. Transverse and longitudinal sections were made with a scalpel on frozen specimens. This technique was preferred over histological techniques because the sections could be further dissected under the microscope for the determination of muscle fiber orientations and identification of sheets of connective tissue.

The techniques used for the preparation of the illustrations were selected from various sources (Lucas and Stettenheim 1972; P. R. Stettenheim, pers. comm.; W. J. Bock, pers. comm.). I first outlined the anatomical preparations with the help of a camera lucida. I enlarged this drawing through an AM200 Lucygraph to cover a surface of about 20 × 30 cm. I then compared this enlargement directly with the specimen and added corrections and structural details. This pencil drawing was subsequently transposed into a pen and ink illustration by an illustrator using one of two methods. In one, the pencil drawing was first traced in "photographic blue" onto engineer's paper; the tracing was then inked. In the second method, the pencil drawing was traced directly onto transparent vellum in ink. The second method is more accurate and requires less time. Finally, the inked illustrations were labelled on an acetate overlay by using Leroy lettering equipment. The completed illustrations were assembled into plates that were photostatically reduced to a plate size of 6 × 9 inches. These photostatic reproductions were used as originals for publication.

In illustrating the skeletal parts, I first made an outline drawing with the camera lucida, enlarged this drawing to the appropriate size, and sketched some details into the enlargement. The illustrator then prepared a graphite drawing on engineer's paper on the basis of my drawing and of the object under the stereomicroscope. Finally, I checked the illustrator's rendering by comparing it with the object. These drawings were then labelled as described above and photographed.

ANATOMY

Introduction

The anatomical description of the lingual apparatus covers only mechanically functioning structural elements. Blood and lymph vessels (except cavernous vessels), nerves, and sensory organs are not included. Special care has been devoted to the determination of individual variation. This aspect of descriptive anatomy has generally been neglected in non-medical anatomical descriptions of muscle-bone systems, despite its relevance for evolutionary and taxonomic conclusions based on comparative studies. Indeed, individual variations have been mistaken for interspecific differences because of a typological approach to anatomy (e.g., Mudge 1903).

The sequence in which the various structures are treated follows partly the suggestions given for the construction of a mechanical model and partly practical considerations to minimize references to later sections during the description of particular structures. Some cross-references to the different sections are included to stress the interrelationships of the structures.

The descriptive sections are organized according to the particular nature of the material and are, therefore, not uniform. Most anatomical descriptions include a brief statement on the functional properties of the structures. This functional statement does not duplicate the functional analysis in the analytical part of the study, but describes a structure as a form-function complex, whereas the functional statements in the analytical part of the study are part of the mechanical model.

The anatomical nomenclature follows the "Nomina Anatomica Avium" (NAA) (Baumel et al. 1979) whenever possible. However, many morphological structures are described here for the first time or, sometimes, proper homologies with the structures described in the NAA could not be established with certainty. In these cases, the structures were given new names, including Latin ones. For certain structures, names that differ from the NAA, but are better established and better suited, were used. The terms indicating the orientation of structures follow common usage. The words "anterior" and "posterior" are used synonymously with "apical" and "caudal." The reference system for expressions of direction or orientation is given by the mandible, the longitudinal axis of which is considered to lie horizontally.

SKELETON

INTRODUCTION

The skeletal basis of the lingual apparatus consists of the hyoid skeleton, which is an assemblage of several distinct elements. The hyoid skeleton interacts with other skeletal elements that are part of different systems, namely, the mandible as a part of the jaw apparatus, and the cricoid and the tracheal rings as parts of the laryngeal apparatus. The mandible serves as an attachment site for all but one of the extrinsic lingual muscles, and the hyoid skeleton must move within the framework provided by the mandible. The laryngeal skeleton is attached to the hyoid skeleton through ligaments and fasciae, and the majority of its extrinsic muscles originate from the hyoid skeleton. Thus, the movements of the larynx are confined by the framework provided by the hyoid skeleton (Homberger 1979b).

The following description of the skeleton concentrates on the hyoid skeleton and its articulations, whereas only those parts of the mandible and laryngeal skeleton needed to understand the structure and function of the lingual apparatus are described.

HYOID SKELETON

The hyoid skeleton consists of several bony elements and one sesamoid bone: paired paraglossalia, which combine to form a single element; an unpaired basihyale and urohyale, which completely fuse; paired ceratobranchialia; paired epibranchialia; and the Nodulus, which is a sesamoid bone ventral to the urohyale

(Figs. 1A–2A). Cartilaginous processes extend from the caudal ends of the epibranchialia and urohyale. The ceratobranchialia and epibranchialia form the paired hyoid horns. The paraglossalia as a unit and the basihyale, ceratobranchialia, and epibranchialia are interconnected through diarthroses, i.e., the Articulatio paraglosso-basihyalis, Art. cerato-basihyalis, and Art. epi-ceratobranchialis. The other joints among hyoid skeletal elements are a synchondrosis between the anterior processes of the paraglossalia, a syndesmosis between the central portions of the paraglossalia, and a synostosis between the embryonically separate basihyale and urohyale. The Nodulus is held against the ventral surface of the urohyale by the Lig. nodulo-ceratobranchiale. Various articular ligaments strengthen the diarthroses and often limit the extent of their excursions.

The hyoid skeleton is suspended by extrinsic lingual muscles within the framework of the lower jaw. Caudally, the hyoid horns lie along the medial surface of the pterygoid musculature covering the medial side of the mandibular ramus. Apically, they lie along the medial surface of the pterygoid muscles attaching to the medial sides of the palatina because the palatina of parrots form almost vertically oriented blades that adhere to the medial sides of the Rami mandibulares and reach the ventral mandibular edges when the bill is closed (Homberger 1981). The anterior ends of the hyoid horns embrace the laryngeal skeleton (Homberger 1979b), whereas the posterior parts of the hyoid horns embrace the neck, i.e., the cervical vertebral column with the neck musculature. Posteriorly, the cartilaginous processes of the epibranchialia curve dorsally around the medial portion of the M. depressor mandibulae (Figs. 17A, B). The longitudinal axis of the ceratobranchiale is more or less horizontal, with its anterior end tilting upwards or downwards depending on the position of the hyoid apparatus with respect to the lower jaw and the position of the jaw with respect to the skull. The anterior ends of the ceratobranchialia converge towards the unpaired basihyale. The longitudinal axes of the epibranchialia are in line with the ceratobranchialia but can rotate depending on the position of the hyoid skeleton relative to the lower jaw. The anterior end of the longitudinal axis of the basihyale points upwards to a varying degree depending on the position of the lingual apparatus within the lower jaw. The longitudinal axis of the paraglossale is to a varying degree bent downwards with respect to the basihyale.

The nomenclature of the hyoid skeleton follows Bock (1972, 1978, pers. comm.) and Richards and Bock (1973). The synonymies have been compiled (Appendix II A) from the literature on hyoid skeletons of parrots and compared with the nomenclature suggested by Baumel (1979) and Zweers (1982). Additional synonymies can be found in Gadow (1891), Fürbringer (1922), and Zweers (1982).

Os epibranchiale.—The epibranchiale is a short and bilaterally compressed skeletal element. Its bulbous anterior end bears the articular facet for the Art. epi-ceratobranchialis (Facies articularis epi-ceratobranchialis, F. art. ec; Figs. 3A, B). Caudally, the epibranchiale flattens into a plate (Fig. 3B) the broad sides of which face dorso-laterally and ventro-medially and whose edges point dorso-medially (Margo dorso-medialis, Ma. d-m) and ventro-laterally (Margo ventro-lateralis, Ma. v-l). Usually, the epibranchiale ends caudally in a small, flat, ovoid surface (A. CP; Fig. 3B) that serves as the attachment site for the cartilaginous process of the epibranchiale. This cartilaginous process (Cartilago epibranchialis, CP) is roughly twice as long as the epibranchiale (Figs. 33B, C). It is always present but usually missing in skeletal preparations (e.g., Figs. 1A–3B).

The ventro-lateral edge of the epibranchiale curves dorso-medially at about mid-length (e.g., LSUMZ 94491), or after the anterior two-thirds (e.g., AMNH 1968) or one-third of its length (e.g., AMNH 2354, AMNH 3448, AMNH 9029; Fig. 3A). The dorso-medial edge also curves dorso-medially, usually continuing into a prominent crest (Crista dorsalis epibranchialis, Cr. d. eb) (e.g., AMNH 2354, AMNH 3448, AMNH 9029). Sometimes, however, the dorso-medial edge does not form a crest but continues directly to the attachment site for the cartilaginous process (e.g., AMNH 1968, LSUMZ 94491). In these cases the cartilaginous process extends its attachment anteriorly over the medio-dorsal edge of the epibranchiale and forms a cartilaginous crest replacing the Crista dorsalis epibranchialis. The top of the dorsal crest bends slightly laterally so that it points dorsally.

The dorsal crest serves as the insertion site for the M. branchiomandibularis posterior (Figs. 12A, C, 13A, 33B). The larger part of the remaining surface of the epibranchiale, the dorsal surface of the anterior end, and the ventro-lateral edge cranial to the curvature serve as the insertion sites for the M. branchiomandibularis ant. lat. (Figs. 12A, C, 13A, 33B, C).

The size of the epibranchiale varies greatly within particular pairs and among individuals, thus, partly accounting for the individual variation in shape and proportion described above. For five pairs of epibranchialia the total length varied between 3.7–6.4 mm ($\bar{X}$ = 4.77 mm; s.d. = 0.79 mm), the width at the anterior end varied between 1.1–1.5 mm ($\bar{X}$ = 1.3 mm; s.d. = 0.13 mm), and the height at the anterior end varied between 1.3–1.9 mm ($\bar{X}$ = 1.6 mm; s.d. = 0.12 mm).

Os ceratobranchiale.—The slender ceratobranchiale is rod shaped. It is straight except for the anterior one-third which is bent slightly medially (Figs. 1A, B, 4A, B). The cross-section of this anterior one-third is roughly triangular with one side of the triangle forming the dorsal surface of the ceratobranchiale. Caudally, the cross-section becomes smaller and more circular or laterally compressed. The anterior end is the widest part of the ceratobranchiale and bears the articular facet for the Art. cerato-basihyalis. For descriptive purposes, this articular facet (Facies articularis cerato-basihyalis, F. art. cb) can be divided into two parts. The medial part (F. art. cb med) is bulbous and faces apically; the lateral part (F. art. cb lat) is flat and faces dorsally (Figs. 6A–7A). Immediately behind the lateral part of the articular facet, the ceratobranchiale is usually constricted (Figs. 4B, 7A) and continues to narrow for about the anterior one-third of its length. Thereafter, the ceratobranchiale maintains a fairly constant width. The posterior end of the ceratobranchiale, however, is bulbous and bears the articular facet of the Art. epi-ceratobranchialis (Facies articularis epi-ceratobranchialis, F. art. ec; Figs. 4A, B). The medial side of the bulbous caudal end extends somewhat more caudally than the lateral side, so that the articular facet faces caudo-latero-ventrally.

The dorsal surface of the anterior one-third of the ceratobranchiale is molded into a deep depression (Fossa dorsalis ceratobranchialis, F. d. cb; Fig. 4A). Part of this fossa serves as the site of origin for the M. supraglossus ceratobranchialis (Figs. 12A, B) and occasionally also as the insertion site for the M. tracheohyoideus ceratobranchialis (Fig. 12B). The lateral border of the Fossa dorsalis is formed by a sharp crest (Crista lateralis ceratobranchialis, Cr. l. cb), which points dorsally. This lateral crest extends from behind the lateral part of the articular facet of the Art. cerato-basihyalis to the end of the anterior one-third of the ceratobranchiale (Fig. 4A). When viewed laterally, this crest describes a downward curve (Figs. 2A, 6A). The Crista lateralis serves as the site of origin for the M. hypoglossus obliquus lateralis (Figs. 12A, B, 29B, 32A). Occasionally, the M. tracheohyoideus ceratobranchialis inserts on the anterior end of the lateral crest (Fig. 12A). The medial border of the Fossa dorsalis is formed by a blunt ridge pointing medially (Crista dorso-medialis, Cr. d-m; Figs. 4A, 5A, B, 7A, B). The broadened base of the Crista dorso-medialis near the articular facet of the Art. cerato-basihyalis forms the attachment site for a strong articular ligament (Lig. interarticulare cerato-basihyale; Figs. 5A, 9A). This crest also forms the dorsal border of the Sulcus ventro-medialis (see below) and continues backwards as a dorsal or dorso-medial ridge along the apical two-thirds of the ceratobranchiale (Figs. 5A, B). The Crista dorso-medialis usually does not serve as an attachment site for muscles, but the origin of the M. hypoglossus obliquus lateralis occasionally spreads over its dorsal side (Fig. 12B). Caudally, the Crista dorso-medialis acts as a dividing line between the origins of the M. ceratoglossus lateralis and M. ceratoglossus medialis (Figs. 12A, B).

The underside of the ceratobranchiale has a more complex structure than the dorsal side due to the twisting and shifting of its surface structures along the ceratobranchiale. The underside of the anterior end of the ceratobranchiale can be divided into two areas, namely a medial surface facing medially (Facies medialis, Fac. med; Figs. 5A, B, 7A, B) and a ventro-lateral surface behind the ventral border of the articular facet of the Art. cerato-basihyalis (Margo ventralis faciei articularis cerato-basihyalis, M. v. F. art. cb; Fig. 4B). This ventro-lateral surface is usually much broader than the rest of the ceratobranchiale (e.g., AMNH 1968, AMNH 2354, AMNH 9029; Fig. 4B) but is sometimes only slightly broadened as compared to the rest of the anterior part of the ceratobranchiale (e.g., AMNH 3448, LSUMZ 94491; Fig. 1B). A shallow but distinct groove (Sulcus ventro-medialis, S. v-m) is molded on the medial part of the ventro-lateral surface (Figs. 1B, 4B). This sulcus curves almost immediately to the medial side of the ceratobranchiale where it merges with the Facies medialis (Figs. 5A, B). The main part of this sulcus serves as the site of origin for the M. ceratoglossus medialis (Fig. 33A), whereas its anteriormost part serves usually as a channel in which the muscle body of the M. ceratoglossus medialis lies (Fig. 28A). The Sulcus ventro-medialis is bordered apico-medially by the thickened ventral margin of the articular facet of the Art. cerato-basihyalis. Laterally, it is bordered by a low mound which occupies the entire lateral part of the ventro-lateral surface (Figs. 1B, 4B). The greater part of this lateral mound serves as the attachment site for the M. serpihyoideus and M. ceratoglossus profundus (Figs. 13A, B, 29B). The medial border of the mound forms a distinct crest, which turns medially and becomes the ventral crest (Crista ventralis ceratobranchialis, Cr. v. cb; Figs. 4B, 5A, B). At the point where the ventral crest curves medially, the lateral mound usually bears a small tuberosity (Tuberositas ceratobranchialis, Tub. cb) where the Lig. nodulo-ceratobranchiale attaches (Figs. 4B, 5A, B, 13A, B). The ventral crest points medially but usually flattens out before reaching the caudal one-third of the ceratobranchiale. Rarely, it continues to the caudal end of the ceratobranchiale (e.g., AMNH 1968). The ventral slope of the ventral crest serves, together with a greater part of the ventral surface of the ceratobranchiale, as the site of origin for the M. ceratohyoideus (Figs. 13A, B, 33A).

The lateral surface of the ceratobranchiale serves to a varying extent as the site of origin for the M. ceratoglossus lateralis (Figs. 12A–13B, 29B).

The bulbous caudal end of the ceratobranchiale serves as the attachment site for the articular capsule of the Art. epi-ceratobranchialis and as the insertion site for the M. branchiomandibularis anterior medialis. Rarely, a short, but sharp crest (Crista branchiomandibularis, Cr. bm; Figs. 5A, B) may be found on the ventro-medial surface. This crest points ventro-medially and arises between the borders of the almost circular origin of the M. branchiomandibularis anterior medialis (e.g., AMNH 2354; Fig. 13A).

The total length of the ceratobranchiale varies individually much less than its width and than the measurements of the articular facets. For five pairs of ceratobranchialia the total length varies between 21.2–23.5 mm ($\bar{X} = 22.56$ mm; s.d. = 0.83 mm), the maximum width of the anterior end bearing the facet of the cerato-basihyal joint varies between 2.6–3.2 mm ($\bar{X} = 2.96$ mm; s.d. = 0.17 mm), and the maximum width at the posterior end varies from 1.4–1.9 mm ($\bar{X} = 1.63$ mm; s.d. = 0.15 mm). Mivart (1895) stated that the ceratobranchiale of *Psittacus* is twice as long as the basihyale, which is in agreement with my present data (see next section).

Os basihyale, Os urohyale, and Nodulus.—The basihyale and urohyale are fused and form a single, unpaired skeletal element. For descriptive purposes, this element can be divided into the body of the basihyale (Corpus basihyale, C. bh), the central process of the basihyale (Processus centralis basihyalis, P. centr. bh), and the urohyale (UH).

The Corpus basihyale is a roughly rectangular or square plate (Figs. 1A, B). Usually, its lateral edges curve gently upwards, thus making the plate ventrally convex. Often, the Corpus basihyale narrows apically (e.g., AMNH 1968, AMNH 9029). A pair of processes (Proc. parahyalis, PH) project anteriorly from the latero-apical corners of the body of the basihyale. The parahyal processes are slender bony bars of varying but equal length with a dorso-ventrally flattened oval cross-section. They are oriented obliquely with respect to the basihyale (Figs. 2A, B). Usually, the tip of the parahyal process forms a knob that often curves medially (Fig. 1A).

The central process of the basihyale arises between the bases of the parahyal processes. The width of its base is approximately one-third of the width of the basihyal body. The central process becomes laterally compressed apically. It ends in a saddle-shaped articular facet for the articulation with the paraglossale (Facies articularis paraglosso-basihyalis, F. art. pb; Fig. 2B). The dorsal surface of the tip of the Processus centralis is broadened and flattened and forms a round platform (Facies dorsalis, Fac. d; Figs. 1A, 8A, 32A). On the ventral surface, the tip of the central process projects ventro-apically and usually tapers into a sharp, elongated point that projects anteriorly from the articular facet of the Art. paraglosso-basihyalis (e.g., AMNH 1968, AMNH 2354, LSUMZ 94491). Sometimes, this ventro-apical tip is more flattened and knoblike (e.g., AMNH 3448, AMNH 9029; Figs. 1B, 2B, 8A).

The urohyale arises from the caudal end of the Corpus basihyale. The base of the urohyale is about as broad as one-third of the width of the body of the basihyale, but it soon narrows into a laterally compressed bar (Figs. 1A, B). Usually, the urohyale forms a straight bar (e.g., AMNH 1968, AMNH 3448, LSUMZ 94491; Figs. 2A, B), but sometimes its caudal end curves slightly upwards (e.g., AMNH 2354, AMNH 9029). Often, the height of the urohyale is markedly reduced in the caudal one-third (e.g., AMNH 3448, AMNH 9029; Figs. 2A, B). The narrow dorsal surface of this bar is flattened along its entire length, whereas the ventral surface forms a rounded ridge. The bony urohyale ends caudally in a knob with a concave caudal facet (Area cartilaginis urohyalis, A. CUH). This facet serves as the attachment point for the cartilaginous process of the urohyale (Cartilago urohyalis, CUH). This cartilaginous process is always present, but of variable length and shape and usually missing in skeletal preparations. Often, its tip curves laterally to accommodate the curvature of the trachea (e.g., Z2; Fig. 28B).

The articular facet for the articulation with the ceratobranchiale (Art. cerato-basihyalis) is concave and covers the lateral one-third of the caudal margin of the Corpus basihyalis and the lateral surface of the base of the urohyale, forming an arch of roughly 90° (Figs. 1B, 2B).

The ventral, crest-like surface of the urohyale is somewhat broadened where it faces the Nodulus. The Nodulus is an unpaired sesamoid bone embedded in the Lig. nodulo-ceratobranchiale. It is irregularly shaped and individually variable (Figs. 1B, 2A). Cartilaginous extensions at its anterior and posterior ends give the structure an oval shape (Figs. 23A, B). Its dorsal surface is concave and faces the Crista ventralis of the urohyale. The Nodulus and its cartilaginous extensions are cushioned against the urohyale by a bursa (Bursa synovialis urohyalis, BTuh). The Nodulus serves as an attachment site for the M. serpihyoideus, M. cerato-hyoideus, M. mylohyoideus posterior, and M. ceratoglossus superficialis noduli.

The cartilaginous process of the urohyale serves as the insertion site for a small slip of the M. tracheolateralis (Figs. 25A, B; Homberger 1979b). The flattened dorsal surface of the urohyale supports the laryngeal apparatus (Homberger 1979b). On the lateral side of the base of the urohyale, adjacent to the medial corner of the articular facet of the Art. cerato-basihyalis, is a broad area with a rough surface (A. lig.; Figs. 2B, 14C), which is the attachment site of a thick articular ligament, the Lig. articulare cerato-basihyale. The dorsal surface of the transition zone between the urohyale and basihyale may be smooth (e.g., AMNH 2354, AMNH 9029, LSUMZ 94491) or structured with ridges and crests (Fig. 1A) that serve as the insertion site for the M. tracheohyoideus basihyalis (Figs. 24A, B).

The dorsal surface of the Corpus basihyale is divided into a pair of broad depressions (Fossa basihyalis, F. bh) by a sharp, midsagittal crest (Crista dorsalis basihyalis, Cr. d. bh). This crest extends forward to the anterior one-third of the central process (Figs. 1A, 2B). Part of the Crista dorsalis serves as the attachment site for the Lig. crico-basihyale (Lcrb) of the laryngeal apparatus (Homberger 1979b; Figs. 26A, B) and for the Lig. interparahyale. Each Fossa basihyalis is subdivided into a larger medial and smaller lateral part by a faint crest (Crista lateralis basihyalis, Cr. l. bh) that continues onto the parahyal process to its tip (Fig. 1A). The medial part of the Fossa basihyalis serves as the site of origin for the M. supraglossus basihyalis and M. cricohyoideus ventralis (Homberger 1979b), whereas the lateral part of the fossa serves as the insertion site for the M. stylo-hyoideus and M. tracheohyoideus parahyalis (Fig. 12A, B).

The dorsal surface of the Processus centralis serves only occasionally as an extension of the site of origin for the M. supraglossus basihyalis and then only up to its midlength (Fig. 12B). The anterior half of the central process and its Facies dorsalis do not serve as attachment places for any ligaments or muscles. The apical border of the Facies dorsalis is surrounded by a narrow, curved rim which faces dorso-laterally to dorso-apically (Area ligamentosa, A. lig; Fig. 8A). This narrow rim serves as the attachment site for the Lig. articulare paraglosso-basihyale dorsale.

The ventral surface of the Corpus basihyale is less structured than the dorsal surface due to the smaller number of muscles and ligaments attaching to it. Sometimes, a sharp midsagittal ridge (Crista ventralis basihyalis, Cr. v. bh) extends forward from the ventral crest of the urohyale and flattens out midway along the central process of the basihyale (e.g., AMNH 1968). Usually, however, the Crista ventralis is blunter and is not continuous with the ventral crest of the urohyale (Fig. 1B). The M. hypoglossus obliquus medialis originates on both sides of the Crista ventralis basihyalis (Figs. 13A, B).

The ventral surface of the Processus centralis serves as the site of origin for the M. hypoglossus obliquus medialis (Figs. 12A, B). The ventro-apical tip of the central process serves as the attachment site for the aponeurosis of the M. hypoglossus anterior and the Fascia paraglossalis dorsalis (Figs. 8A, 30B, 44B) and for the paired Lig. articulare paraglosso-basihyale ventrale (Figs. 30B, 39B–40B).

Wide ranges of individual variation are found in various measurements of the basihyale and urohyale. Least variable are the length of the basihyale and its central process and the size of the articular facet of the Art. paraglosso-basihyalis, but all other measurements show much individual variation. The total length of the basihyale and urohyale varies between 18.8–21.0 mm (N = 5; $\bar{X}$ = 19.9 mm; s.d. = 0.82 mm), the length of the urohyale excluding its cartilaginous process varies between 7.7–9.2 mm (N = 5; $\bar{X}$ = 8.36 mm; s.d. = 0.51 mm), the length of the body of the basihyale varies between 3.6–4.1 mm (N = 5; $\bar{X}$ = 3.84 mm; s.d. = 0.21 mm), the length of the dorsal side of the central process of the basihyale varies between 6.5–7.2 mm (N = 5; $\bar{X}$ = 7.02 mm; s.d. = 0.17 mm), and the length of the parahyal processes varies between 1.6–2.4 mm (N = 4; $\bar{X}$ = 1.92 mm; s.d. = 0.4 mm). The maximum width of the body of the basihyale varies between 5.7–6.9 mm (N = 5; $\bar{X}$ = 6.42 mm; s.d. = 0.46 mm), the width of the central process of the basihyale at midlength varies between 1.5–2.0 mm (N = 5; $\bar{X}$ = 1.7 mm; s.d. = 0.17 mm), the height of the urohyale at midlength varies between 1.4–1.8 mm (N = 5; $\bar{X}$ = 1.54 mm; s.d. = 0.15 mm), and the height of the central process of the basihyale at midlength varies between 1.9–2.5 mm (N = 5; $\bar{X}$ = 2.05 mm; s.d. = 0.25 mm). The length of the facet of the Art. paraglosso-basihyalis varies between 2.8–3.1 mm (N = 3; $\bar{X}$ = 2.96 mm; s.d. = 0.25 mm), the maximum width of the articular facet varies between 2.1–2.2 mm (N = 5; $\bar{X}$ = 2.12 mm; s.d. = 0.07 mm), and the width at the anterior end of the articular facet varies between 1.3–1.4 mm (N = 5; $\bar{X}$ = 1.34 mm; s.d. = 0.05 mm). Mivart (1895) stated that the urohyale measures three-quarters of the length of the basihyale in *Psittacus,* which is supported by the present data.

Os paraglossale.—The paired paraglossalia form a single skeletal element, the paraglossale, being bound together through a syndesmosis (Syndesmosis paraglossalis) between their central portions and a cartilaginous bridge (Synchondrosis paraglossalis) between their anterior processes. The cartilaginous bridge is always present (Figs. 31B, 32A, 36B, 37A) but is usually not preserved in skeletal preparations (e.g., Figs. 1A, B, 9A–10A).

A single unit of the paired paraglossalia is roughly wing-shaped with the broadest portion at midlength (Fig. 10B). Its outer surface faces from laterally to latero-ventrally and its inner surface faces from medially to medio-dorsally. It can be divided into three portions: a central portion, an anterior process, and a posterior process (Fig. 9A).

The central portion (Pars centralis paraglossalis, P. centr. pg) has a roughly triangular cross-section (Figs. 40A–41A) and is characterized by the slightly convex, elongated articular facet of the Art. paraglosso-basihyalis on its medio-ventral surface (Fig. 10B). The two other surfaces of the central portion face medio-dorsally and latero-ventrally, respectively, and are continuous with the corresponding surfaces of the anterior and posterior processes.

The posterior process (Proc. posterior paraglossalis, P. post. pg) is more laterally compressed than the central portion and becomes increasingly vertically oriented caudally (Figs. 41B, 42A), ending in a knob-like structure (Figs. 9A–10A, 42B). The medial surfaces of the paired posterior processes form an acute angle where they meet (Figs. 9A–10A). Their sides are either more or less parallel (Fig. 9A) or diverge caudally (Figs. 9B, 10A).

The anterior process (Proc. anterior paraglossalis, P. ant. pg) is longer than the posterior process. Its cross-section near the central portion is oval-shaped and becomes increasingly flattened apically (Figs. 36B–39A). The anterior process is obliquely oriented so that its broad surfaces face medio-dorsally and latero-ventrally. Distally, the anterior processes are connected by a cartilaginous bridge (Synchondrosis paraglossalis, SPG) that attaches along their medio-ventral edges (Figs. 9A–10B, 36B, 37A). The anterior edge of the synchondrosis forms a pair of transversally elongated, flat mounds (Figs. 30A, 32A). These mounds underlie the pair of Bursae synoviales paraglossales cushioning the common aponeurosis of the M. hypoglossus anterior. Proximally, the anterior processes enclose the Foramen interparaglossale. Caudally, this foramen is bordered by the Syndesmosis paraglossalis and, apically, by the Synchondrosis paraglossalis (Figs. 30B, 32A, 37B–40A, 43).

The surface structures of the paraglossale usually indicate attachment sites for muscles, tendons, or ligaments. On the outer surface of the paraglossale, the most prominent surface structure is the Tuberositas paraglossalis (Tub. pg), a usually well-defined, elongated tuberosity with a rough surface (e.g., AMNH 9029, AMNH 1968, AMNH 3448, LSUMZ 94491; Fig. 9B). Sometimes, this tuberosity is rather flat and smooth (e.g., AMNH 2354; Fig. 9A). It serves as an insertion site for the tendon of the M. ceratoglossus (Fig. 39A). Usually, a low crest (Crista lateralis, Cr. l. pg) extends posteriorly from the Tuberositas paraglossalis, although it may be absent (e.g., AMNH 9029). It marks the dorsal border of the origin of the M. hypoglossus anterior and leads either directly to the terminal knob of the posterior process of the paraglossale (e.g., AMNH 3448, AMNH 2354; Figs. 2A, 9A, B) or joins the Crista dorso-lateralis at its caudal one-third or midlength (e.g., AMNH 1968, LSUMZ 94491). The Crista dorso-lateralis (Cr. d-l) projects laterally from the dorsal margin of the paraglossale (Margo dorsalis). It extends from the terminal knob of the Proc. posterior paraglossalis to the midlength of the Margo dorsalis and serves as an insertion site for the M. genioglossus (Figs. 2A, 9A, B, 12C, 40A–42A). The dorsal surface of the terminal knob of the posterior process serves as an insertion site for a portion of the M. hypoglossus obliquus and occasionally for the posteriormost fibers of the M. hypoglossus anterior (Figs. 12A, 13A, 32B). The posterior half of the Margo dorsalis serves as an attachment site for the tendinous fibers of the M. hypoglossus obliquus which form the envelope of the lingual salivary gland (Fig. 32A). The apical half of the Margo dorsalis serves mainly as an attachment site for the tendinous fibers of part C of the M. supraglossus (Figs. 12A, 29B, 36B–38A, 45A). The apical end of the Margo dorsalis, i.e., the latero-dorsal corner of the anterior process, serves as an attachment site for the Lig. paraglossale apicale (Figs. 30A, 32A, 45A) and for a portion of the tendinous fibers from the M. hypoglossus anterior (Figs. 12C–13B, 27B, 35B).

The inner surface of the paraglossale is less structured than the outer surface because fewer and smaller muscles attach here (Fig. 10B). The medial surface of the posterior process of the paraglossale serves as an insertion site for the M. hypoglossus obliquus medialis (Figs. 12A, 13A, 32A, 42B). A narrow strip along the medial side of the Margo dorsalis serves as the insertion site for part of the M. cricohyoideus dorsalis superficialis, an extrinsic laryngeal muscle (Figs. 12A, 32A; Homberger 1979b). Directly above the anterior half of the articular facet of the Art. paraglosso-basihyalis is a round area (Area syndesmosis paraglossalis, A. synd. pg) that forms the contact zone with the contralateral half of the paraglossale through a syndesmosis. If the paired paraglossalia are separated at the syndesmosis, this area shows a rough surface (Fig. 10B). This area may be rather small (e.g., AMNH 3448, AMNH 9029, LSUMZ 94491; Figs. 9B, 10A) or more extended (e.g., AMNH 1968, AMNH 2354; Figs. 9A, 10B). A sharp crest (Crista medialis, Cr. m) extends apically from the dorsal border of the Area syndesmosis paraglossalis (Figs. 10A, B). It serves as the attachment site for the Lig. interparaglossale caudale (Figs. 30B, 32A) and as the dorsal border of the origin of the medial portion of the M. hypoglossus anterior (Fig. 39B).

The shape and proportions of the paired paraglossalia show a peculiar individual variability (the morphological variations of the surface structures have been described above). The dimensions of certain parts of the paraglossale are more variable than others. In five specimens, the length of the paraglossale varies between 10.1–11.3 mm ($\bar{X}$ = 10.76 mm; s.d. = 0.17 mm), the width at the posterior end varies between 4.9–6.1 mm ($\bar{X}$ = 5.76 mm; s.d. = 0.53 mm), the width at the apical end varies between 5.2–5.4 mm ($\bar{X}$ = 5.28 mm; s.d. = 0.09 mm), and the width at midlength of the central portion varies between 3.9–4.9 mm ($\bar{X}$ = 4.38 mm; s.d. = 0.39 mm). Fürbringer (1922) indicated that the paraglossale is longer than wide and that the proportions range from 1.22 to 1.95 for most parrots. However, the proportions for the present sample of five specimens range from 1.65 to 2.31. The dimensions of the articular facet for the Art. paraglosso-basihyalis and the height of the paraglossale are only slightly variable. The length of the articular facet varies between 3.2–3.3 mm (N = 5; $\bar{X}$ = 3.25 mm; s.d. = 0.05 mm), the distance between the posterior ends of the articular facets varies between 3.1–3.4 mm (N = 5; $\bar{X}$ = 3.25 mm; s.d. = 0.12 mm), and the distance between the anterior ends of the articular facets varies between 1.7–1.9 mm (N = 4; $\bar{X}$ = 1.8 mm; s.d. = 0.08 mm). The height of the paraglossale at the level of the central part varies between 3.1–3.4 mm (N = 5; $\bar{X}$ = 3.26 mm; s.d. = 0.14

mm). Variation in the angle formed between the posterior processes of the paraglossalia (Figs. 3A, B) allows the distinction between two types of the paraglossale. In one, the angle enclosed between the posterior processes is small, and the difference between the distance between the posterior processes and the width of the paraglossale at the central portion is small, so that the posterior processes appear to be in line with the anterior part of the paraglossale (e.g., AMNH 1968, AMNH 2354; Fig. 9A). In the other type, the angle enclosed between the posterior processes is large as is the difference between the distance between the posterior processes and the width at the central portion, so that the posterior processes appear to spread laterally (e.g., AMNH 3348, AMNH 9029, LSUMZ 94491; Figs. 9B, 10A). A larger sample size might provide intermediate forms.

Discussion.—The hyoid skeleton of parrots differs in many respects from the one in other avian groups, mainly in the shape of the paraglossale and of the basihyale with its parahyal processes (for a review of the avian hyoid skeleton, see Giebel 1858, Fürbringer 1922).

Most of the studies dealing with the psittacine hyoid skeleton provide only a cursory description and illustration of its composition and of the shape of its elements (Kutorga 1832; Duvernoy 1835; Thuet 1838; Giebel 1858; Nitzsch in Giebel 1862; Shufeldt 1886; Gadow 1891; Mudge 1903; Kallius 1906; Denker 1907; Steinbacher 1951; Dubale and Rawal 1965; Evans 1969; Rawal 1970). Some non-anatomical studies treat only the basihyalc, urohyale, and parahyal processes (Hargrave 1970; Holyoak 1973; Smith 1975). The most detailed descriptions of the psittacine hyoid skeleton were made by Mivart (1895, 1896). He, however, did not describe the sites of muscle and ligament attachments or the spatial relationships between skeletal elements and surrounding soft parts. Quantitative data for the psittacine hyoid skeleton are scarce and consist only of relative measurements (e.g., Mivart 1895; Fürbringer 1922).

Articulations of the Hyoid Skeleton: Structure and Function

The articulations of the hyoid skeleton form organs that are distinct from the skeletal elements in composition, structure, and function and are hence described separately. Ligaments are mentioned only insofar as they are a structural part of the articulations and influence movements at the articulations; they are, however, fully described separately.

Movements at articulations have been determined through an analysis of the surface structure of the articular facets and through manipulation of skeletal preparations. As a guideline for the manipulations, it was assumed that the joints in vivo would allow only those movements during which the articular facets are kept adjoined. The movements determined in this way were corroborated by analyses of the influence of articular ligaments on the mobility of the particular joints (pp. 91 ff.).

Articulatio epi-ceratobranchialis.—The articulation between the epibranchiale and ceratobranchiale is a diarthrosis. The articular capsule is thick and fibrous but not reinforced by any particular articular ligaments. A fibrous annular articular disc projects from the articular capsule into the articular cavity.

The articular facet of the ceratobranchiale is an almost flat, minimally concave, elongated surface that faces caudo-latero-ventrally, and its longitudinal axis extends from medio-dorso-caudal to latero-ventro-apical (Figs. 4A, B). The articular surface is slightly raised above the surface of the bulbous caudal end of the ceratobranchiale.

The complementary articular facet of the epibranchiale is an almost flat, minimally convex, elongated surface that faces apico-medio-dorsally (Fig. 3A).

The structures of the articular facets allow only for limited movement. An angular rotation of the epibranchiale is possible about the longitudinal axis of the articulation so that the epibranchiale rotates from caudo-medio-ventral to apico-latero-dorsal and vice versa. Rotations in other planes are more restricted (pp. 113 ff.).

Articulatio cerato-basihyalis.—The articulation between the ceratobranchiale and basihyale is a diarthrosis. The articular capsule is a sleeve-like ring of connective tissue with a longitudinal fibrillar structure. This sleeve is thickened on the dorso-medial and ventro-medial sides of the articulation to form articular ligaments. On the dorso-medial side, the Lig. interarticulare cerato-basihyale (Licb) binds together the dorso-medial corners of the articular facets of the basihyale and ceratobranchiale and, thus, forms the axis of angular rotation at the articulation (see below; Figs. 14C, D, 26A).

On the medio-ventral side, the Lig. articulare cerato-basihyale (Lacb) stretches from the medial corner of the basihyal articular facet to the medio-ventral corner of the ceratobranchial articular facet (Figs. 14C, D, 27A, 28A). On the dorso-lateral and ventro-lateral sides of the articulation, the articular sleeve is not reinforced and is much longer than the rest of the articular capsule. This localized elongation of the articular capsule is necessary because the distance between its attachments on the basihyale and ceratobranchiale increases disproportionately during the extension of the articulation (see below).

The articular facet of the basihyale-urohyale is socket-like, but has a complex structure. It is situated in the right angle formed by the base of the urohyale and the caudal border of the Corpus basihyale. For descriptive purposes, the oblong articular facet can be subdivided into a medial and a lateral part (Fig. 8B). The larger medial part is round and socket-shaped. The ventral rim of this socket protrudes more laterally than the dorsal rim. The perpendicular axis of the socket points caudo-latero-dorsally (assuming a horizontally oriented basihyale and urohyale; if the basihyale is assumed to be in the slightly inclined "normal" position, the perpendicular axis points latero-caudally, i.e., into the direction of the longitudinal axis of the ceratobranchiale). The elongated, but narrow lateral part of the articular facet extends ventro-laterally from the medial part and is distinguished from the latter by the different orientation of its surface. It starts at a very narrow angle from the ventral border of the medial part and curves dorso-laterally. In doing so, it gradually broadens and twists so that it faces increasingly ventrally because it follows the ventral surface of the dorsally overhanging caudal border of the Corpus basihyale to its caudo-lateral corner (Fig. 8B). Except for its lateralmost part, the entire articular facet is surrounded by a bulging rim (Figs. 1A, B, 2B, 8B). Laterally, the dorsal border of the articular facet forms a crest-like edge, but on the ventral side the articular surface is slightly raised over the surface of the basihyale.

The complementary articular facet on the ceratobranchiale is convex and can also be divided into two confluent parts. The larger medial part is strongly curved from medio-ventral to latero-dorsal (Figs. 5A, B, 7A) but is almost flat with a shallow transverse sulcus from dorso-medial to ventro-lateral (Figs. 6A, B, 33A). The elongated lateral part of the articular facet starts at a very narrow angle on the ventral side of the medial part. It curves latero-dorsally (Figs. 5A, 6A–7A) and, dorsally, forms a narrow plate that extends caudally beyond the dorsal border of the medial part (Figs. 4A, 5A, 6A–7A). The entire articular facet is slightly raised above the surface of the ceratobranchiale along its dorsal and medial sides (Figs. 4A, 5A–7A), but along its lateral and ventral sides, it is bordered by a crest-like margin (Margo ventralis faciei articularis cerato-basihyalis, Ma. v. F. art. cb; Figs. 4A–7A).

The articular facets of the ceratobranchiale and basihyale exactly match only when the articulation is flexed. As soon as the articulation is extended, the lateral corners of the articular facets protrude from under their counterparts (e.g., Figs. 25A, 32A) so that they are covered only by the wall of the articular capsule.

The movements of the ceratobranchiale and basihyale at the Art. cerato-basihyalis are guided and restricted in their movement by the Lig. interarticulare cerato-basihyalis and Lig. articulare cerato-basihyalis and by the shape of the curvature of the lateral parts of the articular facets. Yet, for the Art. cerato-basihyalis it is not possible to distinguish strictly a moving skeletal element from a stationary one. Rather, both the ceratobranchiale and basihyale move simultaneously about the axis of articulation. The relative extent of movement of each element is determined by the type of motion of the hyoid (e.g., retraction, protraction, etc.). The freedom of movement at the Art. cerato-basihyalis is further restricted by the fact that the two ceratobranchialia move simultaneously about the single basihyale during movements of the hyoid. For descriptive reasons and to facilitate understanding of the movements at the Art. cerato-basihyalis, the movements of each skeletal element will be described separately by assuming a static position of the other element (complex movements are described on pp. 99 ff.).

If a stationary basihyale is assumed, the movement of the ceratobranchiale can be described, in simplest terms, as angular rotation about the dorso-medial corner of the articular facet of the basihyale (Fig. 14C). Starting from the extended position of the basihyale and ceratobranchiale, at which the dorso-lateral corner of the lateral part of the basihyal articular facet and the caudo-dorsal corner of the lateral part of the ceratobranchial articular facet are exposed (Figs. 1A, 2A), the angular rotation makes the ceratobranchiale describe an arc towards latero-dorso-apical. At the end of this rotation, at which the articulation is flexed, the caudo-dorsal corner of the lateral part of the articular facet of the ceratobranchiale meets the dorso-lateral corner of its counterpart on the basihyale. During this angular rotation, the ceratobranchiale also rotates slightly about its own longitudinal axis, i.e., the dorsal surface of the ceratobranchiale turns laterally so that it faces dorso-laterally. This axial rotation is due to the complex structure of the articular facets. The basihyal articular facet along the dorsally overhanging caudal border of the Corpus basihyale faces ventrally and, therefore, dictates the level of the lateralmost part of the ceratobranchial articular facet, which comes in contact with it during the angular rotation of the ceratobranchiale. Because the lateral part of the ceratobranchial articular facet curves upwards (i.e., dorsally), the caudal end of the lateral part of the ceratobranchial articular facet is, so to speak, pushed down by the caudal border of the Corpus basihyale. In this way, the ceratobranchiale is forced to rotate from dorsal to dorso-lateral about its own longitudinal axis. Conversely, when the articulation is being extended, the ceratobranchiale performs an axial rotation from dorsal to dorso-medial.

The movement of the basihyale mirrors exactly the movement of the ceratobranchiale, but it will also be explained in detail to facilitate the understanding of the complex motions about the Art. cerato-basihyalis explained later (pp. 99 ff.).

If a stationary ceratobranchiale is assumed, the movement of the basihyale can be described as an angular rotation about the medio-dorsal corner of the ceratobranchiale (Figs. 14D, E). The course of the rotation follows the lateral part of the ceratobranchial articular facet. Starting from the extended position of the basihyale and ceratobranchiale, an angular rotation makes the basihyale describe an arc towards latero-dorso-caudal. Concurrently, the tip of the urohyale describes an arc towards contralatero-ventro-apical. The end of this rotation, at which the articulation is flexed, is reached when the dorso-lateral corner of the lateral part of the articular facet of the basihyale meets the caudo-dorsal corner of its counterpart on the ceratobranchiale. During this angular rotation, the basihyale also rotates slightly about its own longitudinal axis, i.e., the dorsal surface of the basihyale turns medially so that it faces dorso-medial. This axial rotation of the basihyale is due to the complex structure of the articular facets. The curvature of the caudo-dorsal section of the lateral part of the ceratobranchial articular facet flattens horizontally towards its end. Therefore, when the dorso-lateral end of the lateral part of the basihyal articular facet reaches the caudal part of its counterpart on the ceratobranchiale, it remains at the same level instead of continuing dorsally by following the latero-dorso-caudal curve of the angular rotation. In this way, the basihyale is forced to rotate from dorsal to dorso-medial about its own longitudinal axis. Conversely, when the articulation is being extended, the basihyale performs an axial rotation from dorsal to dorso-lateral.

An overextension of the Art. cerato-basihyalis is prevented by the Lig. articulare cerato-basihyale. This articular ligament extends from the dorso-medial corner of the basihyal articular facet to the ventro-medial corner of the ceratobranchial articular facet. The ventro-medial corner of the ceratobranchiale glides dorso-laterally when the articulation is extended (Fig. 14C); thus, the articular ligament is stretched (Fig. 28A) and functions as an anchor line for the medio-ventral corner of the ceratobranchiale. In contrast, the flexion of the Art. cerato-basihyalis is controlled by the system of the Lig. nodulo-ceratobranchiale (pp. 99 ff.).

Articulatio paraglosso-basihyalis.—The articulation between the paraglossale and basihyale is a diarthrosis. The articular capsule is especially thick and forms the Lig. articulare paraglosso-basihyale dorsale on its dorsal side (Figs. 41A, 43–44B). A pair of heavy articular ligaments (Lig. articulare paraglosso-basihyale ventrale) reinforces the ventro-apical tip of the articulation (Figs. 30B, 39B–40B). Fibrous articular menisci or pads project from the articular capsule into the articular cavity to make up for incongruences between the articular facets of the basihyale and paraglossale (Figs. 40B–41B, 43–44B).

The articular facet of the basihyale is saddle-shaped. Its midsagittal crest is gently rounded in cross-section (Fig. 8A), and its curved profile leads from dorso-caudal to ventro-apical (Figs. 2B, 43–44B). The transverse curvature of the ventro-apical part of the articular facet has a much smaller radius then the transverse curvature of the caudo-dorsal part (Figs. 8A, 14B). Dorsally, the sides of the saddle embrace the Facies dorsalis of the basihyale (Figs. 1A, 8A). Towards ventro-apical, the sides of the saddle become increasingly shorter (Fig. 2B). The articular facet is slightly raised above the surface of the basihyale along its lateral and ventral border. Along its dorsal edge, it is bordered by an oblique rim-like surface that serves as the attachment site for the thickened articular capsule, i.e., Lig. articulare paraglosso-basihyale dorsale.

The complementary articular facet of the paraglossale is chevron-shaped and formed by the combination of the articular facets on the ventro-medial surface of each half of the paired paraglossalia. The top of the chevron is formed by the Syndesmosis paraglossalis (Figs. 9A, B, 40B), which is shorter than the midsagittal crest of the saddle-shaped articular facet on the basihyale. Apically and caudally from the syndesmosis, the articular facets of the paired paraglossalia diverge, forming an angle between them (Figs. 9A, B). The chevron is much narrower apically than caudally (Figs. 9A, B), and the sides of the chevron become broader from apical to caudal (Fig. 10B). The entire articular surface of the chevron is slightly smaller than that of its counterpart on the basihyale. The entire articular facet is bordered ventrally by a sharp crest (Margo ventralis faciei articularis paraglosso-basihyalis, Ma. v. F. art. pb). The articular facet is slightly raised over the surface of the paraglossale along the dorsal border (Margo dorsalis faciei articularis paraglosso-basihyalis, Ma. d. F. art. pb).

Movements of the paraglossalia are restricted by the articular ligaments and the particular shape of the articular surfaces to an up-down angular rotation, which is guided by the curved midsagittal ridge of the saddle-shaped articular facet on the basihyale, and a limited, but complex axial rotation about the midsagittal axis of the articulation (Figs. 14A, B). No pure angular lateral rotation of the paraglossalia exists.

During the up-and-down movement, the paraglossale performs an angular rotation in the midsagittal plane, using the top of the chevron-shaped articular facet as a pivot. At the same time, the top of the chevron-shaped articular surface slides back and forth along the midsagittal crest of the saddle-shaped articular facet on the basihyale (Figs. 14A, 53A–54B, 55A, B, 60A–64B).

When the anterior tip of the paraglossale is depressed and the posterior end raised, i.e., when the paraglossale is flexed, the top of the chevron-shaped paraglossal articular facet glides towards the caudo-dorsal end of the midsagittal crest of the basihyal articular facet. In this position, the diverging caudal free ends of the paraglossal articular facet fit across the transversal curvature of the midsagittal crest of the basihyal articular facet. At the same time, the smaller diverging anterior free ends of the paraglossal articular facet are pointing ventro-apically and fit across the horizontally oriented midsagittal crest of the ventro-apical end of the basihyal articular facet. The extreme tip of the midsagittal crest of the basihyal articular facet projects apically underneath the anterior border of the paraglossal articular facet.

When the anterior tip of the paraglossale is raised and the posterior end depressed, i.e., when the paraglossale is extended, the top of the chevron-shaped paraglossal articular facet glides towards the ventro-apical end of the midsagittal crest of the basihyal articular facet. In this position, the caudal rim of the tip of the chevron-shaped paraglossal articular facet abuts against the center of the curved profile of the midsagittal crest of the basihyal articular facet. Here, the diverging caudal free ends of the paraglossal articular facet do not fit exactly over the midsagittal crest of the basihyal articular facet because the radius of the transversal curvature of the crest is smaller ventrally than dorsally. The caudo-dorsal end of the midsagittal crest of the basihyal articular facet projects dorsally above the paraglossal articular facet. At the same time, the diverging anterior free ends of the paraglossal articular facet are directed apically and project beyond the ventro-apical tip of the midsagittal crest of the basihyal articular facet.

The paraglossale can rotate axially about the midsagittal axis of the articulation only if it is flexed. In this position, the articular facet of the paraglossale fits over the midsagittal crest of the basihyal articular facet and can glide from side to side across the crest so that the dorsal surface of the paraglossale rotates to face latero-dorsal. But because the radius of the curvature of the midsagittal crest of the basihyal articular facet is much larger in its caudo-dorsal part than in its apical part, the axial rotation movement of the paraglossale is tied to an angular rotational movement about the ventro-apical tip of the articulation (Figs. 14B, 56A, B). Thus, the anterior and posterior tips of the paraglossale are moved sideways simultaneously with the axial rotation of the paraglossale. For example, when the anterior tip of the paraglossale rotates to the left from the straight position, the right anterior tip of the paraglossale describes an arc from centro-dorsal to dextro-ventral, whereas the right posterior tip of the paraglossale describes an arc from dextro-dorsal to medio-ventral. Simultaneously, the left anterior tip of the paraglossale describes an arc from sinistro-ventral to medio-dorsal, whereas the left posterior tip describes an arc from medio-ventral to sinistro-dorsal.

The paraglossale cannot perform an axial rotation when its anterior tip is raised, i.e., when it is extended. In this position the longitudinal rotational axis of the paraglossale is almost horizontally oriented, and an axial rotation about this axis is prevented by the more vertically oriented caudo-dorsal part of the midsagittal crest of the basihyal articular surface that acts like a peg between the diverging caudal free ends of the paraglossal articular facet. A pure angular rotation from side to side by the paraglossale about the ventro-apical end of the Art. paraglosso-basihyalis is also prevented by the vertically oriented caudo-dorsal part of the midsagittal crest of the basihyal articular surface. Even a pure angular rotation from side to side about the caudo-dorsal end of the Art. paraglosso-basihyalis would not be possible without luxation of the anterior end of the articulation.

A lateral luxation and an excessive raising of the anterior part of the Art. paraglosso-basihyalis during normal movements is prevented by the paired Lig. articulare paraglosso-basihyalis ventralis, whereas an overflexion is resisted by the thick Lig. articulare paraglosso-basihyalis dorsalis.

Discussion.—The anatomy of articulations has only rarely been carefully described in studies of muscle-bone systems of the avian feeding apparatus (e.g., Bühler 1970; Bock and Morioka 1971; Zweers 1974; Zweers et al. 1981). The articulations of the psittacine hyoid apparatus have previously been described only briefly and incompletely. Mivart (1895) described the saddle-shaped Art. paraglosso-basihyalis in *Psittacus,* but his description of the Art. cerato-basihyalis does not agree with my observations; moreover, he did not analyze the movements at the joints. Steinbacher (1951) described the saddle-shaped Art. paraglosso-basihyale in *Lorius,* a parrot that is not closely related to *Psittacus* (Homberger 1980a), and described the movements of the paraglossale as a pure up-down angular rotation.

Mandible

The mandible is a U-shaped skeletal element in which the limbs of the U are formed by the Rami mandibulares ("Unterkieferäste" of Homberger 1980a) and in which the base of the U is represented by the symphyseal portion of the mandible ("Unterkiefersymphyse" of Homberger 1980a). The rami are bilaterally flattened, deep bars whose height increases slightly towards apical. Caudally, an articular process bearing the elongated socket for the condyle of the quadrate bone projects from the medial surface at mid-height (Homberger 1981). The symphyseal portion of the mandible is U-shaped in frontal view. The bottom of the U is elongated and its profile ("Gonys" of Homberger 1980a) curves from caudally to dorso-apically. The symphyseal portion is covered by the rhamphotheca on its outer and inner surfaces. Only the posterior one-quarter of the inner surface remains free ("rhamphothekenfreier Abschnitt" of Homberger 1980a).

Much of the Ramus mandibularis is covered with surface structures indicating the insertions of jaw muscles; only a few places serve as sites of origin for extrinsic lingual muscles (Figs. 11A, B). The lateral surfaces of the caudal edge and the lateral and medial surfaces of the caudo-ventral corner of the Ramus mandibularis serve as the site of origin for the M. stylohyoideus and M. serpihyoideus. The dorso-apical corner on the medial surface of the ramus serves as the site of origin for the M. mylohyoideus and M. branchiomandibularis anterior. The ventral edge near the anterior one-quarter of the mandibular ramus serves as the site of origin for a portion of the M. branchiomandibularis posterior, whereas the rest of the muscle originates from the ventral one-quarter of the lateral surface of the ramus close to the border of the rhamphotheca. The M. genioglossus originates from a circular depression in the center of the free part of inner surface of the symphyseal portion of the mandible.

Larynx and Trachea

This part of the skeleton is generally regarded as a part of the respiratory system, although it is, in many respects, also a part of the lingual apparatus to which it is attached by extrinsic muscles and ligaments (pp. 67 ff.; Homberger 1979b). The laryngeal skeleton consists of a ring-like cricoid, an unpaired procricoid, and a pair of arytaenoids. The arytaenoids form the lateral borders of the glottis. The tracheal skeleton is built from bony rings which form a tube. Shortly below its upper end, that tube increases its diameter and curves apically. A series of tracheal half-rings are added apically to the uppermost complete tracheal ring. The cricoid lies on top of the upper end of the trachea and is supported by the caudal half of the uppermost tracheal ring and by the lateral and apical edges of the tracheal half-rings. The laryngeal apparatus is laterally embraced by the anterior ends of the ceratobranchialia. The ventro-apical plate of the cricoid lies on the borderline between the basihyale and urohyale, and the tracheal half-rings rest on the dorsal surface of the urohyale.

Ligaments and Fasciae (Syndesmology)

Introduction

Syndesmology has traditionally been the least explored part in anatomical studies of skeletomuscular systems, and the syndesmology of the avian lingual apparatus in particular has only rarely been described (e.g., Zweers 1974, 1982; Zweers et al. 1977).

Ligaments and fasciae do not take an active role in the movements of a biomechanical apparatus, but rather react passively to forces placed upon them (pp. 82 f., 91 ff.). Ligaments play an important role in holding together skeletal elements, in guiding and coordinating movements of the hyoid, or in serving as attachment sites for muscles, whereas fasciae, in general, serve to limit dislocation of various structures during movements of the hyoid.

Several types of ligaments can be distinguished following the classification of Bühler (1981) based on functional criteria (see also Bock 1974):

(a) Intraosseal ligaments connect two different areas of the same skeletal elements. In the psittacine hyoid apparatus, such ligaments have various functions.

(b) Articular ligaments connect two different skeletal elements that articulate with each other. Special forms of articular ligaments are syndesmoses, articular capsules, and intracapsular ligaments that are ligamentous thickenings of the articular capsules. They limit the extent of movements of the skeletal elements, thus preventing the disruption of the articulations. (The articular capsules are not treated on pp. 22 f., but see pp. 14 ff.).

(c) Linkage ligaments span more than one articulation and are most important for the coordination of the movements of the various skeletal elements of a system.

(d) Fasciae are sheets of connective tissue enveloping or suspending organs or connecting different structures.

Intraosseal Ligaments

1. Ligamentum paraglossale apicale (Lpga).—This intraosseal ligament surrounds the dorsal part of the apical surface of the M. mesoglossus anterior and is firmly attached to the latero-apical tips of the Procc. anteriores paraglossales (Figs. 27A, B, 29A–30B). Sometimes (e.g., Z7), the ligament forms a broad sheet covering the entire apical surface of the M. mesoglossus anterior. The ligament is bound strongly to the M. mesoglossus ant. on one side and to the concentrically arranged connective tissue passing between the cavernous veins of the Corpus cavernosum apicale on the other side (Figs. 27A, B). Also connected to the ligament is the apical end of the main tendon of the tendinous part D of the M. supraglossus. The main function of the ligament is to connect the tendinous fibers of the M. supraglossus and M. mesoglossus ant. to the Corpus cavernosum apicale so that the muscle fibers of the M. mesoglossus ant. can be extended and the tendinous fibers of the M. supraglossus can be pulled forward by an inflation of the Corpus cavernosum apicale.

2. Ligamentum interparaglossale apicale (Lipga).—This intraosseal ligament attaches on the ventro-medial edges of the paired Procc. anteriores paraglossales caudally to the Synchondrosis paraglossalis and spans the apical part of the Foramen interparaglossale (Figs. 30B, 32A, 37B, 38A, 43, 44A). It forms a somewhat yielding, but resilient edge over which the Fascia dorsalis paraglossalis and the aponeurosis of the M. hypoglossus anterior glide when passing through the Foramen interparaglossale (Figs. 30B, 43, 44A). It also reinforces the Synchondrosis paraglossalis in holding the two halves of the paraglossale together.

3. Ligamentum interparaglossale caudale (Lipgc).—This intraosseal ligament attaches along the Cristae mediales of the Procc. anteriores paraglossales and

spans the Foramen interparaglossale just in front of the Syndesmosis paraglossalis with which it is confluent (Figs. 30B, 32A, 39B, 40A). It serves as a site of origin for the medial portion of the M. hypoglossus anterior as well as a resilient edge over which the Fascia dorsalis paraglossalis passes before going through the Foramen interparaglossale (Figs. 30B, 32A). It also reinforces the Syndesmosis paraglossalis in tying the two halves of the paraglossale together.

4. Syndesmosis paraglossalis (Synd. pg).—This syndesmosis is very variable in size and structure. It is formed by very short fibers of connective tissue binding together the two halves of the paraglossale at their central portions (Fig. 40B). It also forms the top of the chevron-shaped articular facet of the paraglossale (Figs. 9A, B). On the dorsal surface of the paraglossale, the connective tissue fibers form a cruciate flat ligament over the syndesmosis between the two halves of the paraglossale (Fig. 32A). This ligament strengthens the syndesmosis against ventrolateral forces exerted on the two halves of the paraglossale by the M. genioglossus and M. ceratoglossus.

5. Ligamentum interparahyale (Liph).—This intraosseal ligament was observed in only one specimen (Z3). It spans the space between the anterior tips of the paired parahyal processes and is fastened to the Crista dorsalis basihyalis through strands of connective tissue. It serves as the insertion site for the M. cricohyoideus dorsalis intermedialis, an extrinsic laryngeal muscle (Fig. 22; Homberger 1979b).

6. Ligamentum parahyale.—This intraosseal ligament consists of a thin sheet of connective tissue bridging the gap between the Proc. parahyalis and the Proc. centralis basihyalis. It attaches along the medial edge of the parahyal process and along the lateral surface of the central process of the basihyale. The ligament helps the parahyal process withstand the lateral components of the ventro-caudo-laterally acting forces exerted by the M. stylohyoideus and the M. tracheohyoideus parahyalis.

Articular Ligaments

1. Ligamentum articulare paraglosso-basihyale ventrale (Lapbv).—This articular ligament is paired and runs from the lateral surface of the ventro-apical tip of the basihyale to the ventro-lateral surface of the paraglossale close to the transition zone between the Pars centralis and the Proc. anterior paraglossalis (Figs. 30B, 39B–40B). The pair of ligaments prevents lateral luxation of the anterior end of the Art. paraglosso-basihyalis as well as excessive protraction of the paraglossale during its extension (pp. 115 f.).

2. Ligamentum articulare paraglosso-basihyale dorsale (Lapbd).—This ligament is an intracapsular ligament; i.e., it is an especially thickened part of the articular capsule on the dorsal side of the Art. paraglosso-basihyalis. It attaches on the rim-like surface of the dorsal edge of the anterior end of the basihyale (Fig. 8A) and on the Margo dorsalis of the articular facet of the paraglossale (Figs. 10A, 41A, B, 43, 44A). It prevents overflexion of the paraglossale (pp. 116 f.).

3. Ligamentum interarticulare cerato-basihyale (Licb).—This intracapsular ligament is an especially thickened part of the dorsal side of the articular capsule of the Art. cerato-basihyalis. It connects the dorso-medial border of the articular facet of the ceratobranchiale with the dorso-medial border of the transition zone between the urohyale and basihyale (Figs. 14C, D, 26A). Most fibers of this ligament cross over the dorsal surface of the transition zone between the urohyale

and basihyale to form a cruciate ligament. The function of this ligament is to stabilize the axis of angular rotation of the articulation at this point by binding the ceratobranchiale and basihyale especially strongly together and by limiting movements of these skeletal elements with respect to each other at this particular place.

4. Ligamentum articulare cerato-basihyale (Lacb).—This articular ligament is an especially thickened part of the articular capsule of the Art. cerato-basihyalis. It is a very thick band attaching near the medial corner of the articular facet on the basihyale and running to the medio-ventral corner of the articular facet of the ceratobranchiale (A. lig; Figs. 2B, 7A–8B, 14C, D, 27A, 28A). It is relaxed when the articulation is flexed and stretched taut when the articulation is extended. Thus, its main function is the prevention of an overextension of the Art. cerato-basihyalis.

Linkage Ligaments

Ligamentum nodulo-ceratobranchiale (Lnc).—This is a long, strap-like linkage ligament attaching on the Tuberositas ceratobranchialis on the ventral side of the ceratobranchiale and crossing the ventral surface of the urohyale (Figs. 16A–19B, 21B). At the point where it crosses the urohyale it contains a sesamoid bone, the Nodulus (Fig. 16B, 23A). The ligament serves as an insertion site for the M. serpihyoideus and, rarely, also for the M. ceratohyoideus and as the site of origin for the M. ceratoglossus superficialis (Figs. 16A–19B, 23A, B). It plays an important role in the coordination of the movements of the ceratobranchialia and of the urohyale and basihyale about the Art. cerato-basihyalis and prevents overflexion of the Art. cerato-basihyalis (pp. 99 ff.). Mudge (1903) called this ligament the "uro-hypobranchial tendon."

Fasciae

1. Ligamentum crico-basihyale (Lcrb).—This is not a ligament of the lingual apparatus, but of the laryngeal apparatus. It forms a broad sheet and is, therefore, considered a fascia. It fastens the cricoid of the larynx to the caudal half of the Crista dorsalis basihyalis (Figs. 26A, B, 32A). Its function is to prevent excessive caudal dislocation of the larynx during the lingual movements (Homberger 1979b).

2. Fascia paraglossalis dorsalis (Fpgd, vFpgd).—This is a broad, thick sheet of connective tissue that is firmly attached to the dorsal surface of the paraglossale along the caudal border of the paired Bursae synoviales paraglossales and along the Margo dorsalis paraglossalis (Fig. 32A). In the region of the Foramen interparaglossale, it is perforated by the aponeurosis of the M. hypoglossus anterior and passes through the foramen to attach to the ventral surface of the paraglossale (Figs. 30A, B, 37B, 38A, 43–44B) and, where the medial surface is covered by the origin of the medial portion of the M. hypoglossus anterior, to the connective tissue underlying the lingual nail (Figs. 38B, 39A). The fascia serves to anchor the aponeurosis of the M. hypoglossus anterior firmly, but not rigidly, to the dorsal surface of the paraglossale. When the M. hypoglossus ant. contracts, the fascia is compressed, and when the M. supraglossus contracts to lift and flatten the epithelial surface of the tip of the tongue, it expands through elastic recoil (pp. 123 ff.).

3. Fascia uro-trachealis.—This broad sheet of connective tissue attaches to the midsagittal raphe of the M. ceratohyoideus and to the Nodulus and passes over

the lateral side of the urohyale and the Bursa synovialis urohyalis to attach to the superficial connective tissue on the dorso-lateral side of the trachea. This fascia covers the entire cartilaginous process of the urohyale reaching caudally far beyond its caudal tip and enveloping the trachea. It holds the trachea and the urohyale together, while still allowing for some back-and-forth movement of the laryngeal apparatus (Homberger 1979b).

4. Fascia nodulo-lingualis.—This fascia attaches along the apical edge of the central portion of the Lig. nodulo-ceratobranchiale and passes forward between the Glandula sublingualis and the M. ceratoglossus and M. hypoglossus obliquus medialis to anchor itself on the connective tissue covering the two above-mentioned muscles and binding the anterior edge of the M. mylohyoideus anterior to the posterior surface of the M. genioglossus. It prevents an excessive caudal dislocation of the Nodulus, especially when the trachea is pulled caudally, and, therefore, prevents also an overextension of the Art. cerato-basihyalis (pp. 101 f.).

5. Fascia facialis.—This fascia covers the lateral and ventral sides of the head and consists of at least two layers. All the dermal muscles, e.g., M. dermotemporalis, and a part of the M. tracheohyoideus are embedded in it. Laterally, it attaches to the connective tissue covering the surface of the superficial jaw muscles, and apico-ventrally, it attaches to the caudal borders of the M. mylohyoideus anterior and of the Glandula sublingualis. It provides the lingual apparatus with ventral support and prevents apical slippage of the M. mylohyoideus anterior during a combined protraction-elevation of the hyoid (pp. 103 f.).

6. Fascia vaginalis hyoidei.—This fascia forms a sleeve-like sheath of connective tissue in which the hyoid horns glide back and forth during movements of the hyoid. The sheath lies along the medial surface of the pterygoid musculature covering the medial side of the mandibular ramus, starting at about the posterior one-third of the mandibular ramus, and passes caudally between the neck musculature and the caudo-ventral corner of the mandibular ramus to curve dorsally along the caudal border of the M. depressor mandibulae to the insertion of the neck musculature on the skull (Figs. 15A, B). The Fascia vaginalis hyoidei is attached to the surrounding tissue and especially to the surface of the neck musculature through connective tissue.

The fascia guides the hyoid horns into the right position during the retraction of the lingual apparatus. This function is necessary because the retractors of the lingual apparatus consist only of muscles that push the hyoid horns backwards (M. serpihyoideus, M. stylohyoideus) and not of muscles that pull the hyoid horns into the correct position (pp. 91 ff.).

The Fascia vaginalis hyoidei was described, but not named, in woodpeckers by Leiber (1907).

Discussion

The ligaments and fasciae of *Psittacus* are very different from those described for *Anas* and *Columba* (Zweers 1974, 1982; Zweers et al. 1977) and cannot be homologized at this time with them. Given the strictly functional context in which ligaments and fasciae are found and the well-known easy remodelling of connective tissue, great differences in the syndesmology among the lingual systems of different species are to be expected. This explains at least partially why homologies have traditionally been difficult to establish for ligaments and fasciae in comparative anatomical work. Noncontinuous variations are found even among individuals of the same species (e.g., Lig. interparahyale of *Psittacus*) and within the psittacine group (e.g., the Lig. nodulo-ceratobranchiale is missing in *Agapornis*; pers. obs.).

Functional analysis of the articular ligaments has shown that they are located at strategic points of the hyoid skeleton so that they can withstand muscular forces acting on particular articulations and, thus, limit or guide the movements initiated by muscular contractions. The evolutionary origin of these ligaments lies probably in a rearrangement of connective tissue or differentiation of articular capsules to meet the exigencies of the biomechanical system.

The intraosseal ligaments have various functions and, thus, probably also different evolutionary origins. The functions include: withstanding tensile stresses (e.g., Lig. parahyale, Lig. interparaglossale apicale, Lig. interparaglossale caudale), serving as a resilient border (e.g., Lig. interparaglossale apicale and Lig. interparaglossale caudale), connecting two different tissue masses (e.g., Lig. paraglossale apicale), or serving as a site for muscle attachment (e.g., Lig. interparahyale). Whereas most of these intraosseal ligaments probably evolved as ligamentous extensions of bone (e.g., Lig. parahyale), cartilage (e.g., Lig. interparaglossale apicale), or connective tissue (e.g., Lig. paraglossale apicale, Lig. interparaglossale caudale), the Lig. interparahyale probably was derived from a tendinous insertion of a muscle.

The only linkage ligament of the lingual apparatus, the Lig. nodulo-ceratobranchiale, was probably also derived from a tendon which may have served to increase the insertion surface for the M. serpihyoideus and, at the same time, may have coupled the movements of the ceratobranchialia and basihyale.

All the fasciae have more or less the same function, namely to limit the dislocation or movement of certain structures. They probably arose through reorganization of loose connective tissue within the system.

Musculature

Introduction

For descriptive purposes, the 13 paired and unpaired muscles of the psittacine lingual apparatus have been divided into two functional groups, namely into the extrinsic and the intrinsic lingual muscles (Table 1). The extrinsic lingual muscles originate from the lower jaw, whereas the intrinsic lingual muscles move the elements of the hyoid skeleton with respect to one another. This division of the lingual muscles remains consistent within the avian class and is especially useful for functional antomical studies (see also Nitzsch in Giebel 1862; Zweers 1974, 1982; Zweers et al. 1977) in contrast to a classification based on ontogenetic development (e.g., Kallius 1906; Edgeworth 1935; Engels 1938; Richards and Bock 1973). The muscles connecting the larynx to the hyoid apparatus and trachea have usually been described as lingual muscles (e.g., Zweers 1982). However, in parrots, at least, they are actually extrinsic laryngeal muscles (Table 1; Homberger 1979b).

The muscle nomenclature follows that introduced by Bock (1972). This nomenclature is based on names that are most commonly used in the literature. I have not adopted the recently advocated nomenclature based on the attachment site of the muscles (McLelland 1968; Vanden Berge 1975; Zweers 1982) because

TABLE 1

CLASSIFICATION OF THE LINGUAL AND LARYNGEAL MUSCLES[a]

Homberger (1979b, present study)		Zweers (1981, 1982)
		External hyoid muscles:
		M. serpihyoideus
Extrinsic lingual muscles:		M. mylohyoideus
M. serpihyoideus		M. genioglossus
M. mylohyoideus		
M. genioglossus		
M. stylohyoideus		Extrinsic hyoid muscles:
M. branchiomandibularis		M. stylohyoideus
M. tracheohyoideus		M. branchiomandibularis
		M. tracheohyoideus
		M. ceratohyoideus
Intrinsic lingual muscles:		Intrinsic hyoid muscles:
M. ceratohyoideus		M. ceratoglossus
M. ceratoglossus		M. hypoglossus obliquus
M. hypoglossus obliquus		M. hypoglossus anterior
M. hypoglossus anterior		
M. hypoglossus transversus	}	
M. supraglossus	} only in parrots	
M. mesoglossus	}	
Extrinsic laryngeal muscles:		Extrinsic hyoid muscles:
M. cricohyoideus		M. cricohyoideus
M. tracheolateralis		M. tracheolateralis
Intrinsic laryngeal muscles:		Laryngeal muscles:
M. dilator laryngis		M. dilator laryngis
M. constrictor laryngis		M. constrictor laryngis

[a] All specific muscle names follow Homberger.

the muscle attachments can be quite variable among different individuals and avian species. The synonymies for the avian lingual muscles have been listed for each muscle in separate tables in Appendix II. Additional synonymies can be found in Kutorga (1832), Shufeldt (1890), Edgeworth (1935), Fisher and Goodman (1955), George and Berger (1966), Vanden Berge (1979), and Zweers (1982).

Innervation of the lingual muscles has not been traced for several reasons. First, nerves perform no mechanical functions. Secondly, innervation of avian lingual muscles has already been described (e.g., Gadow 1891; Edgeworth 1935; Baumel 1975). Thirdly, the idea that a muscle keeps its original innervation through any evolutionary change (e.g., Fürbringer 1888; Barnikol 1951; Starck 1959) has repeatedly been challenged (e.g., Edgeworth 1935; Straus 1946) and appears, therefore, to be of uncertain value for the establishment of muscle homologies. Furthermore, it was possible to establish homologies between the lingual muscles of *Psittacus* and those of other bird species on morphological grounds (see also Richards and Bock 1973:90; Bock 1974:229 ff.).

Descriptions of the psittacine lingual muscles are organized as follows:

(a) Diagnosis and differentiation: a brief overview of the organization and subdivision of the muscle to allow for quick identification and homologization with muscles of other avian species.

(b) Origin and insertion: the origin is determined as the muscle end attached to the most stable skeletal element (in the broad sense) and the insertion as the muscle end attached to the more mobile skeletal element.

(c) Description: information on muscle architecture and on the spatial arrangement of muscles and their relationships with other structures.

(d) Measurements: measurements of muscles are included here as a basis for future comparative work.

(e) Articulations: those affected by the contraction of the particular muscle are listed; this information is crucial for determining whether the muscle is a one-joint or multi-joint muscle and for the analysis of the muscle function.

(f) Function: a brief statement of the function of a muscle in terms allowing easy comparison with functional interpretations of other workers, with references to the sections of the mechanical analysis in which the particular muscle is treated.

(g) Antagonists: muscles that counteract some aspect of the function of the treated muscle; antagonistic muscles are listed separately for each affected articulation.

(h) Synergists: muscles that support some aspect of the function of the treated muscle; synergistic actions are stated separately for each affected articulation.

(i) Discussion: a comparison of my results with those of earlier publications on the psittacine lingual apparatus. (The pre-1832 publications cited by Kutorga 1832, Duvernoy 1835, and Shufeldt 1890 were not available before the completion of the present study.) Functional interpretations by various authors were compared for each muscle, but a detailed discussion and comparison of psittacine muscles with non-psittacine muscles is not included.

EXTRINSIC LINGUAL MUSCULATURE

***M. serpihyoideus* (M. sh., Msh, Fig. 17B).—**

Diagnosis and differentiation: Originates from the caudo-ventral end of the mandibular ramus and inserts on the anterior end of the ceratobranchiale, on the Lig. nodulo-ceratobranchiale, on the Nodulus, and on a midventral raphe. Not subdivided.

Origin: Mandibula. Tendinous. The M. sh. originates along the caudal margin of the Ramus mandibularis from the ventral half of the lateral side of the ramus around the ventral edge to the Crista articularis on the medial surface of the ramus (Fig. 11A). On its lateral side, the origin of the M. sh. merges with the origin of the M. stylohyoideus (Figs. 15A, B).

Insertion: Lig. nodulo-ceratobranchiale, Fascia urohyalis, including the Nodulus, and ceratobranchiale. Fleshy. The M. sh. inserts medially along a midsagittal raphe that is part of the connective tissue attaching on the Nodulus and enveloping the urohyale (Fascia urohyalis) and that serves also as the insertion site for the M. mylohoideus posterior and M. ceratohyoideus. The insertion on the midsagittal raphe extends to the caudal one-fifth of the urohyale (Figs. 15B, 16A, 17A, B). Occasionally, a few caudal muscle fibers attach on the connective tissue enveloping the trachea and the M. tracheolateralis (e.g., AMNH 7203 and Z4). Apically, muscle fibers proceed to insert along the entire length of the ventral surface of the Lig. nodulo-ceratobranchiale (Figs. 17A, B). Laterally, fibers insert directly on the ceratobranchiale. This insertion may cover a relatively large area laterally and apically to the Tuberositas ceratobranchialis (e.g., Z4; Fig. 13B) or even caudally to the Tuberositas (e.g., Z2). Part of this insertion on the ceratobranchiale is ventrally subtended by a superficial slip of the Lig. nodulo-ceratobranchiale which serves as origin for the M. ceratoglossus profundus (Fig. 18). Sometimes, such as in Z3, only the deep fibers of the M. sh. insert directly on the ceratobranchiale and their insertions are concentrated at the caudal border of the Tuberositas ceratobranchialis. These deep fibers are ventrally subtended by the Lig. nodulo-ceratobranchiale. The superficial fibers cross the Lig. nodulo-ceratobranchiale ventrally and merge rostrally with some fibers of the M. ceratoglossus profundus (Fig. 16A).

In rare cases, continuity of the insertion along the Lig. nodulo-ceratobranchiale is interrupted by a gap (e.g., Z4 left side; Fig. 17A). In other cases (e.g., Z2 right side), fibers inserting on the central portion of the Lig. nodulo-ceratobranchiale almost merge with the originating muscle fibers of the M. ceratoglossus superficialis ligamenti superficialis (Fig. 17B).

Description: Parallel-fibered; forms a fan-shaped sheet in a bipolar fashion. Starting from the origin, the muscle fibers converge slightly towards the insertion of the muscle (Figs. 15B, 16A). The medial border of the M. sh. is slightly thickened. The lateral edge merges with the M. stylohyoideus at the level between midlength of the M. sh. and the tip of the urohyale (Figs. 15B, 16A). Generally, the muscle portion inserting on the central part of the Lig. nodulo-ceratobranchiale is thinner than the rest of the muscle. In Z2, this part is especially thin (Fig. 17B). Sometimes, it is missing altogether, thus, giving the impression that the M. sh. inserts with two separate heads (e.g., Z4 left; Fig. 17A; see Discussion).

The main variations of the M. sh. are found in the extent of the insertion on the ceratobranchiale, on the Lig. nodulo-ceratobranchiale and, to a lesser degree, on the Fascia urohyalis.

Measurements: The muscle fiber length in one specimen ranged between 27–29 mm, the muscle width at the origin (including the origin of the M. stylohyoideus) varied between 2–5 mm ($N = 7$; $\bar{X} = 3.5$ mm; s.d. = 0.96 mm), and the width at the insertion varied between 9–14 mm ($N = 7$; $\bar{X} = 11$ mm; s.d. = 1.82 mm). The thickness of the muscle at midlength was between 0.5–1.0 mm.

Articulations: As an extrinsic muscle, the M. sh. does not pass directly over an articulation but moves the hyoid apparatus with respect to the mandible. It acts indirectly on the Art. cerato-basihyalis.

Functions: The M. sh. retracts the hyoid and, to a certain degree, also lowers it. Indirectly, with the help of the M. stylohyoideus, it simultaneously flexes the basihyale (for details, see pp. 99 ff.). It also pulls the Nodulus back.

Antagonists: For retraction and lowering of hyoid: M. branchiomandibularis, M. mylohyoideus; for flexion of basihyale: M. ceratohyoideus, M. ceratoglossus medialis; for pulling the Nodulus back: M. stylohyoideus, M. tracheohyoideus, M. mylohyoideus posterior, M. ceratoglossus superficialis.

Synergists: For retraction and lowering of hyoid: M. stylohyoideus, M. tracheohyoideus; for flexion of the basihyale: M. stylohyoideus, M. tracheohyoideus, M. ceratoglossus lateralis, M. ceratoglossus superficialis, M. hypoglossus obliquus lateralis, M. supraglossus ceratobranchialis; for pulling the Nodulus back: M. ceratohyoideus.

Discussion: All authors agree, with varying degrees of accuracy, that the psittacine M. sh. originates on the posterior end of the mandibular ramus, but considerable disagreement exists regarding its insertion and function. Kutorga (1832),

Nitzsch (in Giebel 1862), Gadow (1891), and Denker (1907) described the insertion on the urohyale. Kallius (1906) described it on the Nodulus in *Melopsittacus*. Dubale and Rawal (1965) stated that the M. sh. in *Psittacula* inserts on the Art. cerato-basihyalis and on the sheath investing the urohyale. Rawal (1970) maintained that the M. sh. inserts on the basihyale up to the paraglossale. Lubosch (1933:606) stated that the M. sh. is missing in parrots. It is evident from his text and figures, however, that he misidentified the M. sh. as the M. stylohyoideus, although he described the origin on the lower jaw and the insertion on the connective tissue of the floor of the mouth. Duvernoy (1835), Kasai (1957), and Burton (1974c) did not describe the M. sh. but showed it in their figures, whereas Chaine (1905) and Steinbacher (1951) did not mention this muscle at all.

The description of Mudge (1903:253, and table facing p. 264) is the most detailed and accurate one and agrees with my observations. Mudge also described the division of the muscle into an outer and inner portion through retrogression of the central muscle fibers at the insertion of the muscle. He listed the various degrees of subdivision of the M. sh. found in parrots and mentioned that the M. sh. is clearly divided in *Psittacus,* although he does not use this character for classificatory purposes. My own data for *Psittacus* suggest that these various degrees of subdivision of the M. sh. represent individual, rather than interspecific, variations.

Various authors also disagree about the function of the M. sh. (Table 2). The functional misinterpretations of the earlier workers were probably due to a lack of proper functional analytical methods and also to the misdetermination of the insertion site of the M. sh.

Synonymies of the avian M. serpihyoideus are listed in Appendix II B.

M. stylohyoideus (M. st., Mst; Figs. 15B, 24A) —

Diagnosis and differentiation: Originates from the caudo-lateral border of the mandibular ramus and inserts on the basihyale and its parahyal processes. Not subdivided.

Origin: Mandibula. Tendinous. The M. st. originates along the caudal border on the lateral surface of the mandibular ramus. The ventral border of the origin merges with the origin of the M. serpihyoideus (Figs. 11A, 15A, B).

Insertion: Basihyale. Fleshy. The M. st. inserts on the dorsal side and along the lateral border of the basihyal body and parahyal process. The insertion extends from the tip of the parahyal process just behind the origin of the M. cricohyoideus ventralis to the Art. cerato-basihyalis. The closeness of the insertion to the Art. cerato-basihyalis varies individually. Medially the insertion of the M. st. is flanked by the insertion of the M. tracheohyoideus parahyalis (Figs. 12A, B, 25A).

Description: Parallel-fibered, ribbon-like. From the insertion, the muscle rises vertically between the insertion of the M. tracheohyoideus parahyalis and the belly of the M. hypoglossus obliquus lateralis (Figs. 24A, 32A). It then turns laterally around the ceratobranchiale and the surrounding muscles, e.g., the M. hypoglossus obliquus lateralis and M. ceratoglossus lateralis, and continues towards the ventral side of the lingual apparatus by passing medially to the Glandula mandibularis and the M. branchiomandibularis anterior (Figs. 16B, 20, 24A, B). Here, the M. st. joins the lateral border of the M. serpihyoideus with which it merges towards their origins on the mandible (Figs. 15B, 16A). The fibers of the M. st. do not remain concentrated along the lateral portion of the combined muscles, but spread to the dorsal side of the M. serpihyoideus.

TABLE 2

COMPARISON OF FUNCTIONAL INTERPRETATIONS OF THE PSITTACINE M. SERPIHYOIDEUS

Author	Function
Kutorga (1832), Duvernoy (1835)	retracts the tongue
Nitzsch (in Giebel 1862)	raises the tongue apparatus
Gadow (1891)	retracts and raises the tongue and larynx
Dubale and Rawal (1965), Rawal (1970)	lowers the tongue
Homberger, this study	retracts and lowers the hyoid

Measurements: The muscle fiber length varied between 36–39 mm (N = 3; $\bar{X}$ = 37.6 mm; s.d. = 1.53 mm), and the muscle width at the insertion varied between 2–4 mm (N = 3; $\bar{X}$ = 3 mm; s.d. = 1 mm). The thickness of the muscle at mid-length was <1 mm. For the muscle width at the origin, see the measurements for the M. serpihyoideus.

Articulations: As an extrinsic muscle, the M. st. moves the hyoid mainly with respect to the mandible. Therefore, it does not act directly on an articulation for this movement, but it has some effect on the Art. cerato-basihyalis because it passes dorso-laterally over it (pp. 99 ff.).

Functions: The M. st. retracts and lowers the hyoid and, at the same time, flexes the basihyale.

Antagonists: For retraction and lowering of hyoid: M. branchiomandibularis, M. mylohyoideus; for flexion of basihyale: M. ceratohyoideus, M. ceratoglossus medialis.

Synergists: For retraction and lowering of hyoid: M. serpihyoideus, M. tracheohyoideus; for flexion of basihyale: M. serpihyoideus, M. tracheohyoideus, M. ceratoglossus lateralis, M. ceratoglossus superficialis, M. hypoglossus obliquus lateralis, M. supraglossus ceratobranchialis.

Discussion: Most authors agree that the psittacine M. st. originates from the posterior extremity of the mandibular ramus (Nitzsch in Giebel 1862; Mudge 1903; Kallius 1906; Denker 1907; Dubale and Rawal 1965; Rawal 1970) and inserts on the parahyal process (Nitzsch in Giebel 1862; Denker 1907; Dubale and Rawal 1965; Rawal 1970; Burton 1974c). Mudge (1903) observed a tendency towards reduction of the M. st. among parrots, of which the Loriinae (*Lorius, Vini, Eos*) have lost this muscle. Mudge (1903) also mentioned a trend towards increasing separation between the M. st. and M. serpihyoideus among parrots. He found this to be the least pronounced in *Psittacus* and some other, not closely related species (sensu Homberger 1980a) in which the two muscles are confluent for the caudal half of the M. st. My observations confirm this for *Psittacus*. Lubosch (1933) stated that the M. st. is missing in parrots (p. 29). In some publications, the description of the M. st. consists only of a figure (Duvernoy 1835; Gadow 1891), whereas Kutorga (1832), Chaine (1905), and Steinbacher (1951) did not mention this muscle. I could not read Kasai's (1957) Japanese description.

Functional interpretations of the M. st. conflict among earlier publications (Table 3).

Synonymies of the avian M. st. are listed in Appendix II C.

TABLE 3

COMPARISON OF FUNCTIONAL INTERPRETATIONS OF THE PSITTACINE M. STYLOHYOIDEUS

Author	Function
Duvernoy (1835)	retracts the hyoid
Nitzsch (in Giebel 1862)	raises the hyoid
Gadow (1891)	retracts and raises the hyoid and larynx
Dubale and Rawal (1965), Rawal (1970)	lowers the hyoid
Homberger, this study	retracts and lowers the hyoid, flexes the basihyale

M. branchiomandibularis (M. bm., Mbm; Figs. 15B, 23A).—

Diagnosis and differentiation: Originates from the anterior end of the mandibular ramus and inserts on the epibranchiale and caudal end of the ceratobranchiale. Two main portions can be distinguished: M. branchiomandibularis posterior and M. branchiomandibularis anterior. The latter is subdivided into a medial (M. bm. anterior medialis) and lateral portion (M. bm. anterior lateralis).

A. M. branchiomandibularis posterior (M. bm. post., Mbmp).—*Diagnosis:* Originates from the ventro-lateral surface of the mandibular ramus and inserts on the epibranchiale. Not subdivided.

Origin: Mandible and jaw musculature. Fleshy, partially tendinous. The M. bm. post. originates with muscular fibers from the lateral surface of the mandibular ramus close to the border of the rhamphotheca and tendinously from the surface of the M. adductor mandibularis externus ventralis extending over the ventral one-quarter of the mandibular ramus (e.g., Z3; Figs. 11A, 15A, D). Sometimes, the M. bm. post. has an additional fleshy origin on the ventral edge of the mandible (e.g., Z4; Fig. 11A).

Insertion: Epibranchiale. Fleshy. The M. bm. post. inserts on the dorsal crest of the dorso-medial edge of the posterior half of the epibranchiale (Figs. 12A, B, 13A, 33B). The insertion borders dorsally and medio-ventrally to the insertion of the M. bm. ant. lat. (Figs. 12A, B, 13A, 33A, B).

Description: Parallel-fibered, for the most part ribbon-like (Figs. 21A, B). From its insertion, the M. bm. post. runs forward enveloping the dorsal half of the epibranchiale and Art. epi-ceratobranchialis (Figs. 21B, 33A). Apically from this articulation, the M. bm. post. turns laterally in such a fashion that its dorsal surface faces laterally (Fig. 21B). Farther ahead, the muscle turns around the ventral edge of the mandibular ramus so that its formerly lateral surface now faces the lateral surface of the mandibular ramus and the surface of the M. adductor mandibularis externus ventralis (Figs. 15A, B, 21B). The variability of the M. bm. post. is minimal and involves mainly the extent of the origin and the presence or absence of the partial origin on the ventral edge of the mandible.

Measurements: For three specimens, the muscle fiber length varied between 24–27 mm ($\bar{X}$ = 25.3 mm; s.d. = 1.5 mm) and the muscle width at the origin varied between 3–5 mm ($\bar{X}$ = 4 mm; s.d. = 1 mm).

Articulations: As an extrinsic muscle, the M. bm. post. mainly moves the hyoid with respect to the mandible. It does not act on an articulation for this movement but has some effect on the Art. epi-ceratobranchialis (pp. 113 ff.).

Functions: The M. bm. post. protracts the hyoid along the Ramus mandibularis and the pterygoid muscles of the palate without raising it and tends to rotate the epibranchiale dorsally. Together with the M. bm. ant. lat., it stabilizes the Art. epi-ceratobranchialis. Indirectly, with the help of the M. ceratohyoideus, it extends the basihyale.

Antagonists: For protraction of hyoid: M. serpihyoideus, M. stylohyoideus, M. tracheohyoideus; for dorsal rotation of epibranchiale: M. bm. ant. lat.; for extension of basihyale: M. stylohyoideus, M. tracheohyoideus, M. ceratoglossus lateralis, M. ceratoglossus superficialis, M. hypoglossus obliquus lateralis, M. supraglossus ceratobranchialis.

Synergists: For protraction of hyoid: M. bm. ant., M. mylohyoideus; for dorsal rotation of epibranchiale: none; for extension of basihyale: M. ceratohyoideus, M. ceratoglossus medialis.

B. M. branchiomandibularis anterior (M. bm. ant., Mbma).—*Diagnosis and differentiation:* Originates on the antero-dorsal part of the medial surface of the mandibular ramus. Two portions can be distinguished: a lateral portion, M. bm. ant. lateralis, which inserts mainly on the epibranchiale, and a medial portion, M. bm. ant. medialis, which inserts on the ceratobranchiale.

Origin: Common for both the lateral and medial portions. Mandible. Fleshy. The M. bm. ant. originates from a large area on the anterior part of the medial surface of the Ramus mandibularis (Figs. 11A, B). The origin borders dorsally on the origin of the M. mylohyoideus and caudally on the insertion of the M. ethmomandibularis (Figs. 11A, B). Dorsally, some fibers originate tendinously from the cushion of connective tissue forming the angle of the mouth.

Insertion: 1. M. branchiomandibularis anterior lateralis (M. bm. ant. lat., Mbmal): Epibranchiale and capsule of Art. epi-ceratobranchialis. Fleshy. At the insertion, the muscle wraps itself around the apical half of the epibranchiale (Figs. 12A, C, 13A, 33C) so that its lateral borders meet on the dorso-medial side of the epibranchiale (Fig. 33B). Ventrally, some fibers insert on the capsule of the Art. epi-ceratobranchialis. Here, the insertion is often not clearly separated from the insertion of the M. bm. ant. med. (Figs. 12A, 33B).

2. M. branchiomandibularis anterior medialis (M. bm. ant. med., Mbmam): Ceratobranchiale. Fleshy. The M. bm. ant. med. inserts mainly on the caudal end of the dorsal side of the ceratobranchiale along the Crista branchiomandibularis and close to the Art. epi-ceratobranchialis (Figs. 12A, C). Occasionally, the insertion completely surrounds the caudal end of the ceratobranchiale (e.g., Z3 right side; Fig. 13A). In such a case, the ventral part of the insertion of the M. bm. ant. med. cannot be clearly separated from the insertion of the M. bm. ant. lat. if the latter spreads over the articular capsule.

Description: 1. M. branchiomandibularis anterior lateralis (M. bm. ant. lat.): Parallel-fibered, ribbon-like. Starting from its insertion on the epibranchiale, the M. bm. ant. lat. runs forward and envelops the ventral side of the epibranchiale and Art. epi-ceratobranchialis (Figs. 33A, B). The fibers inserting latero-dorsally pass over the insertion of the M. bm. ant. med. (Fig. 33B). Anterior to the articulation, the M. bm. ant. lat. turns laterally in such a way that its former ventral surface now faces laterally (Figs. 23A, B, 33A) and runs straight forward to its origin (Fig. 15B).

2. M. branchiomandibularis anterior medialis (M. bm. ant. med.): Parallel-

fibered, ribbon-like. Starting from its insertion on the ceratobranchiale, the M. bm. ant. med. envelops the dorsal side of the ceratobranchiale up to the point where the M. trachyeohyoideus crosses the ceratobranchiale (Fig. 24A). It then turns towards the lateral side of the ceratobranchiale and flattens laterally so that the former midsagittal fibers of the dorsal side of the muscle form the dorsal border of the now vertically oriented ribbon-like muscle. The M. bm. ant. med. adheres to the medial surface of the M. bm. ant. lat., with which it may exchange fibers, and follows it to their common origin (Figs. 15B, 24A, 33A).

Variations for both portions of the M. bm. ant. involve the extent of their insertions and the number of muscle fibers they exchange.

Measurements: For the M. bm. ant. lat. the muscle fiber length varied between 29–37 mm ($N = 3$; $\bar{X} = 32$ mm; s.d. = 3.04 mm), the muscle width at the origin varied between 3–4 mm ($N = 2$; $\bar{X} = 3.5$ mm), and the muscle width at the insertion varied between 3–6 mm ($N = 3$; $\bar{X} = 4.7$ mm; s.d. = 1.5 mm). The muscle thickness at midlength was between 0.5–1.0 mm for three specimens.

For the M. bm. ant. med. the muscle fiber length varied between 26–32 mm ($N = 3$; $\bar{X} = 27.8$ mm; s.d. = 2.35 mm), and the muscle width at the insertion varied between 3–9 mm ($N = 3$; $\bar{X} = 6.3$ mm; s.d. = 3.06 mm). The muscle thickness at midlength of one specimen was about 1 mm.

Articulations: The M. bm. ant. moves the hyoid mainly with respect to the mandible and, therefore, does not act on an articulation for this movement. However, the M. bm. ant. lat. affects the Art. epi-ceratobranchialis (pp. 113 f.).

Functions: The M. bm. ant. protracts the hyoid along the Ramus mandibularis and the pterygoid musculature of the palate. The M. bm. ant. lat. tends to rotate the epibranchiale ventrally, but together with the M. bm. post., it stabilizes the Art. epi-ceratobranchialis. Indirectly, with the help of the M. ceratohyoideus, it extends the basihyale.

Antagonists: For protraction of hyoid: M. serpihyoideus, M. stylohyoideus, M. tracheohyoideus; for flexion of epibranchiale: M. bm. post.; for extension of basihyale: M. stylohyoideus, M. tracheohyoideus, M. ceratoglossus lateralis, M. supraglossus ceratobranchialis.

Synergists: For protraction of hyoid: M. bm. post., M. mylohyoideus; for extension of epibranchiale: none; for extension of basihyale: M. ceratohyoideus, M. ceratoglossus medialis.

Discussion: The M. bm. of parrots is more complex than that of other avian orders. It consists of three parts instead of two because the M. bm. ant. is divided into a medial and a lateral part. With the exception of Mudge (1903), however, most workers reported that the psittacine M. bm. consists of two parts (Duvernoy 1835; Nitzsch in Giebel 1862; Gadow 1891; Lubosch 1933; Dubale and Rawal 1965) or only of one (Kutorga 1832 and Kallius 1906, both of whom described only the M. bm. post., and Denker 1907, who described only the M. bm. ant. med.). Mudge (1903) recognized that the psittacine M. bm. consists of three parts, but listed some species (including *Psittacus*) in which it consists of only one or two parts. However, the clarity with which the M. bm. is divided into three parts varies individually, and the taxonomic value of this character must, therefore, be questioned.

The different origins of the two parts of the psittacine M. bm. have been described by Nitzsch (in Giebel 1862), Gadow (1891), Mudge (1903), and Dubale

TABLE 4

COMPARISON OF FUNCTIONAL INTERPRETATIONS OF THE PSITTACINE M. BRANCHIOMANDIBULARIS

Author	Function
Kutorga (1832), Duvernoy (1835), Nitzsch (in Giebel 1862), Gadow (1891)	protracts the tongue
Dubale and Rawal (1965), Rawal (1970)	raises the tongue
Homberger, this study	protracts the hyoid and, indirectly, extends the basihyale

and Rawal (1965), but only Mudge (1903) described the M. bm. post. as also originating from the lateral surface of mandible. He interpreted the two different origins of the M. bm. post. as an interspecific variation. My study shows, however, that the origin from the lateral surface of the mandible is always present in *Psittacus* but considerable individual variation exists with respect to the presence of the additional slip of the M. bm. post. originating from the ventral margin of the mandible.

With respect to the insertion of the various parts of the M. bm., Nitzsch (in Giebel 1862) and Gadow (1891) differentiated between the insertion of the M. bm. post. on the epibranchiale and the insertion of the M. bm. ant. on both the epibranchiale and ceratobranchiale. Only Mudge (1903) distinguished the two parts of the M. bm. ant. on the basis of their different insertions, namely the M. bm. ant. lat. inserting on the ventral side of the epibranchiale and the M. bm. ant. med. inserting on the posterior end of the ceratobranchiale. As was shown in my study of *Psittacus,* these two insertions can occasionally become confluent so that the two parts are not clearly separated.

Burton (1974c) did not describe the M. bm., but the muscle can be identified from his figures, as it can be from the figures in Kasai (1957). The M. bm. was not mentioned by Chaine (1905) or Steinbacher (1951).

Little disagreement in the functional interpretations of the M. bm. was found by previous workers (Table 4).

Synonymies of the avian M. bm. are listed in Appendix II D.

***M. mylohyoideus (M. mh., Mmh; Fig. 16A).*—**

Diagnosis and differentiation: Originates from the medial surface of the dorso-anterior corner of the Ramus mandibularis and inserts on the Glandula sublingualis or on the Nodulus. Two portions can be distinguished: M. mylohyoideus anterior and M. mylohyoideus posterior.

A. M. mylohyoideus anterior (M. mh. ant., Mmha).—*Diagnosis:* Originates from the medial surface of the dorso-anterior corner of the Ramus mandibularis and inserts on the sublingual salivary gland (Gl. sublingualis).

Origin: Mandibula. Fleshy. The M. mh. ant. originates on the medial surface of the anterior end of the mandibular ramus and here along the crest that lies parallel to and just below the dorsal margin of the ramus (Figs. 11A, B). Some fibers extend their attachment onto the cushion of connective tissue that forms the angle of the mouth. Ventrally, the origin is framed by the origin of the M. branchiomandibularis anterior. Caudally, the origin of the M. mh. ant. extends to the origin of the M. mh. post. from which it is usually incompletely separated (Figs. 11A, B).

Insertion: Glandula sublingualis. Fleshy. The M. mh. ant. inserts on a midsagittal raphe that is firmly attached to the ventral surface of the Glandula sublingualis along its entire length (Fig. 16A). Some muscle fibers attach along the caudal border of the gland.

Description: Parallel-fibered, slightly fan-shaped. Starting from the origin, the muscle fibers fan out slightly and form a sheet covering the lateral surface and the ventral edge of the anterior portion of the mandibular salivary gland (Gl. mandibularis) (Fig. 20) and the entire ventral surface of the Gl. sublingualis (Fig. 16A). The apical border of the M. mh. ant. abuts against the caudal surface of the M. genioglossus (Figs. 15A, 20). In the center of the ventral side of the tongue, the M. mh. ant. is separated from the M. mh. post. by a gap of about 3 mm (Figs. 15B, 16A). Practically no variations were observed in the M. mh. ant.

Measurements: The muscle fiber length varied between 10–21 mm ($N = 3$; $\bar{X} = 14.6$ mm; s.d. = 4.5 mm), the muscle width at the origin varied between 2.5–3.0 mm ($N = 3$; $\bar{X} = 2.8$ mm; s.d. = 0.3 mm), and the muscle width at the insertion varied between 5–7 mm ($N = 3$; $\bar{X} = 5.7$ mm; s.d. = 1.2 mm). The thickness of the muscle at midlength was between 0.5–1.0 mm in two specimens.

Articulations: The M. mh. ant. moves the hyoid mainly with respect to the mandible and, therefore, does not act directly on an articulation; it has an indirect effect on the Art. cerato-basihyalis (pp. 102 ff.).

Functions: Lifts and protracts the hyoid and indirectly supports the extension of the basihyale. Exerts pressure on the sublingual and mandibular salivary glands, thereby inducing the expulsion of saliva.

Antagonists: For lifting of hyoid: M. genioglossus, M. serpihyoideus, M. stylohyoideus, M. tracheohyoideus; for protraction of hyoid: M. serpihyoideus, M. stylohyoideus, M. tracheohyoideus.

Synergists: For lifting of hyoid: M. mylohyoideus post.; for protraction of hyoid: M. branchiomandibularis.

B. M. mylohyoideus posterior (M. mh. post., Mmhp).—*Diagnosis:* Originates from the medial surface of the dorso-anterior corner of the Ramus mandibularis and inserts on the Fascia urohyalis and the Nodulus.

Origin: Mandibula. Fleshy. The M. mh. post. originates on the medial surface of the anterior portion of the Ramus mandibularis, along a crest that lies parallel to and just below the dorsal margin of the ramus (Fig. 11B). The origin extends caudally from the caudal border of the origin of the M. mh. ant. to the insertion of the M. ethmomandibularis of the jaw musculature (Fig. 11B). Sometimes, the origin extends also forward along the ventral border of the M. mh. ant. (Fig. 11A). The ventral side of the origin of the M. mh. post. is bounded by the origin of the M. branchiomandibularis anterior (Figs. 11A, B).

Insertion: Fascia urohyalis and Nodulus. Fleshy. The M. mh. post. inserts along a midsagittal line on the ventral surface of the Fascia urohyalis and Nodulus (Figs. 15B, 16A). Towards caudal, the insertion often continues onto the Lig. nodulo-ceratobranchiale. The fibers of the collateral portions usually interdigitate at their insertions.

Description: Parallel-fibered. The M. mh. post. is ribbon-like with an indication of a subdivision into two or three bundles (Figs. 15B, 16A). Starting from its insertion, the M. mh. post. passes over the lateral side of the posterior half of the mandibular salivary gland (Gl. mandibularis) and latero-ventrally over the Art. cerato-basihyalis (Fig. 20). Near the origin, the muscle fibers converge and pass ventrally over the M. mh. ant. Variations are minimal, mostly concerning the muscle fiber arrangement at the insertion and origin.

Measurements: The muscle fiber length varied between 16–22 mm (N = 3; $\bar{X}$ = 18.3 mm; s.d. = 2.3 mm), the muscle width at the origin varied between 1–2 mm (N = 3; $\bar{X}$ = 1.7 mm; s.d. = 0.3 mm), and the muscle width at the insertion varied between 2–4 mm (N = 5; $\bar{X}$ = 2.8 mm; s.d. = 0.8 mm). Muscle thickness at midlength was less than 1 mm in two specimens.

Articulations: The M. mh. post. moves the hyoid mainly with respect to the mandible and, therefore, does not act directly on an articulation. It has, however, an indirect effect on the Art. cerato-basihyalis (pp. 103 f.).

Functions: Lifts and protracts the hyoid apparatus. Indirectly, it supports the extension of the basihyale. It also pulls the Nodulus forward. Simultaneously, it exerts pressure on the mandibular salivary gland, thereby inducing the expulsion of saliva.

Antagonists: For lifting of hyoid: M. genioglossus, M. serpihyoideus, M. stylohyoideus, M. tracheohyoideus; for protraction of hyoid: M. serpihyoideus, M. stylohyoideus, M. tracheohyoideus; for protraction of Nodulus: M. serpihyoideus, M. ceratohyoideus.

Synergists: For lifting of hyoid: M. mylohyoideus anterior; for protraction of hyoid: M. branchiomandibularis; for protraction of Nodulus: M. ceratoglossus superficialis.

Discussion: The psittacine M. mylohyoideus consists of two clearly distinct portions in contrast to the usually undivided M. mh. of most other birds. Anatomists generally agree on the bipartite nature of the psittacine M. mh. and on the location of its origin, but reports concerning the insertions and functions of the M. mh. conflict. The M. mh. ant. is generally described as simply merging with its collateral part and forming a sling under the ventral side of the tongue (Duvernoy 1835; Nitzsch in Giebel 1862; Gadow 1891; Denker 1907; Dubale and Rawal 1965; Rawal 1970). Only Mudge (1903) and Kallius (1906) mentioned the attachment of the M. mh. ant. to the sublingual salivary gland. The M. mh. post. has been described as inserting on the base of the urohyale (Duvernoy 1835; Nitzsch in Giebel 1862; Gadow 1891) or on the Fascia urohyalis (Dubale and Rawal 1965; Rawal 1970). Mudge (1903) presented the most detailed description and stated that the M. mh. post. inserts on the Nodulus or, if the Nodulus is absent, on the Fascia urohyalis or the cartilaginous process of the urohyale. Mudge (1903) also stated that the M. mh. post. varies from a state equally well developed as the M. mh. ant. (e.g., in *Psittacus*) to a state of total reduction (e.g., in *Probosciger, Nestor*). In addition, he reported that the gap between the two portions of the M. mh. varies from very small (e.g., in *Psittacus*) to very large in other species and that the caudal border of the M. mh. ant. is attached to the Fascia facialis.

Lubosch (1933) mentioned only that the M. mh. is present in parrots and that the M. serpihyoideus extends apically onto its dorsal side. The M. mh. can be identified only from a figure in Kasai (1957). It is not mentioned by Kutorga (1832), Chaine (1905), Steinbacher (1951), or Burton (1974c).

TABLE 5

COMPARISON OF FUNCTIONAL INTERPRETATIONS OF THE PSITTACINE M. MYLOHYOIDEUS

Author	Function
Duvernoy (1835)	M. mh. ant.: supports the floor of the mouth and lifts the hyoid
Gadow (1891)	pushes the tongue and larynx against the palate (together with M. serpihyoideus)
Dubale and Rawal (1965), Rawal (1970)	depresses the floor of the mouth
Homberger, this study	protracts and lifts the hyoid apparatus; supports the expulsion of saliva

The function of the M. mh. has generally been related to the floor of the mouth (Table 5). Parrots, however, do not possess a membraneous floor of the mouth like other birds (Homberger 1980a).

Synonymies of the M. mh. are summarized in Appendix II E.

M. genioglossus (M. gg., Mgg; Figs. 15A, 21B).—

Diagnosis and differentiation: Originates from the mandibular symphysis and inserts on the paraglossale. Not subdivided.

Origin: Mandible. Fleshy. M. gg. originates from a round depression on the dorsal surface of the caudal one-quarter of the mandibular symphysis (Figs. 11B, 15A). The collateral parts originate jointly but are clearly separated by a sheet of connective tissue. Rostrally, the origin borders the base of the rhamphotheca on the mandible; caudally, some fibers extend occasionally almost to the caudal edge of the mandibular symphysis.

Insertion: Paraglossale. Fleshy. M. gg. inserts along the Crista dorso-lateralis of the paraglossale (Figs. 39B–42A, 45B). The insertion extends from the tip of the Proc. posterior paraglossalis to the level of the caudal border of the lingual nail, i.e., to the level of the middle of the insertion of the M. ceratoglossus (Fig. 12C). Some superficial muscle fibers insert on the tendinous fibers of the portions B and C of the M. supraglossus (Fig. 21B).

Description: Parallel-fibered. For the most part, the muscle is strap-like (Figs. 16A, B) but the muscle fibers fan out slightly towards the insertion. The collateral parts arise from the origin with a circular cross-section (Fig. 15B) and flatten gradually into transversally oriented ribbons. They are firmly tied together by connective tissue. The two collateral parts separate when they reach the ventral surface of the body of the tongue, i.e., the apical border of the M. mylohyoideus anterior (Fig. 15A). There, each individual ribbon-like muscle rotates approximately 90° about its longitudinal axis in such a way that its formerly apically facing side is now facing laterally (Figs. 15A, 21B). The muscle fibers diverge slightly towards the insertion on the paraglossale (Figs. 15A, B, 21A, B). The free portion of M. gg., i.e., the portion that is not in immediate contact with the body of the tongue, is covered on the apical side by the mucosa of the underside of the tongue (Fig. 22) and on the caudal side by the bare skin of the chin region ("Kinnhöhle" of Homberger 1980a). The upper portion of M. gg., i.e., the portion that is in immediate contact with the lingual body, passes laterally over the inserting tendon of the M. ceratoglossus.

Variations observed in the M. gg. are minimal and involve mostly the extent of the origin and insertion.

Measurements: The muscle fiber length varied between 12–17 mm (N = 3; $\bar{X}$ = 13.8 mm; s.d. = 1.4 mm), the muscle width at the origin varied between 2–4 mm (N = 3; $\bar{X}$ = 2.7 mm; s.d. = 1.2 mm), and the muscle width at the insertion varied between 5.5–6 mm (N = 3; $\bar{X}$ = 5.7 mm; s.d. = 0.4 mm). In two specimens the muscle thickness was 1.5 mm at the origin and about 0.5–1.0 mm at the insertion.

Articulations: M. gg. moves the hyoid mainly with respect to the mandible and does not directly affect any articulation of the hyoid itself (pp. 104 f.).

Functions: Lowers and protracts the hyoid.

Antagonists: For lowering of hyoid: M. mylohyoideus; for protraction of hyoid: M. serpihyoideus, M. stylohyoideus, M. tracheohyoideus.

Synergists: For lowering the hyoid: none; for protraction of hyoid: M. branchiomandibularis.

Discussion: M. genioglossus is not always present in birds. If present, it usually originates from the caudal border of the mandibular symphysis and runs caudally under the floor of the mouth cavity to insert on the hyoid apparatus. In parrots, however, the M. gg. originates from the dorsal surface of the mandibular symphysis and runs dorsally to insert on the hyoid apparatus. In most birds, the floor of the mouth is long and membraneous and extends caudally from the mandibular symphysis between the mandibular rami (Homberger 1980a). In parrots, in contrast, the short floor of the mouth cavity is formed entirely by the rhamphotheca-covered part of the mandibular symphysis (Fig. 15A).

Most authors agreed, with varying accuracy, on the sites of origin and insertion, that the psittacine M. gg. originates from the dorsal surface of the posterior part of the mandibular symphysis and inserts on the lateral edge of the paraglossale (Kutorga 1832; Duvernoy 1835; Nitzsch in Giebel 1862; Gadow 1891; Mudge 1903; Kallius 1906; Denker 1907; Steinbacher 1951; Dubale and Rawal 1965). Lubosch (1933) erroneously stated that the M. genioglossus is missing in *Platycercus.* Burton (1974c) showed the M. gg. only in a figure, Kasai (1957) did not mention it, and Chaine (1905) showed it in a figure and stated only that it lies directly under the lingual epithelium. Mudge (1903) stated that the M. gg. is similar in all parrots, except *Nestor.* This observation concurs with my observation of little individual variation found in *Psittacus.*

In contrast to the anatomical descriptions, the functional interpretations of the M. gg. vary greatly (Table 6).

The synonymies of the avian M. gg. are found in Appendix II F.

M. tracheohyoideus (M. th., Mth; Figs. 16B, 24A).—

Diagnosis and differentiation: Inserts on the basihyale and, occasionally, on the ceratobranchiale, and originates from the sternum. Up to three different slips can be distinguished, depending on their insertion on the Processus parahyalis (M. th. parahyalis, Mthp), the basihyale (M. th. basihyalis, Mthb), or the ceratobranchiale (M. th. ceratobranchialis, Mthc).

Origin: Sternum. Fleshy. The main portion of the M. th. originates from the antero-ventral corner of the Carina sterni ("Apex carinae" of Vanden Berge 1979) and extends its origin to a varying degree onto the tendinous surface of the M. pectoralis. Occasionally, some muscle fibers originate from the tough connective tissue lying between the pectoral musculature and the skin. The origins of the

TABLE 6

COMPARISON OF FUNCTIONAL INTERPRETATIONS OF THE PSITTACINE M. GENIOGLOSSUS

Author	Function
Kutorga (1832)	lifts the tongue to the palate
Duvernoy (1835)	pulls the tongue down and retracts the protracted tongue
Nitzsch (in Giebel 1862), Gadow (1891)	protracts and lifts the tongue
Steinbacher (1951)	pulls the tongue ventrally and pulls back the protracted tongue
Dubale and Rawal (1965)	raises the tongue
Rawal (1970)	raises the tongue and makes the upper surface of the tongue concave
Homberger, this study	lowers the anterior portion of the lingual apparatus

collateral parts of the M. th. are separated by a raphe of connective tissue. The collateral parts, however, are not always symmetrical; i.e., the origin of one collateral part can extend to the contralateral side of the Apex carinae (e.g., LSUMZ 1287).

Insertion: Basihyale and ceratobranchiale. Fleshy. The main insertion of the M. th. (M. th. parahyalis), which is always present and is, in some specimens, even the only insertion of the M. th. (e.g., AMNH 7203), is located on the dorsal surface of the basihyale. It extends from the Proc. parahyalis, caudally to the origin of the M. cricohyoideus dorsalis profundus, to the border of the Art. cerato-basihyalis (Figs. 12A, B, 25A–26A, 32A). Occasionally, the insertion is shorter (e.g., Z1; Fig. 26B). The insertion of the M. th. parahyalis is flanked laterally by the insertion of the M. stylohyoideus and medially by the origin of the M. cricohyoideus ventralis (Figs. 12A, B, 24A, 25A). Usually, additional slips of the M. th. insert on the transitional region between the basihyale and urohyale or on the ceratobranchiale. A fairly large insertion (M. th. basihyalis) may be found on the dorsal surface of the caudal end of the basihyale close to the base of the urohyale (e.g., Z3; Figs. 12A, 24A, B). Occasionally (e.g., Z1 left side), this insertion extends to the dorsal surface of the urohyale and to the caudo-medial corner of the articular capsule of the Art. cerato-basihyalis (Fig. 26B), but the insertion of the M. th. basihyalis may also be restricted to the caudo-medial corner of the articular capsule (e.g., Z1 right side, Z6; Fig. 26B). The slip inserting on the ceratobranchiale (M. th. ceratobranchialis) is often missing. This insertion can vary from a distinct patch along the medial to caudo-medial margin of the lateral part of the articular facet of the Art. cerato-basihyalis (Figs. 12B, 26A, B) to a small spot on the caudo-lateral margin of that articulation (e.g., AMNH 7203; Fig. 12A).

Description: Parallel-fibered, ribbon-like. Starting from the main insertion along the lateral border of the basihyale and Proc. parahyalis, the anteriormost muscle fibers of the M. th. parahyalis rise as a vertical sheet between the M. stylohyoideus and M. cricohyoideus ventralis, whereas the muscle fibers inserting more caudally run immediately laterally (Figs. 24A–26A, 32A). The M. th. basihyalis, which inserts on the basihyal-urohyal region, is ribbon-like and joins the M. th. parahyalis at the level of the caudal end of the origin of the M. supraglossus ceratobranchialis (Figs. 24A, B, 26B). The M. th. ceratobranchialis inserts on the ceratobranchiale (if present) and joins the two other portions of the M. th. on their ventral surface (Figs. 25B–26B) after emerging from beneath the M. supraglossus ceratobranchialis (Figs. 25B, 26B). The portions of the M. th. described thus far lie directly under the laryngeal apparatus. After having merged to a single ribbon-like muscle, the M. th. emerges from beneath the laryngeal apparatus (Figs. 21A–22) and turns laterally around the ceratobranchiale and the surrounding caudal portions of the M. hypoglossus obliquus lateralis and M. ceratoglossus lateralis to the ventral side of the tongue apparatus (Figs. 20, 21B, 24A, B, 25B). (The M. stylohyoideus, which also turns around the ceratobranchiale, lies anteriorly and laterally to the M. th.) On the ventral side of the tongue apparatus, the ribbon-like M. th. crosses obliquely over the ventral surface of the M. ceratohyoideus and is covered ventrally by the M. serpihyoideus (Figs. 16A, B). When turning to the ventral side of the tongue, the muscle fibers of the M. th. are rearranged in such a way that the originally dorso-medial fibers of the ribbon-like muscle now face the ceratobranchiale and the ventral side of the tongue apparatus. At the same time, the fibers of the former apico-lateral border of the muscle form at first a crest on the dorsal surface of the muscle and then form the caudo-lateral border of the muscle on the ventral side of the tongue. When the M. th. has emerged from underneath the caudal border of the M. serpihyoideus, it descends along the neck, closely following the latero-ventral surface of the trachea. At the end of the uppermost one-fifth of the total distance between the origin of the M. th. and the caudal border of the M. serpihyoideus, the muscle fibers of the M. th. merge with those of the M. dermotemporalis of the neck. At the same time, the two collateral parts of the M. th. merge with each other and firmly attach to the skin of the neck. For the next one-fifth of the M. th. (down to where it meets the crop), the muscle fibers are interrupted by interspersed irregular transverse bands of connective tissue. (The more lateral muscle fibers coming from the M. dermotemporalis are not interrupted.) For the remaining three-fifths of the M. th., the muscle fibers are continuous down to the origin. In its lowest one-fifth the M. th. is separated from the skin. Here, it adheres loosely to the wall of the crop (Ingluvies) and is divided by a midsagittal raphe into two collateral, often unequal parts. The last one-tenth of the M. th. passes over the surface of the M. pectoralis from where it originates in addition to the sternum.

Measurements: In one specimen, the muscle fibers were about 92–100 mm long, the muscle width at the origin was 18 mm, the muscle width at the insertion 5 mm, and the muscle thickness varied between less than 0.5 mm and 1.0 mm.

Articulations: The M. th. moves the hyoid mainly relative to the mandible and, therefore, does not act directly on an articulation. It has, however, some effect on the Art. cerato-basihyalis because it passes over the dorso-lateral side of this joint.

Functions: Pulls down and retracts the hyoid. Simultaneously, it rotates the basihyale into a more vertical position.

Antagonists: For pulling down and retraction of hyoid: M. mylohyoideus; for retraction of hyoid: M. branchiomandibularis; for flexion of basihyale: M. ceratohyoideus, M. ceratoglossus medialis.

Synergists: For pulling down and retraction of hyoid: M. serpihyoideus, M. stylohyoideus; for flexion of basihyale: M. serpihyoideus, M. stylohyoideus, M.

ceratoglossus lateralis, M. ceratoglossus superficialis, M. hypoglossus obliquus lateralis, M. supraglossus ceratobranchialis.

Discussion: A great deal of confusion surrounds the M. tracheohyoideus. This is at least partly due to the high interspecific variability of the muscle, which has often been confounded with the neighboring M. tracheolateralis. The confusion has frequently been exacerbated by the failure of many authors to recognize the M. sternotrachealis and M. cricohyoideus as separate additional muscles associated with the trachea. Recent attempts to resolve the confusion concerning the "tracheal" muscles, namely the M. tracheohyoideus and M. tracheolateralis, have not been completely successful (e.g., George and Berger 1966; Vanden Berge 1979). To clarify the situation it is necessary to include briefly the M. tracheolateralis in the discussion of the M. tracheohyoideus although it will not be treated here because it is not a part of the lingual muscles (see Table 1).

The present study on *Psittacus,* as well as previous and ongoing studies on passerine birds by Bock (e.g., 1972, 1978, pers. comm.) and his associates, all show that avian tracheal muscles can be divided into two distinct parts. One part, the M. tracheohyoideus, is characterized by its association with the skin of the neck, whereas the other part, the M. tracheolateralis, is characterized by its close association with the trachea. The origins and insertions of both muscles, however, are quite variable among birds.

The psittacine M. tracheolateralis originates from the base of the trachea and inserts with several heads on the cricoid of the larynx (Homberger 1979b).

As shown in the present study on *Psittacus,* the psittacine M. tracheohyoideus originates from the sternum, attaches to the skin of the neck and inserts on the basihyale and, occasionally, on the ceratobranchiale. Mudge (1903) stated that the M. th. originates invariably from the sternum in parrots. Nitzsch (in Giebel 1862) mentioned an alternative origin on the clavicle. Gadow (1891) and Dubale and Rawal (1965) mistook the attachment of the M. th. onto the skin of the neck for the origin of the muscle (see below). The insertion of the M. th. along the Proc. parahyalis has been described by several authors (Nitzsch in Giebel 1862; Mudge 1903; Lubosch 1933; Dubale and Rawal 1965). Mudge (1903) and Lubosch (1933) reported that the insertion extends along the lateral margin of the Corpus basihyalis, although Mudge (1903) thought that this is peculiar to the genus *Prosopeia.* Kallius (1906) described the insertion to be on the basihyale "where it articulates with the ceratobranchiale," but from his description it is unclear whether he described the basihyal or parahyal head of the M. th. Mudge (1903) maintained that an insertion on the ceratobranchiale had been found in only a few genera (*Cacatua, Probosciger, Coracopsis, Eclectus*). However, the present study shows that a ceratobranchial head of the M. th. may be present in certain individuals also of *Psittacus.* Denker (1907) described the M. th. following the trachea and inserting on the dorsal surface of the basihyale but apparently mistook the parahyal and ceratobranchial heads of the M. th. for a separate muscle (i.e., "M. keratohyoideus III").

In all parrots (except *Strigops*), the M. th. attaches to the skin of the neck on its way to the hyoid apparatus (Mudge 1903). The midcervical portion of the M. th., which is attached to the skin, undergoes various stages of reduction of its muscle fibers to tendinous fibers, so that the muscle may become divided into an

TABLE 7

COMPARISON OF FUNCTIONAL INTERPRETATIONS OF THE PSITTACINE M. TRACHEOHYOIDEUS

Author	Function
Duvernoy (1835), Nitzsch (in Giebel 1862)	retracts the tongue
Dubale and Rawal (1965)	depresses the tongue
Homberger, this study	retracts and pulls down the hyoid apparatus

upper and a lower muscular portion by a tendinous middle segment (Mudge 1903). Apart from this division, Mudge (1903) described various stages of reduction of the lower portion of the M. th. between the sternum and the attachment to the skin of the neck. Mudge (1903) stated that *Psittacus* is one of only six parrot species with an undivided M. th., which concurs with my own observations on *Psittacus,* although I have no data on interspecific variations of the muscle.

With regard to the functional interpretations of the M. th., previous authors are in general agreement (see Table 7).

The synonymies for the M. th. and the M. tracheolateralis are presented and annotated in Appendix II G.

INTRINSIC LINGUAL MUSCULATURE

M. ceratohyoideus (M. ch., Mch; Figs. 16B, 19A).—

Diagnosis and differentiation: Originates from the ceratobranchiale and inserts on the Nodulus and surrounding structures. Not subdivided.

Origin: Ceratobranchiale. Fleshy. The M. ch. originates along the ventral surface of the ceratobranchiale, extending from near the caudal end of the Tuberositas ceratobranchialis (Figs. 13A, B) to the caudal one-fourth of the ceratobranchiale (e.g., AMNH 7203, Z3; Figs. 12C, 13A, 33A) or even close to the Art. epiceratobranchialis (e.g., Z3, Z4; Fig. 13B). Apically, the origin lies on the ventro-medial surface of the ceratobranchiale, but soon spreads to the ventro-lateral edge and shrinks again to the ventro-medial edge near the caudal end of the ceratobranchiale (Figs. 12C–13B). Laterally, the origin of the M. ch. is flanked for one-half of its length (e.g., AMNH 7203; Figs. 12C, 13A, 19A) or for one-fourth of its length (e.g., Z3; Figs. 16B, 23A) by the origin of the M. ceratoglossus lateralis. Medially, almost the entire origin of the M. ch. is bordered by the origin of the M. ceratoglossus medialis (Figs. 13A, B, 23A, 24B). The muscle fibers originating most caudally are often not clearly separated from the inserting fibers of the M. branchiomandibularis anterior medialis (Fig. 23B).

Insertion: Nodulus and Fascia urohyalis, rarely also Lig. nodulo-ceratobranchiale. Fleshy. The M. ch. inserts along the Nodulus and along a midsagittal raphe on the Fascia urohyalis. Rarely (e.g., Z4 right side), the insertion not only spreads onto the Lig. nodulo-ceratobranchiale, but the M. ch. is also divided into a superficial and a deep layer with separate insertions. In this case, the superficial (ventral) layer of the M. ch. inserts along the Fascia urohyalis, almost as far caudally as to the level of the caudal border of the insertion of the M. serpihyoideus, and along the caudal edge of nearly the entire length of the Lig. nodulo-ceratobranchiale (Fig. 19A), whereas the deep (dorsal) layer inserts in the more usual fashion along the Fascia urohyalis and the Nodulus (Fig. 19B).

Description: Parallel-fibered, sheet-like and slightly fan-shaped. Starting from the broad origin on the ceratobranchiale, the muscle fibers converge towards the insertion in such a way that the most caudal fibers are the longest and the most obliquely oriented (Figs. 16B, 19A, B, 23A–24B). The ventral surface of the muscle sheet is crossed by the Lig. nodulo-ceratobranchiale and by the M. tracheohyoideus (Fig. 16B) and is almost completely covered by the M. serpihyoideus, except for a small triangle rostro-medially to the Lig. nodulo-ceratobranchiale (Figs. 16A, 17A–18). This free triangle is covered ventrally by the M. mylohyoideus posterior (Fig. 16A). The observed variations of the M. ch. involve mainly the extent of its origin and the extent and location of its insertion.

Measurements: The length of the shortest muscle fibers varied between 2–7 mm ($N = 6$; $\bar{X} = 4.3$ mm; s.d. = 1.7 mm), and the length of the longest muscle fibers varied between 15–17 mm ($N = 6$; $\bar{X} = 15.8$ mm; s.d. = 0.7 mm). The muscle width at the origin varied between 14–18.5 mm ($N = 7$; $\bar{X} = 15.8$ mm; s.d. = 1.5 mm), and the muscle width at the insertion varied between 4.5–9 mm ($N = 7$; $\bar{X} = 5.7$ mm; s.d. = 1.9 mm). The muscle thickness at midlength was about 1 mm in two specimens.

Articulation: As a one-joint muscle, the M. ch. acts directly on the Art. cerato-basihyalis.

Functions: Retracts the Nodulus and, simultaneously, extends the basihyale. It also pulls the ceratobranchiale towards the urohyale.

Antagonists: For extension of basihyale: M. stylohyoideus, M. tracheohyoideus, M. ceratoglossus lateralis, M. ceratoglossus superficialis, M. hypoglossus obliquus lateralis, M. supraglossus ceratobranchialis; for retraction of Nodulus: M. mylohyoideus posterior, M. ceratoglossus superficialis.

Synergists: For extension of basihyale: M. ceratoglossus medialis; for retraction of Nodulus: M. serpihyoideus.

Discussion: The psittacine M. ceratohyoideus does not differ very much from the condition in other birds, but it is more massive and shows a certain variability in its origin and insertion. Most authors agree, with more or less accuracy, that the psittacine M. ch. originates from the medial surface of the apical part of the ceratobranchiale (Kutorga 1832; Duvernoy 1835; Nitzsch in Giebel 1862; Gadow 1891; Mudge 1903; Kallius 1906; Denker 1907; Dubale and Rawal 1965; Rawal 1970). With respect to the insertion, however, conflicting statements are found on whether it is on the urohyale (Kutorga 1832; Duvernoy 1835; Nitzsch in Giebel 1862; Gadow 1891), on the basihyale (Denker 1907), on the Fascia urohyalis (Rawal 1970), on the Fascia urohyalis and Lig. nodulo-ceratobranchiale (Dubale and Rawal 1965), or on a midsagittal raphe ventral to the urohyale (Kallius 1906). Mudge (1903) distinguished four types of insertions: (1) on the Fascia urohyalis only, (2) on the Nodulus and Lig. nodulo-ceratobranchiale, (3) on the Fascia urohyalis and Nodulus, and (4) on the urohyale and basihyale. The insertion types (1) and (4) appear to be doubtful for functional reasons and because Mudge (1903) claimed to have found type (1) in *Psittacus,* whereas my observations do not confirm this. Mudge's (1903) insertion types (2) and (3) have been observed in the present study as individual variants in *Psittacus.*

TABLE 8

COMPARISON OF FUNCTIONAL INTERPRETATIONS OF THE PSITTACINE M. CERATOHYOIDEUS

Author	Function
Kutorga (1832)	unilateral muscle pulls the urohyale towards the ceratobranchiale and the tongue turns toward the opposite side
Duvernoy (1835)	pulls the hyoid horns toward the midline
Nitzsch (in Giebel 1862)	pulls the hyoid horns together
Dubale and Rawal (1965), Rawal (1970)	lowers the hyoid apparatus
Homberger, this study	pulls the ceratobranchialia together and extends the basihyale by raising the urohyale

The functional interpretations of different authors vary widely (see Table 8). The synonymies are assembled in Appendix II H.

M. ceratoglossus (M. cg., Mcg; Fig. 27A).—

Diagnosis and differentiation: Five main divisions with different origins can be distinguished: the M. cg. superficialis originating from the Lig. nodulo-ceratobranchiale and sometimes from the Nodulus, the M. cg. lateralis originating from the lateral surface of the ceratobranchiale, the M. cg. profundus originating from the ventral surface of the ceratobranchiale close to its articulation with the basihyale, and the M. cg. medialis originating from the medial surface of the ceratobranchiale. The fifth portion, the M. cg. centralis originating from the urohyale, is not present in *Psittacus* but was found in *Cacatua* (pers. obs.).

A. M. ceratoglossus superficialis (M. cg. supf., Mcgs; Figs. 16B, 23B).—*Diagnosis and differentiation:* The most superficial (ventral) portion of the M. cg. originating from the Lig. nodulo-ceratobranchiale and, rarely, from the Nodulus. Depending on the individual, up to three portions, occasionally each one with several heads, can be distinguished by their specific origin: the M. cg. supf. ligamenti, which may be subdivided into a superficial and a deep layer, and the M. cg. supf. noduli, which is only occasionally present.

Origin: 1. M. cg. supf. ligamenti (M. cg. supf. lig., Mcgsl): Lig. nodulo-ceratobranchiale. Fleshy. If the M. cg. supf. is undivided (e.g., Z3, AMNH 7203), it originates from the apical margin of the Lig. nodulo-ceratobranchiale extending from the attachment of the ligament on the Tuberositas ceratobranchialis to the point where the ligament reaches the apical border of the M. ceratohyoideus (Fig. 16B). Medially, the origin spreads onto the dorsal surface of the ligament (Fig. 23B). Often, the M. cg. supf. lig. is divided into a superficial (Mcgss) and a deep muscle layer (Mcgsp) (e.g., Z2, Z4) if the Lig. nodulo-ceratobranchiale is split into a superficial and a deep layer to serve as origin to the respective parts of the M. cg. supf. lig. (Figs. 17B–19B). The origin of the deep portion often merges laterally with the origin of the M. cg. prof. (Fig. 19B). Rarely, the M. cg. supf. lig. is absent (e.g., Z6).

2. M. cg. supf. noduli (M. cg. supf. nod., Mcgsn): Fascia urohyalis and Nodulus. Fleshy. The M. cg. supf. nod. originates from the apical end of the Nodulus (Figs. 16B, 23A). It is only rarely present (e.g., Z3).

Insertion: Paraglossale. Tendinous. The muscle fibers of the M. cg. supf. converge and insert on the ventral side of the tendon of the M. cg. medialis (see below), which merges more apically with the common tendon of the entire M. cg. (Figs. 16B–18, 19B, 21B, 23A, B, 27A). The common tendon of the M. cg. inserts on the latero-ventral surface of the paraglossale on the Tuberositas paraglossalis (Figs. 12C, 13A, 27A, B, 29A, B, 32B, 38B, 39A). The insertion of the common tendon lies just behind the base of the lingual nail (Figs. 31B, 38B) or is partially covered by the lingual nail (Figs. 15B–16B, 27B).

Description: 1. M. cg. supf. lig.: Unipinnate, fan-shaped, sheet-like or consisting of separate, more or less parallel-fibered heads. Starting from the origin, the muscle fibers of the M. cg. supf. lig. converge towards the insertion, merging medially with the muscle fibers of the M. cg. supf. nod. (if present) and laterally with the muscle fibers of the M. cg. lateralis (Figs. 16A, B, 23A). If the M. cg. supf. lig. is divided into a superficial and a deep layer, the superficial layer (Mcgss) is sheet-like and its muscle fibers converge towards the insertion. Its lateral fibers merge with the fibers of the M. cg. lateralis. The deep layer (Mcgsp) may consist of several slips (e.g., Z2; Fig. 17B), which are more ribbon-like and whose fibers converge less strongly towards the insertion than those of the superficial layer (Figs. 17B, 19B). Laterally, the deep layer often merges with the M. cg. profundus.

2. M. cg. supf. nod.: Unipinnate, ribbon-like. Starting from the origin, the M. cg. supf. nod. soon merges with the medial border of the M. cg. supf. lig. (Figs. 16B, 23A, 27A). This part of the M. cg. supf. was found only in one out of seven specimens (Z3).

The M. cg. supf. lies between the M. mylohyoideus and Gl. lingualis on its ventral side, the M. cg. medialis and M. cg. profundus on its dorsal side, and the M. cg. lateralis on its lateral side. Variations of the M. cg. supf. are found not only in the number of its subdivisions and layers, but also in the presence or absence of the entire muscle.

Measurements: In undivided muscles, the muscle fiber length varied between 13.5–18 mm (N = 3; $\bar{X}$ = 16.2 mm; s.d. = 1.6 mm), and the muscle width at the origin varied between 1.5–3 mm (N = 3; $\bar{X}$ = 2.5 mm; s.d. = 0.87 mm). The muscle thickness at midlength was less than 0.5 mm in two specimens.

Articulations: The M. cg. supf. is a two-joint muscle. It passes over the ventro-lateral side of the Art. paraglosso-basihyalis and attaches directly or indirectly through the Lig. nodulo-ceratobranchiale to the apical end of the Nodulus, therefore, acting also on the Art. cerato-basihyalis (pp. 114 ff.).

Functions: Flexes the paraglossale and also pulls the Lig. nodulo-ceratobranchiale and the Nodulus forward, thereby assisting in rotating the basihyale into a more vertical position.

Antagonists: For flexion of paraglossale: M. hypoglossus obliquus, M. supraglossus; for protraction of Nodulus: M. serpihyoideus, M. ceratohyoideus; for flexion of basihyale: M. ceratohyoideus, M. ceratoglossus medialis.

Synergists: For flexion of paraglossale: M. cg. lateralis, M. cg. medialis, M. cg. profundus; for protraction of Nodulus: M. mylohyoideus posterior; for flexion of basihyale: M. serpihyoideus, M. stylohyoideus, M. tracheohyoideus, M. cg. lateralis, M. hypoglossus obliquus lateralis, M. supraglossus ceratobranchialis.

B. M. ceratoglossus medialis (M. cg. med., Mcgm; Figs. 23B, 27A).—*Diagnosis and differentiation:* The portion of the M. cg. that originates from the medial surface of the ceratobranchiale. Occasionally, two to three subdivisions can be distinguished: Pars dorsalis (M. cg. med. dors., Mcgmd); Pars ventralis (M. cg. med. ventr., Mcgmv); and Pars accessoria (M. cg. med. acc., Mcgma).

Origin: Ceratobranchiale. Fleshy, occasionally partially tendinous. The M. cg. med. originates mainly on the medial surface of the ceratobranchiale, i.e., within the Sulcus ventro-medialis (Figs. 5A, B). The origin extends from the caudal end of the Tuberositas ceratobranchialis (e.g., Z6), or slightly more caudally to it, to about the caudal one-quarter or even the posterior end of the ceratobranchiale (Figs. 12B, 13B). Ventro-laterally, the origin of the M. cg. med. borders the origin of the M. ceratohyoideus (Figs. 13A, B, 33A). Dorso-laterally, the anterior part of the origin sometimes borders the origin of the M. cg. lat. for a short distance (Figs. 12B, 24B–25B). If the M. cg. med. is subdivided into a dorsal and ventral portion (e.g., Z3 left side, Z4), the M. cg. med. dorsalis originates from the caudal half of the common origin (Fig. 13A). Some fibers of the M. cg. med. ventralis may originate tendinously (Fig. 28A). The M. cg. med. accessorius originates tendinously along the latero-dorsal border of the origin of the main part of the M. cg. med. The origin of the M. cg. med. acc. varies from the apical end of the origin of the main part of the M. cg. med. (e.g., Z4 left side) to opposite the attachment of the Lig. nodulo-ceratobranchiale on the Tuberositas ceratobranchialis (e.g., Z4 right side; Fig. 19B) or to approximately the midlength of the origin of the main part of the M. cg. med. (e.g., Z3; Figs. 23A, B, 27A). If the origin of the M. cg. med. extends far apically, it merges with the origin of the M. cg. profundus (e.g., Z6).

Insertion: Paraglossale. Tendinous. The muscle fibers of the M. cg. med. dors. and M. cg. med. ventr. insert mainly on the ventral surface of a distinct tendon that merges anteriorly with the common tendon of the M. cg. (for the insertion of the common tendon, see M. cg. supf.). The M. cg. med. acc. inserts more apically on the ventral surface of the tendon of the M. cg. med., often in close contact with the M. cg. supf. (Figs. 19B, 23B, 27A, 28B).

Description: Pinnate. The M. cg. med. is the most clearly defined portion of the M. cg. and forms a distinct bundle running over the ventral surface of the apical end of the ceratobranchiale within the Sulcus ventro-medialis (Figs. 23B, 27A, 28A, 33A). If the M. cg. med. is undivided (e.g., Z3 right side, AMNH 7203), the fibers converge from the origin towards a central tendon which starts about at the point where the M. cg. med. reaches the caudal edge of the Lig. nodulo-ceratobranchiale. Apically, this tendon shifts to the dorso-medial surface of the M. cg. med. so that the muscle fibers insert mainly along the ventral and lateral surfaces of the tendon (Figs. 17B, 23B, 27A). The angle of pinnation of the M. cg. med. is about 5°.

If a ventral and dorsal portion can be distinguished (e.g., Z3 left side, Z4), the M. cg. med. dors. originates from the caudal half of the origin and passes along the dorsal side of the M. cg. med. ventr. to insert on the tendon, whereas the M. cg. med. ventr. originates on the apical half of the origin and passes along the ventral side of the M. cg. med. dors. to insert more apically onto the tendon (Figs. 25B–26B, 30A).

The M. cg. med. acc. is usually a thin, ribbon-like slip of variable length. It originates tendinously and adheres to the ventral side of the M. cg. med. (Figs. 17B, 19B, 23A, B, 27A). It inserts most cranially on the tendon of the M. cg. med., often closely associated with the inserting fibers of the M. cg. supf. Despite its association with the M. cg. supf., the M. cg. med. acc. is a part of the M. cg. med. because of its origin on the medial surface of the ceratobranchiale.

The M. cg. med., especially its lateral border, is not always clearly separated from the M. cg. prof. if the origin of the M. cg. med. extends far apically. The variations found in the M. cg. med. are considerable, especially with respect to the number of its subdivisions and the extent of its origin.

On its ventral side, the M. cg. med. is crossed by the Lig. nodulo-ceratobranchiale and almost completely covered by the M. ceratohyoideus and M. cg. supf. (Figs. 16B, 19A, B, 23B).

Measurements: For six specimens, the length of the shortest muscle fibers varied between 5.5–9 mm ($\bar{X}$ = 7.4 mm; s.d. = 1.2 mm), and the length of the longest muscle fibers varied between 15–24 mm ($\bar{X}$ = 19.6 mm; s.d. = 3.4 mm). The muscle width at the origin varied between 11–16 mm (N = 6; $\bar{X}$ = 13.7 mm; s.d. = 2.1 mm). The muscle thickness was about 1 mm in two specimens.

Articulations: The M. cg. med. is a two-joint muscle. It passes over the ventro-lateral side of the Art. paraglosso-basihyalis and over the ventro-medial side of the Art. cerato-basihyalis.

Functions: Flexes the paraglossale and, simultaneously, tends to rotate the basihyale into a more horizontal position.

Antagonists: For flexion of paraglossale: M. hypoglossus obliquus, M. supraglossus; for extension of basihyale: M. serpihyoideus, M. stylohyoideus, M. tracheolateralis, M. cg. lateralis, M. cg. superficialis, M. hypoglossus obliquus lateralis, M. supraglossus ceratobranchialis.

Synergists: For flexion of paraglossale: M. cg. profundus, M. cg. lateralis, M. cg. superficialis; for extension of basihyale: M. ceratohyoideus.

C. M. ceratoglossus lateralis (M. cg. lat., Mcgl; Figs. 23A, 28A).—*Diagnosis:* The most lateral portion of the M. cg. which originates from the lateral surface of the ceratobranchiale. Not subdivided.

Origin: Ceratobranchiale. Fleshy, sometimes partly tendinous. The M. cg. lat. originates mainly on the lateral surface of the ceratobranchiale, extending from the caudal border of the Tuberositas ceratobranchialis (Figs. 12C–13B, 18, 29B) or from the apical end of the tuberosity (e.g., Z3 right side; Fig. 13A) to about the midlength of the ceratobranchiale (Figs. 12A, 12C–13B) or the caudal one-quarter of the ceratobranchiale (e.g., Z3 right side; Fig. 12A). Apically, the latero-dorsal border of the origin follows the lateral crest of the ceratobranchiale (Fig. 4A) and flanks the origin of the M. hypoglossus obliquus lateralis (Figs. 12A, 29B). The fibers of the M. cg. lat. originating along the lateral crest are usually tendinous. At its apical end, the origin of the M. cg. lat. merges often with the origin of the M. cg. prof. Caudally, the latero-dorsal border of the origin of the M. cg. lat. extends onto the dorsal surface of the ceratobranchiale where it may come in contact with the origin of the M. cg. med. (Figs. 12B, 24B, 25A). The ventro-lateral border of the origin of the M. cg. lat., except for its anteriormost part, flanks the origin of the M. ceratohyoideus (Figs. 12C–13B, 16B, 19A, B, 23A).

Insertion: Paraglossale. Tendinous. The muscle fibers of the M. cg. lat. become increasingly tendinous and merge with the other divisions of the M. cg. to form a common tendon (Figs. 21B, 27A). (For the insertion of the common tendon, see M. cg. supf.)

Description: Pinnate. The M. cg. lat. forms a thick, fan-shaped sheet that covers the lateral surface of the ceratobranchiale and continues forward to join the common tendon of the M. cg. The muscle fibers of the M. cg. lat. remain concentrated in the latero-dorsal part of the common tendon. The caudal part of the M. cg. lat. is covered on its latero-ventral surface by the M. stylohyoideus and M. tracheohyoideus (Figs. 15B, 16A, 20). The apical part of the M. cg. lat. is covered laterally by the Glandula mandibularis and ventrally for the greater part by the M. mylohyoideus (Figs. 15B, 16A, 20). On its medial side, the M. cg. lat. is not always clearly separated from the M. cg. prof. (Figs. 28A, B). The observed variations involve mainly the extent of the origin.

Measurements: In two specimens, the shortest muscle fibers were between 11–12 mm long, and the longest muscle fibers measured between 18–19 mm; the muscle width at the origin was 7 mm and 10 mm. The muscle width at the insertion of one specimen was 4 mm. The muscle thickness at midlength was about 0.5 mm in two specimens.

Articulations: The M. cg. lat. is a two-joint muscle that passes over the latero-ventral side of the Art. paraglosso-basihyalis and over the lateral side of the Art. cerato-basihyalis.

Functions: Flexes the paraglossale and, simultaneously, tends to rotate the basihyale into a more vertical position.

Antagonists: For flexion of paraglossale: M. hypoglossus obliquus, M. supraglossus; for flexion of basihyale: M. ceratohyoideus, M. cg. medialis.

Synergists: For flexion of paraglossale: M. cg. profundus, M. cg. lateralis, M. cg. superficialis; for flexion of basihyale: M. serpihyoideus, M. stylohyoideus, M. tracheohyoideus, M. cg. superficialis, M. hypoglossus obliquus lateralis, M. supraglossus ceratobranchialis.

D. M. ceratoglossus profundus (M. cg. prof., Mcgp; Figs. 27A, 28A).—*Diagnosis:* The portion of the M. cg. that originates on the ventral surface of the ceratobranchiale and frequently also on the capsule of the Art. cerato-basihyalis (but not on the basihyale) and lies deepest of all the portions of the M. cg.

Origin: Ceratobranchiale. Fleshy, partly tendinous. Often the anteriormost fibers of the M. cg. prof. originate tendinously. The origin extends from an area lateral and apical of the Tuberositas ceratobranchialis to the capsule of the Art. cerato-basihyalis (Fig. 13B). Rarely (e.g., Z2 left side), a slip of the origin spreads onto the medial surface of the ceratobranchiale (Figs. 13B, 28B). Laterally, the origin borders the origin of the M. cg. lat. from which it is usually not clearly separated (Figs. 12C, 13A). Caudally, the origin borders the attachment of the Lig. nodulo-ceratobranchiale on the Tuberositas ceratobranchialis (Figs. 12C, 13A, 28A, B, 29B) or the insertion of the M. serpihyoideus (Figs. 13A, B). If the origin spreads over the articular capsule, it borders the origin of the M. hypoglossus obliquus medialis (Fig. 13B). Rarely (e.g., Z6), the originating fibers of the M. cg. prof. cannot be clearly separated from the origin of the M. hypoglossus obliquus lateralis if the latter spreads to the ventro-lateral surface of the ceratobranchiale. If a slip of the Lig. nodulo-ceratobranchiale passes ventrally over the portion of the M. serpihyoideus that inserts directly on the ceratobranchiale (e.g., Z4; Figs. 13B, 18), the origin of the M. cg. prof. cannot be clearly separated from the origin of the deep layer of the M. cg. supf. lig. (Mcgsp), which originates on the same ligamentous slip (Figs. 18–19B). Some superficial fibers of the M. cg. prof. often merge with the superficial fibers of the M. serpihyoideus if these two muscles meet by attaching to the same structure, e.g., to the Lig. nodulo-ceratobranchiale or to the ceratobranchiale itself (Fig. 16A). The origin of the M. cg. prof. is not clearly separated from the origin of the M. cg. med. if the latter extends far forward (e.g., Z6).

Insertion: Paraglossale. Tendinous. The muscle fibers converge and their tendinous fibers form the medial part of the common tendon of the M. cg. (For the insertion of the common tendon, see M. cg. supf.)

Description: Pinnate. Forms a thick, fan-shaped sheet. Starting from the origin, the M. cg. prof. converges towards the common inserting tendon of the M. cg. The M. cg. prof. is not always clearly defined against the M. cg. lat. (Figs. 17A–18, 28A, B). The M. cg. prof. passes over the ventral surface of the lateral part of the M. hypoglossus obliquus medialis (Figs. 23A, B, 27A, 28A, B). The ventral surface of the M. cg. prof. itself is completely covered by the M. cg. med. and M. cg. supf. The observed variations involve mainly the extent of the origin.

Measurements: The muscle fiber length varied between 5–10 mm for the shorter fibers and between 7.5–13 mm for the longer fibers (N = 3). In three specimens, the muscle width at the origin varied between 2–4 mm. The muscle thickness at midlength was 0.5 mm in one specimen.

Articulations: Morphologically, the M. cg. prof. is a two-joint muscle passing over the latero-ventral side of the Art. paraglosso-basihyalis and over the ventral side of the Art. cerato-basihyalis, but functionally it is a one-joint muscle (pp. 94 f., 117).

Function: Flexes the paraglossale.

Antagonists: For flexion of paraglossale: M. hypoglossus obliquus, M. supraglossus.

Synergists: For flexion of paraglossale. M. cg. lat., M. cg. supf., M. cg. med.

E. M. ceratoglossus centralis.—This muscle portion has not been found in *Psittacus,* but it is present in *Cacatua roseicapilla* (pers. obs.). It originates from the lateral surfaces of the urohyale and inserts through the common tendon of the M. cg. on the Tuberositas paraglossalis. It is the only portion of the M. cg. that is genuinely a one-joint muscle, passing only over the latero-ventral side of the Art. paraglosso-basihyalis. The M. cg. centralis is mentioned here only for the sake of clarity in the following discussion.

Discussion: The psittacine M. cg. differs considerably from that of other birds by having formed four to five subdivisions with different origins and partly differing functions. Apparently because of the complexity of the psittacine M. cg., many conflicting descriptions of the muscle are found in the literature. Kutorga (1832) and Chaine (1905) described the M. cg. lat. as the only portion of an undivided muscle, originating from the antero-lateral surface of the ceratobranchiale and inserting on the paraglossale. The description of Kallius (1906) is vague. Denker (1907) correctly described the origins of the M. cg. med. and M. cg. lat., but interpreted the two portions as separate muscles inserting on the ventral surface of the basihyale and on the ventro-lateral surface of the tongue respectively. Most authors (Nitzsch in Giebel 1862; Mudge 1903; Steinbacher 1951; Dubale and Rawal 1965; Rawal 1970) have recognized the complexity of the psittacine M. cg., although their descriptions vary greatly in accuracy and detail. (An attempt at homologizing the main subdivisions of the M. cg. as described by different authors is presented in Table 9.)

TABLE 9

Homologies of the Subdivisions of the Psittacine M. ceratoglossus

Homberger	Mudge (1903)	Nitzsch (in Giebel 1862)	Denker (1907)	Steinbacher (1951)	Dubale and Rawal (1965)
M. cg. lateralis	M. cg. lateralis	M. cg. lateralis (sometimes: M. cg. inferior)	M. keratoglossus II	M. cg. lateralis	M. hypoglossus lateralis externus
M. cg. medialis	M. cg. inferior posticus	unidentified muscle marked by * in Fig. 2, Plate V	M. keratohyoideus IV	—	M. hyoglossus inferior
M. cg. profundus	M. cg. inferior anticus, outer portion: "M. cg. i. a." (occasionally: inner portion: "M. cg. i. a^1.")	M. cg. inferior (M. basioglossus)	—	—	M. hyoglossus lateralis internus
M. cg. superficialis	M. cg. inferior anticus accessorius: "M. cg. i^1." (occasionally: "M. cg. i. a^1.")	—	—	—	—
M. cg. centralis	M. cg. inferior anticus, inner portion: "M. cg. i. a^1.", "M. cg. i. a^2."	—	—	M. cg. inferior (?)	—

Mudge (1903) has given the most detailed description of the M. ceratoglossus complex in parrots. He distinguished three main portions: (1) the "M. cg. inferior" (i.e., M. cg. med., M. cg. supf., M. cg. prof., and M. cg. centralis), (2) the "M. cg. lateralis" (i.e., M. cg. lat.), and (3) the "M. cg. superior" (i.e., M. supraglossus). Unfortunately, in his comparative study, he devoted his greatest attention to characters that have been shown to be individually variable, such as the size of the particular divisions, the extent of fusion between the muscle divisions, and the extent and location of fasciae and tendons. At the same time, Mudge did not characterize the various divisions of the M. cg. as functional units and, therefore, had problems homologizing the subdivisions, especially with respect to the M. cg. profundus and M. cg. centralis (see Table 9). A re-examination of Mudge's work allows some comparisons to be made with the present data on *Psittacus*. The M. cg. supf. lig. seems to be present in most species, being reported as missing only in *Nestor* and the Loriinae by Mudge (1903). However, the study of *Psittacus* shows that the M. cg. supf. lig. may occasionally also be absent in other species. A M. cg. supf. nod. was reported by Mudge (1903) only in *Cacatua roseicapilla*, but it was also found in one specimen of *Psittacus* in the present study. According to Mudge (1903), the extent of the origins of the M. cg. med. and M. cg. lat. ranges among parrots from just the anterior one-third of the ceratobranchiale to up to its entire length. For *Psittacus*, Mudge (1903) described the origins of the M. cg. med. and M. cg. lat. as extending only along the anterior half of the ceratobranchiale. However, in the present study of *Psittacus* these origins were shown to vary considerably within the species. Mudge (1903) did not always clearly distinguish between the M. cg. centralis and the M. cg. prof. (his "inner and outer portions of the M. cg. inferior anticus"). Furthermore, he often described the M. cg. prof. as originating from the basihyale. Apparently, Mudge (1903) misinterpreted the origin of the M. cg. prof. on the articular capsule of the Art. ceratobasihyalis (as observed in the present study of *Psittacus*) as being on the basihyale.

A M. cg. centralis that originates along the sides of the urohyale (as described by Mudge 1903) was not found in *Psittacus*. This muscle portion was also absent in *Agapornis personata* but was present in *Cacatua roseicapilla* (pers. obs.). Mudge's data can be interpreted to suggest that the M. cg. centralis is present only in cockatoos and *Strigops*. Steinbacher (1951), however, described in a lory a muscle portion that originates from the urohyale and may be interpreted as homologous to the M. cg. centralis (see Table 9).

Duvernoy (1835), Lubosch (1933), Kasai (1957), and Burton (1974c) showed the M. cg. only in figures, whereas Gadow (1891) just quoted the work of Nitzsch (in Giebel 1862).

The functional interpretations of the M. cg. given by various authors are vague (see Table 10), probably because they did not distinguish between actions of extrinsic and intrinsic muscles (differences between extrinsic and intrinsic muscles are discussed on p. 94).

The synonymies of the avian M. cg. are presented in Appendix II I.

M. hypoglossus obliquus (M. ho., Mho; Fig. 29B).—

Diagnosis and differentiation: Inserts on the posterior process of the paraglossale and on the Glandula lingualis by forming a tendinous envelope around the glandular body. Two portions can be distinguished on the basis of their origins: The

TABLE 10

COMPARISON OF FUNCTIONAL INTERPRETATIONS OF THE PSITTACINE M. CERATOGLOSSUS

Author	Function
Kutorga (1832)	flexes the tip of the tongue and tends to make the lingual surface convex; unilaterally, it bends the caudal part of the tongue towards its own side
Nitzsch (in Giebel 1862)	M. cg. prof.: pulls the tip of the tongue down. M. cg. lat.: pulls the tip of the tongue down and spreads the hyoid horns; unilaterally, it pulls the tongue to the side
Gadow (1891)	pulls the tongue strongly down; unilaterally, it moves the tongue sideways
Steinbacher (1951)	lowers the tip of the tongue
Dubale and Rawal (1965), Rawal (1970)	depresses the tongue
Homberger, this study	flexes the paraglossale. M. cg. lat. and M. cg. supf.: supports the flexion of the basihyale; M. cg. med.: supports the extension of the basihyale

M. ho. medialis originating from the ventral surface of the basihyale and the M. ho. lateralis originating from the dorsal surface of the ceratobranchiale.

A. M. hypoglossus obliquus lateralis (M. ho. lat., Mhol).—*Diagnosis:* Originates from the dorsal surface of the ceratobranchiale and inserts on the paraglossale and the tendinous envelope of the Gl. lingualis.

Origin: Ceratobranchiale. Fleshy, partly tendinous. The M. ho. lat. originates along the Crista lateralis ceratobranchialis extending from the caudal end of the lateral part of the articular facet of the Art. cerato-basihyalis to about the caudal end of the Fossa dorsalis (Fig. 12A). Apically, the origin spreads over the lateral part of the Fossa dorsalis (Fig. 12B). Usually, the origin is flanked along its entire medial border by the origin of the M. supraglossus ceratobranchialis and along its lateral border by the origins of the M. ceratoglossus lateralis and M. ceratoglossus profundus (Figs. 12A, B). Along the lateral border, some fibers may have a tendinous origin. Also, these originating fibers are not always clearly separated from the origins of the M. ceratoglossus prof. and M. ceratoglossus lat. (e.g., Z6) if some fibers of the M. ho. lat. originate on the tendinous fibers of the M. ceratoglossus prof.

Insertion: Paraglossale and tendinous envelope of the Glandula lingualis. Tendinous. The M. ho. lat. inserts along the Margo dorsalis of the Proc. posterior paraglossalis extending from the tip of the process to about the level of the Art. paraglosso-basihyalis (Figs. 12A, C). (For details, see "Description.")

Description: Pinnate. From its origin, the M. ho. lat. arises as a bilaterally flattened muscle between the muscle bodies of the M. supraglossus ceratobranchialis and the M. ceratoglossus lat. (Figs. 24B, 25A, 29B). Farther apically, the cross-section of the M. ho. lat. becomes more oval, and the muscle body passes laterally to the Proc. parahyalis (Figs. 24A–25B, 32A, B). On its dorsal side, the M. ho. lat. is crossed by the M. stylohyoideus and, more caudally, also by the M. tracheohyoideus (Figs. 24A–25B, 32A). Latero-ventrally it is covered by the M. ceratoglossus (Figs. 21B, 24A–25A, 29A, 33A). The M. ho. lat. runs towards the tip of the posterior process of the paraglossale. Starting at the level of the tip of the Proc. parahyalis, the muscle fibers become successively tendinous (Figs. 22, 24A, B). The majority of the tendinous fibers forms a tough sheet that lies atop the ends of the muscle fibers (Figs. 33A, 45A, B). Farther apically, the muscle fibers of the M. ho. med., merging with the M. ho. lat., also contribute to the formation of this sheet. This tendinous sheet runs forward to insert on the Margo dorsalis and upper medial surface of the posterior process of the paraglossale; it forms the floor of the envelope of connective tissue surrounding the Gl. lingualis (Figs. 24A–25A, 32A). The tendinous fibers derived from the peripheral muscle fibers of the M. ho. lat., i.e., those fibers on the surface of the muscle body, contribute to the lateral and caudal walls of the tendinous envelope of the Gl. lingualis. The most caudal tendinous fibers, derived from the most dorsal muscle fibers of the M. ho. lat., form a sheet covering the posterior surface of the Gl. lingualis (Figs. 21B, 22, 45B). This caudal tendinous sheet continues onto the dorsal side of the Gl. lingualis and covers the entire glandular body, attaching apically to the Margo dorsalis paraglossalis where it merges with the tendinous fibers that form the floor of the glandular envelope (Figs. 45A, B). The tendinous fibers contributing to the lateral wall of the envelope of the Gl. lingualis are derived not only from the M. ho. lat., but also from the lateralmost fibers of the M. ho. med. and M. hypoglossus anterior (Figs. 21B, 27A, 29A, B, 32B). All these tendinous fibers merge with the connective tissue cushioning the latero-dorsal edge of the body of the Gl. lingualis. Some superficial muscle fibers of the M. genioglossus may also attach to this lateral wall and cushion of connective tissue (Fig. 21B). The medial wall of the tendinous envelope of the Gl. lingualis incorporates tendinous fibers that arise from the dorsal crest of the paraglossale and are not continuous with muscle fibers. These medial fibers merge with the cushion of connective tissue along the medio-dorsal edge of the body of the Gl. lingualis.

In summary, the body of the Gl. lingualis is enclosed in an envelope of connective tissue of which the floor and the lateral, caudal and dorsal walls contain tendinous fibers derived mainly from the M. hypoglossus obliquus. The dorsal side of this envelope and the edges along the lateral, caudal and medial sides are covered with a layer of fibrous connective tissue that contains a large amount of blood vessels and nerves and serves as cushion between the tendinous envelope of the gland formed by the M. ho. and the tendinous sheet formed by the M. supraglossus (Figs. 40B–42B, 45A, B).

Measurements: In two specimens, the muscle fiber length ranged from 7–12 mm, and the muscle width at the origin varied between 5–6 mm. The cross-sectional area of the muscle at the insertion of one specimen measured 1×2 mm^2. The muscle thickness at midlength was between 0.5–1 mm for one specimen.

Articulations: Two-joint muscle. The M. ho. lat. passes over the dorso-lateral side of the Art. paraglosso-basihyalis and over the latero-dorsal side of the Art. cerato-basihyalis.

Functions: Extends the paraglossale and supports the flexion of the basihyale.

Antagonists: For extension of paraglossale: M. ceratoglossus; for flexion of basihyale: M. ceratohyoideus, M. ceratoglossus medialis.

Synergists: For extension of paraglossale: M. ho. med., M. supraglossus; for flexion of basihyale: M. stylohyoideus, M. tracheohyoideus, M. ceratoglossus lateralis, M. ceratoglossus superficialis, M. supraglossus ceratobranchialis.

B. M. hypoglossus obliquus medialis (M. ho. med., Mhom).—*Diagnosis:* Originates from the ventral surface of the basihyale and inserts on the tendinous envelope of the Gl. lingualis and the posterior process of the paraglossale.

Origin: Basihyale. Fleshy. The origin of the M. ho. med. forms a narrow band along a midsagittal line on the ventral surface of the basihyale (Figs. 12C–13B). Towards caudal, about at midlength of the muscle's origin, the origins of the collateral parts of the muscle may diverge slightly (Figs. 13A, 23A, B, 27A, 28A) or only separate but remain parallel (Figs. 17A, 19A, B). When reaching the cranial borders of the Artt. cerato-basihyales, the origins curve laterally and follow the borders of the articulations (Figs. 13A, B). The extent of the origin of the M. ho. med. varies. Apically, it may start at the level of the caudal ends of the posterior processes of the paraglossale (e.g., Z4, Z6; Fig. 13B) or close to the Art. paraglosso-basihyalis (Figs. 12C, 13A, 16B, 23A, B, 27A–28A, 30A, B, 31B, 32B). Caudolaterally, the origin may extend only to the midlength of the border of the articulation (Figs. 13A, 28A) or all the way to the lateral edge of the basihyale (Fig. 13B). Occasionally, the origin may be divided by a gap in its caudal region (Figs. 17A, 18–19B). The most anterior fibers often do not originate directly on the basihyale but form a continuous sling around the ventral surface of the basihyale (Figs. 23A, B, 27A–28B, 30A, B, 42A, B). If the origin of the M. ceratoglossus prof. spreads onto the capsule of the Art. cerato-basihyalis, the origin of the M. ho. med. is not clearly separated from that of the M. cg. prof.

Insertion: Paraglossale. Mainly fleshy, partly tendinous (tendinous envelope of the Gl. lingualis). The deep (dorsal) muscle fibers, which face the basihyale, insert on the medial surface of the posterior process of the paraglossale from its tip to the border of the Art. paraglosso-basihyalis (Figs. 12A, 25A, 32A, 42A, B). Occasionally, the extreme tip of the posterior process is occupied by the insertion of the M. ho. lat. (Fig. 12A). The caudalmost muscle fibers of the M. ho. med. merge with the insertion of the M. ho. lat. to form the cup-like floor of the tendinous envelope of the Gl. lingualis (Figs. 25A, 32A, 45A, B). The most superficial (ventro-lateral) muscle fibers contribute to the lateral wall of the tendinous envelope of the Gl. lingualis (Figs. 21B, 27A, 29A, B, 32B).

Description: Pinnate, fan-shaped. From their origin on the basihyale, the muscle fibers converge latero-apically towards the caudal tip of the posterior process of the paraglossale. Thus, the caudalmost muscle fibers are the longest and run almost parallel to the midsagittal axis of the basihyale, whereas the most anterior muscle fibers are short and transversally oriented (Figs. 28A, 29B, 32A, B). Superficially, the M. ho. med. is occasionally divided into separate bundles, depending on whether the origin is interrupted by gaps or is continuous. The deepest muscle fibers insert on the medial surface of the posterior process of the paraglossale, whereas the greater part of the M. ho. med. joins the M. ho. lat. to form the tendinous envelope of the Gl. lingualis (see "Description" of the M. ho. lat.). The central part of the M. ho. med. is covered ventrally by the Gl. sublingualis (Fig. 16A). The lateral part is covered on its ventral surface by the M. ceratoglossus (Figs. 16B, 21B, 23A, B, 27A, 28A). Near its insertion, the M. ho. med. is partially covered on its dorsal side by the M. ho. lat. (Figs. 21B, 29A, B, 32A, B). Anteriorly, the M. ho. med. is flanked on its medial side by the M. supraglossus (Fig. 25A).

Measurements: For two specimens, the muscle fiber length ranged from 4–7 mm, the muscle width at the origin varied between 8–9 mm, and the cross-sectional area at the insertion measured 4 × 2 mm^2. The muscle thickness at midlength was about 0.5 mm.

Articulation: One-joint muscle. The M. ho. med. passes over the dorso-lateral side of the Art. paraglosso-basihyalis.

Function: Extends the paraglossale.

Anatagonists: For extension of paraglossale: M. ceratoglossus.

Synergists: For extension of paraglossale: M. ho. lat., M. supraglossus.

Discussion: The psittacine M. hypoglossus obliquus differs considerably from the muscle in other birds through its greater size, its subdivision into two distinct parts, and its formation of a tendinous envelope about the Gl. lingualis. However, many authors described only the M. ho. med. and agreed, more or less accurately, that the origin lies on the latero-ventral surface of the basihyale and that the insertion is located on the Proc. posterior paraglossalis (Kutorga 1832; Nitzsch in Giebel 1862; Gadow 1891; Dubale and Rawal 1965; Rawal 1970). Kallius (1906) disagreed by describing the origin at the base of the ring-shaped cartilage in *Melopsittacus* (homologous to the parahyal processes of other parrots). Steinbacher (1951) noted in the text of his paper that the M. ho. originates only from the basihyale, but in his figure 5 he clearly showed muscle fibers originating also from the apical end of the ceratobranchiale. Only Mudge (1903), Chaine (1905), and Denker (1907) clearly described both portions of the M. ho. Chaine (1905), however, identified the part originating latero-ventrally from the proximal region of the ceratobranchiale and ventrally from the basihyale and inserting on the dorsal crest and inner surface of the Proc. posterior paraglossalis as a portion of the M. ceratoglossus (it actually consists of the M. ho. lat. and the lateral part of the M. ho. med.). Denker (1907) interpreted the two portions of the M. ho. as two different muscles. His description of the muscles and their origins is correct, but he assumed that the insertions were on the paraglossale for the M. ho. med. and in the lateral part of the tongue near the surface for the M. ho. lat. Mudge (1903) distinguished the two parts of the M. ho. on the basis of their origin. In addition, he recognized four different "stages" characterized by differences in the extent of their origin on the ceratobranchiale, in the development of their tendon, and in the clarity of their division into two portions. These stages, nevertheless, appear somewhat arbitrary, especially since their general validity is questionable because of exceptions.

It is remarkable that no previous author has noticed the tendinous envelope of the Gl. lingualis, which is formed by the M. ho. Despite this oversight, many authors were not far from the truth with their respective functional interpretations of the M. ho. (see Table 11), although they did not appreciate the function of the Gl. lingualis as an hydraulic system.

The synonymies of the avian M. ho. are listed in Appendix II J.

M. supraglossus (M. sg., Msg; Fig. 25B).—

Diagnosis and differentiation: Originates from the dorsal surfaces of the ceratobranchiale and basihyale and inserts on the dorsal surface of the paraglossale, on epithelial structures, and on the aponeurosis of the M. hypoglossus anterior. This extremely complex muscle consists of two muscular portions, namely the M. sg. ceratobranchialis originating on the ceratobranchiale and the M. sg. basihyalis originating on the basihyale. Both muscle portions merge to form a central tendon from which tendinous fiber bundles fan out to their respective insertions.

TABLE 11

COMPARISON OF FUNCTIONAL INTERPRETATIONS OF THE PSITTACINE M. HYPOGLOSSUS OBLIQUUS

Author	Function
Kutorga (1832)	extends the tongue after it has been flexed by the M. ceratoglossus
Gadow (1891)	pulls the base of the tongue backwards or points the tip of the tongue upwards; unilaterally, it rotates the tip of the tongue about its longitudinal axis
Steinbacher (1951)	raises the tip of the paraglossale
Dubale and Rawal (1965)	depresses the tongue, flattens the surface of the tongue
Rawal (1970)	helps the lateral movements of the tongue
Homberger, this study	extends the paraglossale; M. ho. lat. also supports the flexion of the basihyale

Origin: 1. M. supraglossus ceratobranchialis (M. sg. cer., Msgc): Ceratobranchiale. Fleshy. The M. sg. cer. originates in the Fossa dorsalis ceratobranchialis, extending from the caudal end of the lateral part of the articular facet of the Art. cerato-basihyalis to the caudal end of the fossa (Fig. 12A). Often, the origin spreads over the capsule of the Art. cerato-basihyalis and to the dorso-medial crest of the ceratobranchiale (Fig. 12B). Occasionally (e.g., Z4 left side), the origin has shifted medially to such a degree that the main portion of the origin is found on the medial surface of the ceratobranchiale instead of in the Fossa dorsalis ceratobranchialis (Fig. 12B). The origin of the M. sg. cer. borders laterally to the origin of the M. hypoglossus obliquus lat. (Fig. 12A) or, if present, to the insertion of the M. tracheohyoideus ceratobranchialis (Fig. 12B).

2. M. supraglossus basihyalis (M. sg. bas., Msgb): Basihyale. Fleshy. The M. sg. bas. originates in the Fossa basihyalis on the dorsal surface of the Corpus basihyale. The origin extends from the capsule of the Art. cerato-basihyalis to the base of the Proc. parahyalis (Fig. 12A) or to the midlength of the Proc. centralis basihyalis (Fig. 12B). Occasionally (e.g., AMNH 7203 left side), some muscle fibers originate laterally on the ventral surface of the basihyale near the base of its central process (Figs. 12C, 13A; shown on the right side). The origin of the M. sg. bas. is laterally flanked by the origin of the M. cricohyoideus ventralis (Fig. 12A). The caudal part of the origin of the M. sg. bas. is usually flanked by the insertion of the M. tracheohyoideus parahyalis if the origin of the M. cricohyoideus ventr. does not extend to the Art. cerato-basihyalis (Figs. 12A, B). The caudal border of the origin is often not clearly separated from the origin of the M. sg. cer.

Insertion: Paraglossale, aponeurosis of the M. hypoglossus anterior, and connective tissue underlying the lingual epithelium. Tendinous. (For details, see next section.)

Description: Starting from their origins, both portions of the M. sg. run straight forward along the dorsal surface of the basihyale. At the level of the angle between the lingual wings, i.e., at about the caudal one-third of the Proc. centralis basihyalis (Figs. 22, 24A–25B), the collateral muscles converge and gradually merge to form a single midsagittal tendon. At about the level of the Art. paraglosso-basihyalis, the last muscle fibers join the tendon (Figs. 31A, 41B, 43, 44A). From this central tendon, a multitude of tendinous fibers branch out to insert on various sites (see below). At their branching point, these tendinous fibers interweave and overcross one another extensively, so that it is impossible to determine whether a particular fiber is derived from the ceratobranchial or basihyal portion of the M. sg.

For descriptive purposes, the tendinous fibers can be divided into five groups (parts A–E). These groups are distinguished mainly by their insertion and the course of their fibers. However, the separations are not always clear-cut, rather, there is a continuous transition from one group of fibers to the other (Fig. 31A). In some cases (e.g., Z7), there is even an actual overcrossing among the fibers of the different parts.

Part A, the caudalmost group of tendinous fibers to emerge from the central tendon, forms a ribbon-like aponeurosis that covers the caudo-medial edge of the tendinous envelope of the Gl. lingualis (Figs. 21A, 22, 44B) and inserts on the connective tissue underlying the epithelium covering the anterior lingual wing.

Part B, the apically following group of tendinous fibers, forms a broad aponeurosis that passes diagonally over the anterior end of the Gl. lingualis (Figs. 22, 42A, B), then continues as a broad ribbon-like tendon along the lateral surface of the Gl. lingualis and along the Margo dorsalis paraglossalis (Figs. 31A, 38A–42A), and finally breaks up into fiber bundles that insert on the Margo dorsalis paraglossalis and within the connective tissue supporting the lingual nail (compare Figs. 36B and 38A).

Part C, the next following group of tendinous fibers, forms an aponeurosis that follows the medio-dorsal surface of the apical end of the Gl. lingualis for a short distance (Figs. 40B, 41A). Apically from the Gl. lingualis, this group of tendinous fibers becomes increasingly thicker because it is joined by fibers branching more apically from the central tendon which is often divided into two portions at this point (Figs. 31A, 40A–41A). The tendinous fibers become also interspersed with cavernous veins, and this mass of tissue forms the caudal part of the V-shaped mounds (Figs. 22, 38B–39B; p. 68). Towards apical, the more caudal fiber bundles pass to the lateral side of the V-shaped mound and tangentially under the lingual nail, then curve medio-ventro-apically to insert on the Margo dorsalis paraglossalis (Figs. 36B–38A). On their way from the lateral side of the V-shaped mound to the Margo dorsalis paraglossalis, the fibers pass through the lateral part of the Corpus cavernosum laterale (Figs. 31A, 36B–38A). The most antero-medial fiber bundles continue beyond the apical end of the paraglossale and either insert on the internal surface of the lingual nail (Figs. 35A–36A) or insert within the connective tissue of the anterior end of the V-shaped mound (Fig. 22). Along the entire medial surface of this group of fiber bundles, tendinous fibers branch off to interdigitate with the mass of muscle bundles of the M. mesoglossus (Fig. 22). These tendinous branches remain near the surface of the tongue where they form a tough network of fibers, among which the muscle bundles of the M. mesoglossus insert, and where they insert within the connective tissue underlying the epithelial surface of the tongue.

Part D of the tendinous fibers lies medio-ventrally of part C and forms a pair of tendons diverging from the apical end of the central tendon (Fig. 31A). This pair of tendons follows the ventro-lateral borders of the mass of the M. mesoglossus (Figs. 31A, 39B). At about the level of the caudal end of the aponeurosis of the M. hypoglossus anterior, the tendons start to send branches into the mass of the M. mesoglossus (Figs. 31A, 36B–38B). These branches break up into smaller units and form a heavy fiber network that firmly adheres to the connective tissue underlying the epithelium of the lingual surface. This network covers the central area of the tip of the tongue above the M. mesoglossus anterior. At about the level of the midlength of the aponeurosis of the M. hypoglossus anterior, the rest of the part D usually passes through the lateral portion of the M. mesoglossus (Fig. 31A) and then follows the periphery of the M. mesoglossus anterior together with the Lig. paraglossale apicale with which it is closely connected. In some specimens (e.g., Z6), the tendons do not pass through the M. mesoglossus, but remain on the outside, i.e., on the latero-ventral surface, of the muscle.

The most apical of the tendinous fiber bundles forming part E arise from the apical end of the central tendon between the paired tendons of part D and radiate into the mass of the M. mesoglossus (Figs. 22, 40A). The most ventral tendinous fiber bundles follow the ventral surface of the M. mesoglossus posterior and attach to the caudal end of the aponeurosis of the M. hypoglossus anterior along the caudo-lateral border of the origin of the M. mesoglossus posterior (Figs. 31A, 43–44B). The majority of the tendinous fibers of the part E passes among the muscle bundles of the M. mesoglossus posterior and inserts on the connective tissue underlying the epithelium covering the caudal part of the central area of the tip of the tongue (Figs. 22, 43–44B).

The muscular part of the M. sg. is surrounded by various other muscles and structures. The caudal part of the dorsal surface of the M. sg. cer. is covered by the M. cricohyoideus ventralis and the laryngeal apparatus (Fig. 22) or, if present, by the M. tracheohyoideus basihyalis (Figs. 24A, B, 26B). More anteriorly, the M. sg. cer. is covered dorsally by the M. cricohyoideus dorsalis intermedius (Fig. 22) and by the M. cricohyoideus dorsalis superficialis (Figs. 44A, B). The M. sg. is framed laterally by the ribbon-like M. cricohyoideus ventralis and by the paraglossal and ceratobranchial heads of the M. tracheohyoideus (Figs. 24A, B, 26A, B). More anteriorly, the M. sg. is flanked by the M. hypoglossus obliquus medialis (Fig. 25A).

Measurements: For the M. sg. cer. of two specimens, the muscle fiber length ranged from 8–13 mm, and the muscle width at the insertion was 1 mm. The cross-sectional area of the muscle at the origin was 4×1.5 mm^2 and 3×2 mm^2, respectively. The muscle thickness at midlength was 1.5 mm. For the M. sg. bas. of two specimens, the muscle fiber length ranged from 6–10 mm, the muscle width at the origin was 4 mm, and the muscle width at the insertion was 2 mm.

Articulations: The M. sg. cer. is a two-joint muscle that passes over the dorsal sides of both the Art. cerato-basihyalis and the Art. paraglosso-basihyalis. In contrast, the M. sg. bas. is a one-joint muscle passing over the dorsal side of only the Art. paraglosso-basihyalis.

Functions: Both portions of the M. sg. probably contract synchronously. The muscle contraction has a multiple effect, namely extending the paraglossale, narrowing the lingual tip and central area of the tip of the tongue by pulling the limbs of the V-shaped mound together, flattening the dorsal surface of the tip of the tongue, and dorsally tilting the part of the lingual tip that is surrounded by the lingual nail. The M. sg. cer. also supports the flexion of the basihyale during the retraction of the hyoid and cooperates with the M. hypoglossus obliquus in the compression of the Gl. lingualis, thus increasing its turgidity (pp. 124 ff., 134 ff.).

Antagonists: For actions at the tip of the tongue: M. mesoglossus, M. hypoglossus anterior, Corpus cavernosum laterale et apicale (pp. 122 ff.); for extension of paraglossale: M. ceratoglossus; for flexion of basihyale: M. ceratohyoideus, M. ceratoglossus medialis.

Synergists: For actions at the tip of the tongue: none; for extension of paraglossale: M. hypoglossus obliquus; for flexion of basihyale: M. stylohyoideus, M. tracheohyoideus, M. ceratoglossus lateralis, M. ceratoglossus superficialis, M. hypoglossus obliquus lateralis.

Discussion: The M. supraglossus is present only in parrots. Duvernoy (1835), Lubosch (1933), and Kasai (1957) did not mention it. All the other authors described the origin of the M. sg. on the ceratobranchiale, although with varying degrees of accuracy; only Mudge (1903) described the additional origin from the basihyale. Mudge also stated that the ceratobranchial origin varies its location from the posterior end (see also Dubale and Rawal 1965) to the proximal end (see also Chaine 1905) of the ceratobranchiale. He also mentioned for certain, not closely related, parrots a third origin on the urohyale, which has not been observed in the present study of *Psittacus.*

The insertion of the M. sg. has generally been described in very vague terms; no author has even hinted at the extremely complex situation that was found in the present study of *Psittacus.* Several authors simply stated that the M. sg. inserts on the dorsal surface of the tongue or in the substance of the tongue (Nitzsch in Giebel 1862; Gadow 1891; Kallius 1906), whereas others mentioned the insertion to be on the paraglossale (Chaine 1905; Dubale and Rawal 1965). Denker (1907) mistook the origin of the M. sg. bas. on the basihyale for the insertion of the M. sg. originating on the ceratobranchiale. Only Kutorga (1832) and Mudge (1903) recognized the connection between the inserting tendon of the M. supraglossus and the M. mesoglossus. Mudge (1903) noted that *Psittacus* differs from the rest of the parrots by having a M. sg. that is continuous with the M. mesoglossus. Other species must be studied before his statement can be properly evaluated.

Steinbacher's (1951) description of the insertion of the M. sg. is the most detailed one, but refers to the special case of the brushtip-tongued parrots, the Loriinae. According to him, the M. sg. inserts partly on the lingual nail, partly on the tendon of the M. hypoglossus anterior, and the M. mesoglossus attaches to the inserting tendons of the M. sg.

Since the M. sg. is unique for parrots, the question of the derivation of this muscle arises. It seems most probable that the M. sg. is derived from the M. hypoglossus obliquus as a subdivision of the bipartite psittacine muscle. How a dorsal shift of the origin of the M. ho. med. could have taken place can be seen with the example of specimen Z4 (Fig. 12B), which has a slip of the M. sg. bas. originating from the ventral surface of the basihyale. A similar case, in which bundles of the M. sg. bas. originate on the ventral surface of the basihyale, has been described by Mudge (1903) for *Amazona.* The close relationship between the M. sg. and M. hypoglossus obliquus is further indicated by the fact that both are extremely similar in their actions on the Art. cerato-basihyalis and Art. paraglosso-basihyalis and in their cooperation in producing certain actions, e.g., those on the Gl. lingualis (pp. 114 ff.).

The M. hypoglossus obliquus changes its insertion to the dorsal side of the paraglossale also in woodpeckers (Piciformes) (W. J. Bock, pers. comm.; see also Leiber 1907: M. ceratoglossus superior).

A derivation of the M. sg. from the M. ceratoglossus (see Appendix II K, for

TABLE 12

COMPARISON OF FUNCTIONAL INTERPRETATIONS OF THE PSITTACINE M. SUPRAGLOSSUS

Author	Function
Kutorga (1832)	raises (the surface of) the apical part of the tongue
Nitzsch (in Giebel 1862)	extends the tongue and pulls the ceratobranchialia together
Steinbacher (1951)	pulls the lingual nail forward
Dubale and Rawal (1965)	depresses the tongue
Homberger, this study	extends the paraglossale, narrows the lingual tip and the central area of the tongue's tip, tilts the lingual nail towards dorsal. M. sg. cer.: supports the flexion of the basihyale during the retraction of the hyoid

the authors) is much less likely because the origin, insertion, function, and relationships to other structures of these two muscles differ completely.

Burton's (1974c) claim of the derivation of the M. sg. from the M. cricohyoideus must be rejected for the same reasons, i.e., the M. cricohyoideus is an extrinsic laryngeal muscle in contrast to the M. sg. which is an intrinsic lingual muscle. The use of the presence of these two muscles as a taxonomic character indicative of close relationship among avian taxa (Burton 1974c) is, therefore, highly questionable.

The functional interpretations of previous authors (see Table 12) differ widely.

The synonymies of the M. supraglossus are compiled in Appendix II K.

M. hypoglossus anterior (M. hg. ant., Mhga; Figs. 27B, 29B, 43).—

Diagnosis and differentiation: Originates from the posterior process of the paraglossale and occasionally from the Lig. interparaglossale, and inserts on the paraglossale, lingual nail, and ventro-apical tip of the central process of the basihyale.

Origin: Paraglossale and Lig. interparaglossale caudale. Fleshy. The main (lateral) portion of the M. hg. ant. originates from the lateral surfaces of the posterior process and central portion of the paraglossale. The origin of this main (lateral) portion of the M. hg. ant. is bordered caudo-ventrally by the Margo ventralis of the paraglossale and apico-dorsally by the insertions of the M. genioglossus and M. ceratoglossus (Figs. 12C, 13A). The posterior part of the origin of the M. hg. ant. follows the insertion of the M. genioglossus more closely than the anterior part (Figs. 12C, 27A, 40B–42A). Occasionally, some muscle fibers originate from the connective tissue of the lateral wall of the Gl. lingualis (Fig. 32B). (For details, see M. ho.) Sometimes, the origin of the main portion of the M. hg. ant. spreads between the Margo ventralis and the insertion of the M. ceratoglossus, almost to the level of the apical end of the insertion of the M. ceratoglossus (e.g., Z1; Figs. 13B, 31B). Occasionally, a smaller (medial) portion of the M. hg. ant. originates from the medial surface of the anterior process of the paraglossale just anteriorly to the Art. paraglosso-basihyalis and from the ventral surface of the Lig. interparaglossale caudale (e.g., Z4, Z6, Z7; Figs. 13B, 39A, B).

Insertion: Paraglossale, basihyale, and connective tissue underlying the lingual nail. Tendinous. (For details, see next section.)

Description: From its origin, the M. hg. ant. runs apico-ventro-medially. It merges with its collateral part at the level of the caudal border of the lingual nail, i.e., at the level of the anterior border of the Foramen interparaglossale (Figs. 16A, B, 23A, B, 38A, B). Shortly before reaching the caudal border of the lingual nail, the superficial muscle fibers on the ventro-lateral surface of the muscle body become tendinous (Figs. 16A, B, 21B, 23A, B, 38A–40B) and are successively joined by the deeper muscle fibers. The last muscle fibers turn into tendinous fibers between the caudal edge of the Synchondrosis paraglossalis (Fig. 37B) and the caudal border of the Bursae synoviales paraglossales (Figs. 43, 44A) (depending on the contractile state of the muscle).

Just apically from the caudal border of the lingual nail, a pair of bundles of superficial (ventral) tendinous fibers branches off the main tendon of the M. hg. ant. and turns laterally (Figs. 23B, 27B). The more lateral tendinous fibers of these bundles run parallel to the caudal border of the lingual nail and insert on the connective tissue underlying the lingual nail (Figs. 35A–37A). The rest of these superficial tendinous fibers insert firmly on the ventro-lateral surface of the apico-lateral tip of the anterior process of the paraglossale (Figs. 27B, 36A, 45B).

The large central tendon of each M. hg. ant. merges completely with its collateral portion and becomes firmly attached to the connective tissue underlying the lingual nail. At midlength of the lingual nail, the tendon turns dorsally around the apical edge of the Synchondrosis paraglossalis against which it is cushioned by the pair of Bursae synoviales paraglossales (Figs. 30A, B, 31A, 43–44B). On the dorsal side of the paraglossale, the tendon flattens into an aponeurosis (Fig. 31A). Along the caudo-lateral edges of the aponeurosis, tendinous fibers branch off the aponeurosis and radiate caudo-latero-dorsally to insert on the connective tissue underlying the epithelium covering the V-shaped mound. The tendinous fibers run parallel to the tendinous fibers of the portion C of the M. supraglossus and pass among the veins of the Corpus cavernosum laterale. Caudally, the aponeurosis becomes increasingly narrow and finally ends in a tendon that passes through the Foramen interparaglossale to insert on the ventro-apical tip of the central process of the basihyale (Figs. 38B–39B, 43–44B). The apico-dorsal surface of the aponeurosis of the M. hg. ant. serves as origin for the M. mesoglossus posterior (Figs. 31A, 36A–39A, 43–45A). The ventral surface of the aponeurosis fuses in its caudal part with the Fascia paraglossalis dorsalis, which separates from the aponeurosis after having passed through the Foramen interparaglossale to attach on the ventral surface of the paraglossale (Figs. 30A, 43–44B). Some tendinous fiber bundles of the portion E of the M. supraglossus attach along the caudo-lateral margin of the aponeurosis of the M. ho. ant. (Fig. 31A).

Some fibers of the central tendon of the M. hg. ant., instead of curving around the apical edge of the Synchondrosis paraglossalis, extend directly apically to insert within the connective tissue underlying the lingual nail. The fibers of this connective tissue are firmly attached ventrally to the inner surface of the lingual nail and lead to the apico-lateral tip of the Processus anterior paraglossalis. The ventral part of these fibers also serves as the site of the origin for the M. mesoglossus anterior (Figs. 35B, 43–44B). Anteriorly, the fibers of this connective tissue are concentrically arranged around the M. mesoglossus anterior and pass among the veins of the Corpus cavernosum apicale.

In summary, the main part of the M. hg. ant. forms a sling originating from the lateral surface of the paraglossale and passing around the apical edge of the Synchondrosis paraglossalis and through the Foramen interparaglossale to insert on the ventro-apical tip of the basihyale. Other parts of the M. hg. ant. attach to the apical end of the paraglossale, the lingual nail, and the connective tissue underlying the epithelium covering the V-shaped mounds after passing among the veins of the Corpora cavernosa lateralia.

Measurements: In two specimens, the muscle fiber length ranged from 4–9 mm, the cross-sectional area of the muscle at the origin measured 2×5 mm^2 and 2×6 mm^2, and the muscle width at the insertion was 1 mm. Muscle thickness at midlength was 1 mm.

Articulation: The M. hg. ant. is a one-joint muscle passing over the ventro-apical side of the Art. paraglosso-basihyalis.

Functions: The M. hg. ant. has multiple functions, but its principal function is to keep its aponeurosis, which serves as the site of origin for the M. mesoglossus posterior, taut during the simultaneous contraction of the M. mesoglossus when the central area of the tip of the tongue is being depressed. It also tilts the lingual nail ventrally and narrows the lingual tip and the central area of the tip of the tongue (pp. 123 f.).

Antagonists: For actions at the tip of the tongue: M. supraglossus, M. hypoglossus transversus, Corpus cavernosum laterale.

Synergist: For depression of the central area of the tip of the tongue: M. mesoglossus.

Discussion: The M. hypoglossus anterior is not present in all birds (Gadow 1891; Bock 1972, 1978), but it is found in all parrots. Most authors agreed that the psittacine M. hg. ant. originates on the lateral surface of the posterior end of the paraglossale (Kutorga 1832; Duvernoy 1835; Mudge 1903; Kallius 1906; Lubosch 1933; Steinbacher 1951; Dubale and Rawal 1965; Rawal 1970). Nitzsch (in Giebel 1862) placed its origin on the apical tip of the basihyale. Only Kallius (1906) described an additional small portion originating from the medial surface of the anterior process of the paraglossale.

In contrast, opinions on the location of the insertion of the M. hg. ant. differ widely. The M. hg. ant. inserts on the ventral surface of the paraglossale according to Kutorga (1832), Nitzsch (in Giebel 1862) and Rawal (1970), on the skin of the tip of the tongue according to Duvernoy (1835), or within the tip of the tongue as vaguely stated by Lubosch (1933) and Dubale and Rawal (1965). Kallius (1906) described the insertion to be on a pair of cartilages lying ventrally to the anterior tip of the paraglossale. These cartilages are actually sesamoid bones of the tendons of the M. hg. ant. and seem to be unique to *Melopsittacus* (see also Evans 1969). Mudge's (1903) description is more detailed, although not accurate. The tendon of the M. hg. ant. inserts partly within the apex of the tongue, partly into the "fascia of the M. mesoglossus." Steinbacher (1951) presented the most accurate description, stating that the M. hg. ant. inserts on the lingual nail but also fuses with the tendons of the M. supraglossus and M. mesoglossus. Nevertheless, even he did not fully appreciate the structural complexity of the M. hg. ant. Gadow (1891) and Kasai (1957) did not describe the M. hg. ant., whereas Chaine (1905) and Burton (1974c) provided only figures of it.

The conflicting views on the functional properties of the psittacine M. hg. ant. reflect the inaccuracy of the anatomical descriptions (see Table 13).

The synonymies of the M. hg. ant. are compiled in Appendix II L. The name "hypoglossus" or "hyoglossus" presumably reflects the opinion that the M. hg.

TABLE 13

COMPARISON OF FUNCTIONAL INTERPRETATIONS OF THE PSITTACINE M. HYPOGLOSSUS ANTERIOR

Author	Function
Kutorga (1832)	stretches the skin of the underside of the tongue
Duvernoy (1835)	flexes and pulls back the fleshy tip of the tongue and also shortens and enlarges it
Nitzsch (in Giebel 1862)	pulls down the tip of the tongue
Mudge (1903)	depresses, together with the M. mesoglossus, the apex of the tongue
Steinbacher (1951)	pulls the lingual nail backwards and erects the papillae at the tip of the tongue (in Loriinae)
Dubale and Rawal (1965)	depresses the tongue, makes the superior surface of the tongue's tip concave
Rawal (1970)	helps the lateral movements of the tongue
Homberger, this study	cooperates with the M. mesoglossus in the depression of the central area of the tip of the tongue, tilts the lingual nail ventrally, narrows the lingual tip and the central area of the tip of the tongue

ant. is derived from the M. hypoglossus obliquus (e.g., Gadow 1891), whereas the name "ceratoglossus anterior" is probably based on the view that the M. hg. ant. is derived from the M. ceratoglossus. The present study on *Psittacus,* however, does not provide any clue as to which opinion may be correct.

M. hypoglossus transversus (M. hg. trans., Mhgt; Fig. 31B).—

Definition and differentiation: Bridges the ventral surfaces of the paired paraglossalia, apically from the Art. paraglosso-basihyalis.

Origin: Paraglossale. Fleshy, partly tendinous. The origin of the M. hg. trans. lies on the ventro-lateral surface of the Proc. anterior paraglossalis apico-medially to the insertion of the M. ceratoglossus on the Tuberositas paraglossalis and to the origin of the main portion of the M. hypoglossus anterior (Figs. 13B, 31B).

Insertion: The M. hg. trans. runs without interruption from origin to origin.

Description: Parallel-fibered, ribbon-like. The M. hg. trans. spans the paraglossalia from origin to origin and covers the anterior half of the Foramen interparaglossale ventrally. On its ventral surface, the M. hg. trans. is completely covered by the M. hypoglossus ant. (Fig. 31B). It was found in only one (Z1) of seven specimens.

Measurements: Not available.

Functions: Counteracts the medio-dorsal pull of the tendinous portion C of the M. supraglossus on the two halves of the paraglossale by exerting a medio-ventral pull on the ventral surface of the two halves of the paraglossale. It also counteracts the medial push exerted on the Proc. anterior paraglossalis by the inflating Corpus cavernosum laterale (pp. 123 f., 126 ff.).

Antagonists: M. supraglossus, M. hypoglossus anterior, Corpus cavernosum laterale.

Synergist: M. ceratoglossus.

Discussion: The M. hg. trans. has not been described previously. It is probably derived from a slip of the M. hypoglossus anterior. The significance of its highly variable presence in *Psittacus* cannot be properly evaluated without more comparative data. In *Agapornis roseicollis,* the M. hg. trans. was found in one of three specimens, whereas it was not present in two specimens of *Cacatua roseicapilla* (pers. obs.).

M. mesoglossus (M. mg., Mmg; Fig. 43).—

Diagnosis and differentiation: A mass of muscle tissue occupying the central area of the tip of the tongue. Depending on its origin, two confluent portions can be distinguished, namely the M. mg. anterior (M. mg. ant.) originating from the lingual nail and the M. mg. posterior (M. mg. post.) originating from the aponeurosis of the M. hypoglossus anterior. The M. mg. inserts on the connective tissue underlying the epithelial cover of the central area of the lingual tip.

Origin: Lingual nail and aponeurosis of M. hypoglossus anterior. Fleshy or by very short tendons. The M. mg. ant. originates from connective tissue that is firmly attached to the ventral inner surface of the ventral part of the lingual nail (Figs. 35B, 43–44B). The area of the origin of the M. mg. ant. is roughly circular (Figs. 30A–31A). The M. mg. post. originates from the apical and dorsal surfaces of the aponeurosis of the M. hypoglossus anterior up to the point where the aponeurosis narrows to a tendon and passes through the Foramen interparaglossale. Thus, the area of origin of the M. mg. post. is roughly triangular (Figs. 31A, 36A–39A, 43–44B; pp. 60 ff.).

Insertion: Connective tissue underlying the epithelial surface of the central area of the lingual tip. Tendinous. The insertion covers the entire triangular central area bordered caudo-laterally by the V-shaped mound and apically by the Corpus cavernosum apicale (Figs. 22, 43–44B).

Description: Consists of a multitude of parallel-fibered muscle bundles interlaced with connective tissue.

The muscle fibers of the M. mg. ant. form a group of muscle bundles which slightly diverge towards their insertions (Figs. 27A, B, 29A–31A, 43–45A). These muscle bundles pass through the network of tendinous fibers formed by the tendinous portions C and D of the M. supraglossus to insert on the connective tissue (Propria mucosae) underlying the epithelium of the lingual surface. This network of tendinous fibers is firmly attached to the underside of the epithelium. The group of muscle bundles of the M. mg. ant. is surrounded apically by the belt-like or sheet-like Lig. paraglossale apicale (Figs. 27A, B, 29A–31A) and by the Corpus cavernosum apicale (Figs. 43–44B).

The muscle fibers of the M. mg. post. follow those of the M. mg. ant. without interruption (Figs. 43–44B). From their origin on the dorsal surface of the aponeurosis of the M. hypoglossus anterior, the muscle fibers diverge dorso-laterally and dorso-caudally (Figs. 31A, 43–45A). Near their origin the muscle bundles are rather closely packed, but toward the surface of the tongue they become more and more interspersed with connective tissue and with the tendinous fibers from the portion E of the M. supraglossus which attach directly to the connective tissue underlying the epithelium of the central area of the lingual surface (Fig. 22). Before the muscle bundles insert on the connective tissue underlying the epithelium, they pass through a network of tendinous fibers formed by the portion C of the M. supraglossus (Fig. 22).

Measurements: Not available.

Articulations: None. The M. mg. does not attach to any skeletal structures in the proper sense.

Function: Pulls the epithelium of the tongue surface ventro-medio-apically, thus depressing the central area of the tip of the tongue (pp. 123 ff.).

Antagonists: M. supraglossus, Corpora cavernosa.

Synergist: M. hypoglossus anterior.

Discussion: The M. mesoglossus is unique to parrots but has been described only by few authors working on the psittacine lingual musculature. It was not mentioned by Duvernoy (1835), Nitzsch (in Giebel 1862), Gadow (1891), Denker (1907), Lubosch (1933), Kasai (1957), Dubale and Rawal (1965), or Rawal (1970). Chaine (1905) provided only a labeled figure as a description of the M. mg. All other students of psittacine lingual muscles agreed that the M. mg. lies within the dorsal depression formed by the two halves of the paraglossale, but every one of them described the origin and insertion of the M. mg. differently. Kutorga (1832) described the M. mg. simply as a pear-shaped muscle with longitudinally arranged muscle fibers of which the medial ones are longest and the peripheral ones are shortest. Mudge (1903) recognized that the M. mg. attaches partly on the tendon of the M. supraglossus and partly on the M. hypoglossus anterior and the epithelium of the lingual tip, but did not provide structural details of these attachments. He also stated that *Nestor* and *Calyptorhynchus* lack a M. mg. and that in the rest of the parrots the muscle varies from only a few muscle fibers within a mass of connective tissue to a well developed muscle mass with only little interspersed connective tissue. He found the latter condition in *Psittacus,* which agrees with my own observation.

The descriptions of Kallius (1906) and Steinbacher (1951) are relatively detailed but refer to special cases within the Psittacidae. The M. mg. of *Melopsittacus* originates from the sesamoid bones of the M. hypoglossus anterior and radiates with a large amount of muscle fibers into the tongue where it interdigitates with fibers of the M. supraglossus (Kallius 1906). In the brushtip-tongued Loriinae, the M. mg. originates from the surface cover of the tongue, from the paraglossale, and from the tendon of the M. supraglossus and inserts on the base of the caudomedial epithelial papillae and on the tendon of the M. supraglossus (Steinbacher 1951).

Dubale and Rawal (1965) did not describe the M. mg. as a separate muscle, but their description of the M. hypoglossus anterior can be interpreted to imply that the authors considered the M. mg. a part of the M. hg. ant. They described the M. hg. ant. as inserting fleshily on the tip of the tongue and functioning in depressing the surface of the tip of the tongue. Both attributes are correct for the M. mg. but not for the M. hg. ant.

Burton (1974c) stated that the M. mg. originates from the anterior tip of the basihyale and inserts broadly on the paraglossale. This description, however, fits no lingual muscle in the parrot tongue. Burton's figures do not help to clarify the situation.

It seems probable that the M. mg. is a unique acquisition of parrots and that it is derived from the M. hg. ant. The origin of the M. mg. on the tendon and aponeurosis of the M. hg. ant. suggests that the M. mg. may originally have been the distal part of the M. hg. ant. that had expanded apically and turned to the dorsal surface of the lingual tip. Further support of the hypothesis that the M. mg. is derived from the M. hg. ant. lies in the fact that both muscles must cooperate to perform the action of depressing the lingual surface.

Burton's (1974c) claim that the M. mg. may be homologous to what he calls the "M. hypoglossus medialis" in certain birds is problematic. Because his description of the M. mg. is incorrect, the resemblance of "his" M. mg. to any other

TABLE 14

COMPARISON OF FUNCTIONAL INTERPRETATIONS OF THE PSITTACINE M. MESOGLOSSUS

Author	Function
Kutorga (1832)	together with the M. hypoglossus obliquus, it makes the fleshy tip of the tongue thicker, rounder, and shorter and bends it
Mudge (1903)	together with the M. hypoglossus anterior, it either acts on the papillae at the tip of the tongue of Loriinae or *Nestor* or depresses the apex of the tongue
Steinbacher (1951)	pulls the papillae of the tongue's tip back and downwards, thereby folding them (only in Loriinae)
Dubale and Rawal (1965)	depresses the surface of the tongue (see Discussion of M. mesoglossus)
Homberger, this study	depresses the central area of the tongue's tip

muscle is uncertain. In addition, it is doubtful that the "M. hypoglossus medialis" is a distinct muscle and different from the M. hg. ant. Burton (1974c) admitted that the two muscles are easily confused, but maintained that the "M. hypoglossus medialis" is characterized by its origin on the basihyale. It is known, however, that the M. hg. ant. does not always originate from the paraglossale, but in certain birds from the basihyale (W. J. Bock, pers. comm.; Boullion and Homberger 1982). For this reason, conclusions on systematic relationships among bird orders based on the presence of the M. mg. and "M. hypoglossus medialis" (Burton 1974c) should be carefully reexamined.

Different functions have been attributed to the M. mg. by various authors (see Table 14). Future comparative studies will have to determine to what degree the different functional interpretations actually reflect morphological differences of the muscle in different taxa.

The synonymies of the M. mg. are compiled in Appendix II M.

BURSAE SYNOVIALES

INTRODUCTION

Two Bursae synoviales are present in the tongue of *Psittacus.* These structures have not been described previously for the avian lingual apparatus.

Bursae synoviales are essentially fluid-filled bags located at pressure points between prominent skeletal ridges on one side and overcrossing tendons on the other. The wall of a bursa has a structure similar to that of an articular capsule; it is pliable, but not distensible due to the orientation of collagen fibers along the lines of greatest tension (Bargmann 1962). A bursa is filled with synovial fluid which, like all liquids, is practically incompressible. In addition, it is viscoelastic, i.e., it resists being squeezed out between two surfaces (Alexander 1968, Fung 1981). A bursa can act as a water cushion ("Wasserkissen"), the function of which is to distribute pressure exerted on it (Bargmann 1962).

The bursal wall is firmly attached to the skeletal ridge on one side and to the surface of the overcrossing tendon on the other. When the tendon is put under tension, either because its muscle contracts or because the skeletal ridge is pushed against the tendon, the underlying bursa is compressed beneath the tendon and bulges out at its periphery. Because of its fluid content and the nondistensibility of its walls, the bursa maintains a fluid-filled space between the tendon and the bony ridge. In this way, the tendon can slide over the bursa and, hence, over the skeletal ridge instead of gliding directly over the skeletal ridge under considerable pressure and friction. In addition, the presence of a bursa between a skeletal ridge and an overcrossing tendon helps to distribute the pressure exerted by the tendon on the skeletal ridge over a broader area. The viscoelastic property of its content also helps a bursa readjust the position of the tendon and the space between the tendon and the skeletal ridge as the tension exerted on the tendon subsides.

Description and Discussion

Bursa synovialis urohyalis (BTuh).—This bursa lies between the ventral surface of the urohyale on one side and the Nodulus and part of the midsagittal raphe separating the collateral parts of the M. ceratohyoideus on the other side (Figs. 23B, 24B, 27A, 28A, B, 29B).

The Nodulus slides sagittally over the bursa when pulled forward during retraction of the hyoid or when pulled backward during protraction of the hyoid. During retraction, the bursa is compressed because the ventro-apically rotating urohyale is pushed against the Nodulus, which is held against the urohyale by the Lig. nodulo-ceratobranchiale and the contracting M. serpihyoideus. During protraction, the bursa is compressed because the contracting M. ceratohyoideus pulls the Nodulus against the urohyale. When the hyoid moves from side to side, the Nodulus is pulled laterally and slides transversally over the bursa.

Bursa synovialis paraglossalis (BTpg).—This bursa is paired and serves as a cushion between the single common tendon of the Mm. hypoglossi anteriores and the paired mounds on the apical edge of the Synchondrosis paraglossalis (Figs. 30A–31A, 32A, 43–44B).

The tendon of the M. hypoglossus anterior slides over the Bursa synovialis paraglossalis after being pulled caudo-ventrally when the epithelial surface of the lingual tip is depressed, being pulled caudo-dorsally when the epithelial surface of the lingual tip is flattened, or when the paraglossale is extended. During depression of the epithelial surface of the lingual tip, the bursa is compressed because the contracting M. hypoglossus anterior pulls its tendon caudally against the edge of the Synchondrosis paraglossalis. In contrast, during flattening of the epithelial surface of the lingual tip, the bursa is compressed because the contracting M. supraglossus pulls the tendon of the M. hypoglossus anterior caudally against the apical edge of the Synchondrosis paraglossalis, especially via its tendinous portion E. During extension of the paraglossale, however, the bursa is compressed because the apical edge of the Synchondrosis paraglossalis is pushed apically against the overcrossing tendon of the M. hypoglossus anterior.

The Tongue; General Shape, Surface Structures, and Relationships to Underlying and Surrounding Structures

The tongue apparatus can be divided into three major portions, namely the free portion of the tongue, the hyoid horns, and the laryngeal area (Homberger 1980a).

The free portion of the tongue is fleshy and elongated. Its cross-section is round in its caudal part and becomes crescent-shaped toward the tip (Figs. 34–42B). It is attached to the mandible along the border of the rhamphotheca ("Radix linguae" of McLelland 1979a, "Rhamphothekenwurzel" of Homberger 1980a; Fig. 15A). The covering epithelium of the entire portion is heavily pigmented.

On the ventral and lateral sides, most of the free portion of the tongue is covered with an epithelium that is loosely attached to the underlying tissues, has a wrinkled appearance and bears only some few blunt papillae. These wrinkles and folds represent tissue reserves so that the epithelium can be stretched when the tongue is protracted or lifted, that is, when the distance between the tip of the tongue and the lingual attachment to the mandible increases. The orifices of the Gll. sublinguales are found on the ventro-lateral sides of the lingual body about midway between the attachment of the tongue to the mandible and the caudal border of the lingual nail (Fig. 20). The epithelium is underlain mainly by connective tissue and muscles, in particular, by the M. genioglossus (Figs. 20, 40A–42B, 44A–45B) and, more apically, by the tendon of the M. ceratoglossus and by the M. hypoglossus anterior (Figs. 38B–41A, 43–44B).

The ventral and lateral surfaces of the tip of the tongue are covered with a broad band of horn-like, very darkly pigmented epithelium, which is the lingual nail ("Cuticula cornea lingualis" of McLelland 1979a, "Nagel" of Homberger 1980a). On the ventral side, the lingual nail is firmly attached to the tendon of the M. hypoglossus anterior and to the connective tissue serving as the site of origin for the M. mesoglossus anterior (Figs. 35A–38A, 43–44B). On the lateral side, the lingual nail is cushioned against the paraglossale by the Corpus cavernosum laterale (Figs. 35A–38A). The caudal border of the lingual nail (Radix cuticulae; "Nagelwurzel" of Homberger 1980a) is clear-cut and somewhat bulging. At the apical border, however, the lingual nail becomes less pigmented and thinner and ends level with the apically adjoining band of epithelium ("Schleimhautrand" of Homberger 1980a), which is about one-third of the length of the lingual nail (Figs. 15A–16B, 20, 21B, 43–45B). This epithelium is especially thick and forms a cap covering the very tip of the tongue (Figs. 43–45B). This cap of thick epithelium is supported by a mass of connective tissue which contains the Corpus cavernosum apicale (Figs. 34, 43, 44A).

On the dorsal side, the free portion of the tongue can be divided into several portions (Fig. 23A). The body of the tongue ("Corpus linguae" of McLelland 1979a, "Zungenkörper" of Homberger 1980a) extends from the apical end of the tongue to the angle between the lingual wings (Angulus alarum linguae; "Zungenflügelwinkel" of Homberger 1980a). The Corpus linguae can be divided into the actual tip of the tongue ("Apex linguae" of McLelland 1979a, "Zungenspitze" of Homberger 1980a, "lingual tip" of Zweers 1982) and the Dorsum linguae (McLelland 1979a, "Zungenrücken" of Homberger 1980a). The tip of the tongue comprises the anterior portion of the tongue that is surrounded by the lingual nail. It is somewhat broader than the dorsum when the Corpora cavernosa are inflated. The surface of its central area bears radially arranged grooves (Sulci radiales; "Radiärrinnen" of Homberger 1980a) which remain clearly visible only to about the level of the Radix cuticulae (Figs. 21A, 34–35B). Laterally, the central area of the tip of the tongue (Area centralis apicis linguae) is framed by the limbs of the V-shaped mound ("lippenartige Wülste" of Homberger 1980a) which converge caudally on the Dorsum linguae midway between the Radix cuticulae and the angle between the lingual wings (Fig. 21A). The epithelium covering the V-shaped mound is firmly attached to the underlying tendinous fibers of the portion C of the M. supraglossus (Figs. 22, 35A–41A). The limbs of the V-shaped mound are medially bordered by deep grooves that mark the boundary line between them and the central area of the Apex linguae (Figs. 36A–38A). The epithelial cover of the central area of the lingual tip is underlaid by and serves as the insertion site for the M. mesoglossus and for some of the tendinous fibers of the portions C, D and E of the M. supraglossus (Figs. 23, 43). The epithelium is pulled back by the contraction of the M. supraglossus, depressed and pulled forward by the contraction of the M. mesoglossus, and lifted by the inflation of the Corpora cavernosa (pp. 123 ff.).

The Dorsum linguae, which extends from the caudal border of the lingual nail to the angle between the lingual wings, has a smooth surface and bears a midsagittal groove along its caudal part. The epithelial cover is firmly attached to the underlying tendinous fibers of the portions B and C of the M. supraglossus which pass over the anterior tip of the Glandula lingualis (Figs. 41B–42B). The caudal tip of the posterior process of the paraglossale does not reach quite as far as the caudal end of the dorsum (Fig. 45B). Caudally, the Dorsum linguae splits into the paired lingual wings (Ala linguae; "Zungenflügel" of Homberger 1980a, "lingual alae" of Zweers 1982). The lingual wings are fleshy lobes that project caudolaterally from the Corpus linguae and overhang the laryngeal area (Figs. 21A, 22, 44B–45B). These lingual wings consist of an anterior and a posterior portion of about equal length. The anterior lingual wing (Ala linguae anterior; "apikaler Abschnitt des Zungenflügels" of Homberger 1980a) is a direct extension of the dorsum and has the same surface structure. Its main supporting structure is the glandular body of the Gl. lingualis which extends apically to slightly beyond the midlength of the dorsum (Figs. 21A, 40B–42B, 45A, B). The epithelial cover is firmly attached to the underlying tendinous fibers of the portions A and B of the M. supraglossus and to the tendinous envelope of the Gl. lingualis formed by the M. hypoglossus obliquus (Figs. 22, 45A). The glandular orifice is found laterally on the anterior lingual wing only slightly caudally from the angle between the lingual wings (Fig. 21A). A row of distinct lingual papillae (Papillae linguales; "Zungenpapillen" of Homberger 1980a) borders the medial and caudal margins of the anterior lingual wing and marks the boundary line towards the posterior lingual wing (Ala linguae posterior; "caudaler Abschnitt des Zungenflügels" of Homberger 1980a) (Figs. 21A, 44B–45B). This posterior lingual wing is narrower than its anterior counterpart and is separated from the latter by a distinct medial notch (Fig. 45A). Its surface is strewn with papillae (Figs. 20, 21A). The posterior lingual wing, unlike the anterior lingual wing, is just an epithelial fold loosely attached to the underlying connective tissue (Figs. 45A, B). It serves as tissue reserve and is stretched when the paraglossale is flexed and, thus, the distance between the preglottal area (see below) and the anterior lingual wing considerably increased (pp. 116 f.).

Laterally, the lingual wing is flanked by an elongated mound formed by the glandular body of the Glandula mandibularis (Figs. 20, 21A). The single glandular orifice is found opposite the row of lingual papillae at the caudal border of the anterior lingual wing (Fig. 21A). The mound is separated from the lingual wings by a shallow groove. Caudally, the mound turns into a skinfold that merges with the skinfold coming from the caudal margin of the posterior lingual wing (Figs. 20, 21A). The skinfold flanks the anterior half of the laryngeal area before making a lateral turn at an angle of 90° to continue as an epithelial crest towards the palate (Homberger 1980a).

The surface of the laryngeal area of the tongue is underlain mainly by the laryngeal apparatus and its extrinsic and intrinsic musculature, but the surface structures of this area are functionally part of the lingual apparatus and, therefore, are treated here. The laryngeal area itself can be divided into two portions, namely a preglottal area followed posteriorly by the laryngeal mound (Figs. 20, 21A).

The preglottal area (Area praeglottalis; "Zungengrund" of Homberger 1980a, "lingual base" of Zweers 1982) is rather short and extends from the Angulus alarum linguae backwards to the apical end of the glottis (Fig. 21A). Laterally, it is bordered by the lingual wings by which it is partially overhung. The area is underlain by extrinsic laryngeal muscles, namely the superficial M. cricohyoideus dorsalis superficialis et intermedialis (Homberger 1979b).

The laryngeal mound ("Mons laryngealis" of McLelland 1979a, King 1979; "Larynxplatte" of Homberger 1980a) extends from the apical end of the glottis to the beginning of the esophagus. It is narrow apically and broadens caudally. The anterior two-thirds of the laryngeal mound is midsagittally split by the glottis. The glottis ("Larynxspalte" of Homberger 1980a, "larynx rostralis" of Zweers 1982) is wider apically than caudally (Fig. 21A). It is framed laterally by thick lips that are formed by the medial edges of the arytenoid cartilages and that converge caudally to form an acute angle. The epithelium covering the laryngeal mound is smooth in the apical part but strewn with large papillae pointing backwards in the caudal part. The laryngeal papillae near the midsagittal line and along the caudal border ("laryngeal flaps" of Bock 1972, 1978; "Larynxpapillenreihe" of Homberger 1980a; "pharyngeal valves" of Zweers 1982) are especially large. Most of the epithelium is directly underlain by the M. cricohyoideus dorsalis superficialis et profundus (Fig. 21A). The strip of epithelium bearing the caudalmost row of papillae serves as the insertion site for the M. cricohyoideus dorsalis superficialis et profundus (Homberger 1979b).

The hyoid horns of the tongue apparatus (Cornua lingualia) are formed by the ceratobranchialia, the epibranchialia, the cartilaginous processes of the epibranchialia, and the surrounding muscles. The posterior part of each hyoid horn (up to the point where the M. branchiomandibularis posterior branches out laterally to attach to the mandibular ramus) is vested in a sleeve-like sheath of connective tissue, the Fascia vaginalis hyoidei, in which it can glide back and forth. When the lingual apparatus is retracted, the distal tip of the cartilaginous process of the epibranchiale reaches the distal end of the sheath. The hyoid horn lies along the medial surface of the jaw (pterygoid) musculature covering the medial side of mandibular ramus. Posteriorly, it passes between the neck musculature and the caudo-ventral corner of the mandibular ramus and then curves dorsally (Figs. 15A, B). The Fascia vaginalis hyoidei is held in place by connective tissue. The distal end of the sheath is attached by strong fasciae and connective tissue to the neck musculature close to its insertion to the skull.

CAVERNOUS VASCULAR TISSUE (CORPORA CAVERNOSA)

INTRODUCTION

Cavernous vascular tissue has been found previously in the tongues of various birds, such as the domestic duck, *Anas platyrhynchos,* and the domestic goose, *Anser anser* (Stadtmüller 1938; Preuss et al. 1969), the parakeet *Melopsittacus undulatus* (Kallius 1906; Feder 1969), various unspecified parrots (Stadtmüller 1938; Portmann 1950), the domestic pigeon, *Columba livia* (Preuss et al. 1969), hummingbirds (Weymouth et al. 1964), and seed-eating oscines (Krulis 1978; Ziswiler 1979, 1980). Functional interpretations of these cavernous vascular tissues have varied from aiding in adjusting the tongue surface to fit the palatal surface (Preuss et al. 1969) to stiffening of the tongue to push seeds against the palate during seed-husking (Krulis 1978; Ziswiler 1979).

The lingual tip of *Psittacus* contains extensive cavernous vascular tissues. For descriptive purposes, an unpaired Corpus cavernosum apicale can be distinguished from the paired Corpora cavernosa lateralia, although all portions of the cavernous vascular tissue are confluent and, therefore, probably inflate and deflate synchronously. The cavernous bodies consist of a mass of more or less coiled, expandable (i.e., "cavernous") veins which can be inflated through an increase in blood volume. The veins are embedded in connective tissue. Additional bundles of tendinous fibers from various intrinsic lingual muscles or bundles of ligamentous fibers pass in specific arrangements among the veins. This mass of cavernous veins and connective tissue fibers forming the cavernous vascular tissue is not surrounded by a special connective tissue sheet ("tunica" of Bargmann 1962) but lies between skeletal elements and muscle masses on one side and epithelium and Tunica propria mucosae on the other.

Neither the detailed histomorphological structure nor the physiology of filling and emptying mechanisms of the cavernous veins have been described for the avian tongue. It can be assumed for purposes of discussion, however, that the filling and emptying of the lingual Corpora cavernosa in birds proceed in a fashion similar to that of human cavernous vascular tissue (cf. Bargmann 1962; Dym 1983; but see McConnell 1982 for a critical review). Therefore, filling of the lingual cavernous bodies is probably effected by an influx of blood into the cavernous veins and subsequent constriction of venous sphincters. In turn, emptying of these cavernous bodies probably starts with a relaxation of the venous sphincters, proceeds with a contraction of the musculature in the wall of the veins and is supported by elastic recoil of the surrounding tissue and by contractions of adjacent muscles.

In contrast to salivary glands and Bursae synoviales, Corpora cavernosa can be termed "active hydrostatic structures" because their internal pressure can be varied by differential fluid content rather than by actions of external forces. Although the histology and physiology of the cavernous bodies in *Psittacus* have not been studied yet, the effect of filling and emptying of the cavernous bodies on the surrounding tissue can be inferred from presently available data. When the cavernous veins inflate, they push apart the tendinous or ligamentous fibers passing among them. Because these fibers are not extensible, the inflating cavernous veins cannot elongate them. This means that the cavernous bodies cannot expand in those directions that would result in elongation of contained tendinous or ligamentous fibers. In this way, the lingual cavernous bodies expand in specific directions and appropriately change the surface shape of the tongue instead of simply inflating it overall.

Increase in turgidity of the cavernous bodies is largely determined by the distensibility of the epithelium and its Tunica propria (see below and pp. 89 ff.). With increasing blood volume and inflation of the cavernous veins, the tension on the epithelial surface covering of the Corpora cavernosa and the turgidity or internal pressure of the cavernous bodies increase.

Description and Discussion

Corpus cavernosum apicale.—The unpaired Corpus cavernosum apicale occupies that part of the lingual tip which lies in front of the M. mesoglossus anterior and is covered by a cap of thickened epithelium (Figs. 34, 43–44B). The cavernous veins are contained among connective tissue fibers that are arranged in concentric layers in front of the M. mesoglossus anterior. These ligamentous fibers form an apical extension of the Lig. paraglossale apicale and attach to the apico-lateral tips of the anterior processes of the paraglossalia (pp. 60 ff.). The Corpus cavernosum apicale is heavily interspersed with touch receptors (Herbst's corpuscles; pers. obs.). The functioning of the Corpus cavernosum apicale and its interactions with surrounding tissues are described on pp. 126 ff.

Corpus cavernosum laterale.—The paired Corpora cavernosa lateralia extend on each side of the tongue from about the apical end of the body of the Gl. lingualis and the caudal end of the V-shaped mound (consult Fig. 21A) to the tip of the tongue. There they join the Corpus cavernosum apicale with which they communicate through large lacunae (Figs. 35A, B). At the caudal end of the Corpus cavernosum laterale, cavernous veins are embedded among the tendinous fibers of the portion C of the M. supraglossus; this tissue mass forms the structural basis of the V-shaped mound (Figs. 39A–40B). From the level of the Radix linguae on forward, the Corpus cavernosum laterale is traversed by those tendinous fibers of the portion C of the M. supraglossus that attach tangentially to the dorsal part of the lingual nail and finally insert on the Margo dorsalis paraglossalis and by those tendinous fibers of the M. hypoglossus anterior (M. hg. ant.) that branch caudo-laterally off the margin of the aponeurosis of the M. hg. ant. and insert on the epithelium covering the lateral border of the V-shaped mound and on the dorsal part of the lingual nail (Figs. 36A–38A). The Corpus cavernosum laterale extends ventrally into the region between the paraglossale and the lingual nail (Figs. 36A–37A). This ventral portion of the Corpus cavernosum laterale is traversed by those tendinous fibers of the M. hg. ant. that branch off its tendon and insert on the lingual nail and on the apico-lateral tip of the Proc. anterior paraglossalis (Figs. 36A–37A). The function of the Corpus cavernosum laterale and its interactions with the surrounding tissues are described on pp. 126 ff.

Salivary Glands

Introduction

Lingual salivary glands in birds have been described or at least mentioned by many authors (e.g., Cholodowsky 1892; Kallius 1906; Leiber 1907; Heidrich 1908; Greschik 1913, 1921, 1928; Antony 1920; Scharnke 1931; Groebbels 1932; Steinbacher 1935; Fahrenholz 1937; Bock 1961; Medway 1962; Fitzgerald 1969; Foelix 1970; Bock and Morioka 1971; Ziswiler and Farner 1972; Bock et al. 1973; Burton 1974a, b; Hodges 1974; McLelland 1975, 1979a, b; Nickel et al. 1977; Zweers et al. 1977; Fährmann 1978; Kilham 1979; Navas et al. 1980a, b; Zweers 1982). Additional references dealing specifically with psittacine salivary glands were reviewed by Homberger (1980a). Fahrenholz (1937) reviewed some of the older literature concerning avian salivary glands, and Zweers (1982) synonymized avian salivary glands which were previously, but independently, described by various authors.

Functional-physiological studies of salivary glands in birds, as with vertebrates in general, have focused on the chemical composition and action of saliva and on the role of saliva in deglutition, thermoregulation, and digestion (e.g., Young and van Lennep 1978). In contrast, the mechanical functions of the salivary glands have received little attention although interactions between secretory glands and surrounding and attaching muscles have been analyzed for venom glands in snakes (e.g., Rosenberg 1967; Jansen and Foehring 1983). Bargmann (1962) referred to the mechanical interactions between salivary glands and surrounding muscles which massage saliva out of the duct system of the gland by contracting and, thus, bulging. Both Leiber (1907) and Fahrenholz (1937) mentioned that in birds lingual muscles often attach to salivary glands, but only Leiber (1907) presented a mechanical interpretation of this structural condition, in particular that it helps to hold the salivary glands in place during lingual movements.

Salivary glands consist of masses of secretory cells that are arranged around a duct system into which part of the content of the secretory cells is released to form saliva. Each cell consists of fluid enclosed in a membrane and, thus, represents a closed hydrostatic system (cf. Bonik et al. 1978). The entire salivary gland is enveloped in a layer of connective tissue (Hodges 1974; Young and van Lennep 1978). Therefore, the entire glandular body can act as an hydrostatic skeletal element if it is compressed by external forces and if the duct system is emptied and collapsed. (The duct system is reopened by the internal hydrostatic pressure of the newly secreted saliva.) Depending on their location and on the nature of the external forces to which they are subjected, the salivary glands can have various specific mechanical functions.

Parrots usually have four salivary glands associated with the lingual apparatus, namely the paired lingual, mandibular and sublingual glands and the unpaired preglottal gland (Homberger 1980a). As an exception among parrots, the Glandula praeglottalis ("unpaare Zungendrüse" of Homberger 1980a) is not present in *Psittacus.* The psittacine lingual salivary glands are mucous glands (pers. obs.) like most avian salivary glands. Unlike the salivary glands of other birds, the psittacine glands form compact glandular bodies. In *Psittacus,* the duct systems of all lingual salivary glands open on the lingual surface, each through a single orifice, i.e., the salivary glands are monostomatic. The lingual salivary glands of *Psittacus* exemplify the wide range of mechanical functions that are possible for glandular structures in general.

Description and Discussion

Glandula lingualis (Gl).—The paired lingual salivary glands ("paarige Zungendrüse" of Homberger 1980a) consist of a pair of compact, ovoid glandular bodies, each with a single duct opening on the dorsal surface of the tongue near the lateral lingual border and at the level of the Angulus alarum (Figs. 20, 21A). Each glandular body is wrapped in an envelope formed by tendinous fibers of the M. hypoglossus obliquus and by collagenous connective tissue (Figs. 41A–42B, 44B–45B). The floor of the caudal half of this envelope serves as the insertion site for the M. hypoglossus obliquus lateralis and for part of the M. hypoglossus obliquus medialis (Figs. 29A, B, 32A, B), whereas the apical half of this envelope is firmly attached to the dorso-medial surface of the posterior process of the paraglossale (Figs. 24A–25A, 29A, B, 32A, B, 40B–42B, 45A, B). The anterior end of the enveloped glandular body is covered by a sheet consisting of the portion C of the tendinous fibers of the M. supraglossus (Figs. 21A, 22). A ribbon-like sheet formed by portion A of the tendinous fibers of the M. supraglossus covers the medio-caudal border of the glandular body (Figs. 21A, 22).

The body of the Gl. lingualis functions as an hydrostatic skeletal element which serves as a caudal extension of the posterior process of the paraglossale. In contrast to massive bony elements, an hydrostatic element, such as a glandular body, yields initially to external forces and, therefore, is able to adjust its surface configuration to fit various molds before external forces render it completely rigid. This mechanical property appears to be of special importance in the Gl. lingualis and the dorsal surface of the anterior lingual wing it supports for the process of swallowing fluid during which the tongue is pressed against the palate (Homberger 1980a).

The glandular body increases its turgidity and assumes functions of an hydrostatic skeletal element when it is compressed; this happens when the M. supraglossus (M. sg.) and M. hypoglossus obliquus (M. ho.) contract and when the paraglossale is flexed.

When the M. ho. contracts, the M. ho. lat. exerts a caudo-ventro-lateral pull on the tendinous envelope of the Gl. lingualis, whereas the M. ho. med. exerts a caudo-ventro-medial pull (Figs. 53A, B). These two forces cooperate in keeping the floor of the tendinous glandular envelope taut and straight. If the M. sg. contracts at the same time (see Tables 17, 18, 19), the glandular body is compressed and its turgidity increases because the tendinous portions B and C of the M. sg. compress the anterior end of the glandular body. If the hyoid is retracted at the same time as the M. ho. and M. sg. contract (see Tables 18, 19), the lateral component of the caudo-ventro-lateral pull exerted by the M. ho. lat. on the floor of the glandular envelope increases at the expense of the caudal component, whereas the medial component of the caudo-ventro-medial pull exerted by the one-joint M. ho. med. does not increase during the retraction of the hyoid (consult Fig. 53B). Therefore, the floor of the glandular envelope is pulled increasingly laterally into an oblique orientation during the retraction of the hyoid. At the same time, however, the caudo-dorsal part of the glandular envelope, which is covered by the tendinous fibers of the portion A of the M. sg., remains at a more or less constant distance from the midsagittal axis of the tongue because the tendinous fibers of the portion A of the M. sg. do not elongate (see pp. 116 f., 124 ff.).

Turgidity of the Gl. lingualis also increases if the paraglossale is flexed. Flexion of the paraglossale always results in an extension of the muscular portion of the M. sg. (see Tables 18, 19). Even if the M. sg. does not contract during flexion of the paraglossale, the Gl. lingualis is compressed during flexion of the paraglossale. Since turgidity of the glandular body increases with flexion of the paraglossale, the glandular body is first compressed by the tendinous fibers of the portions A, B and C of the M. sg. until it is turgid enough to resist the pull of those tendinous fibers; further flexion of the paraglossale will affect the muscular portion of the M. sg. and will extend its fibers. An additional factor is responsible for increased turgidity of the Gl. lingualis during flexion of the paraglossale, namely, that the ventral component of the medio-apico-ventral pull exerted on the dorso-caudal tip of the glandular body by the tendinous portion A of the M. sg. increases during flexion of the paraglossale (compare Figs. 61A and 63B or Figs. 63A and 61B). This increased ventral pull compresses the glandular body and, consequently, increases its turgidity. Increased turgidity is necessary so that the anterior lingual wings, which are underlain by the Gl. lingualis, are raised together with the posterior process of the paraglossale when the paraglossale is flexed. (This mechanism has an important function during swallowing of food when the raised lingual wings are used to push food morsels into the esophagus; see Homberger 1980a.)

Turgidity of the glandular body can be further increased if external forces, such as pressure from food or from the palate, compress it. In this way, the anterior lingual wings, which constitute the caudal part of the free portion of the body, can be used to counteract or resist external forces (see pp. 117 ff., 129 ff.).

Glandula sublingualis (Gs).—The paired sublingual glands ("Unterzungendrüse" of Homberger 1980a) have fused externally into a flattened, roughly pear-shaped body, which is divided internally by a midsagittal septum into separate glandular portions. Apically, a pair of long ducts arise and open on the ventrolateral surface of the Corpus linguae, about midway between the caudal border of the lingual nail and the base of the tongue (Figs. 16A, 20). The glandular body lies between the Fascia nodulo-lingualis and the M. mylohyoideus anterior, which inserts along the midsagittal axis of the connective tissue envelope of the glandular body (Figs. 15B, 16A). It is held in place by connective tissue. The caudal borders of the Gl. sublingualis and M. mylohyoideus ant. are attached to the Fascia facialis, which covers the ventral and caudo-lateral surface of the head. The apical borders of the glandular body and of the M. mylohyoideus ant. are bound to the caudal surface of the M. genioglossus by connective tissue (Fig. 15B). The glandular ducts reach their orifices by passing through the seam line between the M. mylohyoideus ant. and M. genioglossus (Figs. 15A–16A).

The body of the Gl. sublingualis functions as an hydrostatic skeletal element and serves as a site of origin for the M. mylohyoideus anterior. The glandular body is compressed and becomes rigid when the M. mylohyoideus contracts and bulges. The glandular body of the Gl. sublingualis serves also as a cushion between the M. mylohyoideus anterior and M. ceratoglossus (see below).

Glandula mandibularis (Gm).—The paired mandibular glands ("Unterkieferdrüse" of Homberger 1980a) are situated in longitudinal mounds flanking the lingual wings (Figs. 20, 21A). The compact glandular body is lemon-shaped and somewhat bilaterally compressed. It opens with a single orifice on the dorsal surface of the mound at the level of the row of lingual papillae separating the anterior from the posterior lingual wing. The glandular body adheres medially to the lateral surfaces of the M. ceratoglossus, M. hypoglossus obliquus lateralis et medialis, and M. stylohyoideus (Figs. 16A, 20, 21B). The lateral surface of the glandular body is almost completely covered by the M. mylohyoideus anterior et posterior (Figs. 15B, 16A, 20).

The Gl. mandibularis serves as a cushion between the M. ceratoglossus and M. mylohyoideus (Figs. 16A, 20). When these muscles contract and, consequently, bulge, the gland is compressed. Depending on the amount of pressure exerted on the gland, more or less saliva is expelled from the glandular duct system. Furthermore, the gland could act as a cushion between the two muscles. If no compressible cushion existed between the two muscles, the contracting M. mylohyoideus would constrict the underlying M. ceratoglossus because the muscle fibers of the M. mylohyoideus are oriented almost perpendicularly to those of the M. ceratoglossus. Such a constriction could restrict simultaneous contractions of the M. ceratoglossus and could eventually lead to a ring-like atrophy of the muscular body of the M. ceratoglossus where it is covered by the M. mylohyoideus ("wrist-watch effect" of Homberger 1982).

FUNCTIONAL ANALYSIS

Introduction

Complete functional-anatomical studies of structural systems comprise two parts, namely an anatomical description and a functional analysis. Although many anatomical descriptions are not followed by functional analyses (e.g., Mudge 1903), a functional interpretation of the described structures is necessary because every structure has a function, i.e., participates in a form-function complex (Bock and von Wahlert 1965; Bock 1974). In addition, a functional analysis may be used to check the completeness and accuracy of anatomical descriptions (Bühler 1980).

Methods of Functional Analysis

A functional-morphological study permits the establishment of correlations between structure and function of an apparatus, including its constituent parts. The required information for such a study can be divided into two groups: (1) data that can be observed directly and used for the descriptive part of the study, and (2) evidence that is inferred from theoretical considerations of the directly observable data; the latter forms the analytical part of the study. (This distinction is not always clear-cut, but appears to be a useful categorization for the following discussion on methods of functional analysis.) For a skeleto-muscular apparatus, such as the avian lingual apparatus, the directly observable data comprise the movements of the apparatus, its interactions with the environment, and the morphological, physical and physiological properties of its structural parts. The evidence that has to be inferred comprises the functions of the structural parts of the apparatus and the interactions among these parts.

A brief overview of the functional anatomical literature reveals that the descriptions of the directly observable data vary only in the degree of completeness and accuracy. It is the methodological approach (or its absence) to providing the inferred evidence which is responsible for major differences among functional-morphological studies.

There have been two major approaches to the functional analysis of structural parts of an apparatus, namely the "reductionist" and the "holistic" approach.

Reductionist approach.—A "single-structure" reductionist approach has been, and still is, the most commonly used method of functional analysis and has been applied to the discussion of the function of single skeletal elements, bony processes, glands, ligaments, and others. If used for the functional analysis of a muscle-bone system, the function of each muscle is often assessed independently from functional considerations of the rest of the muscles and structures of the apparatus. Several techniques have been developed for this approach:

(a) "Common-sense" estimation.—The movements of the skeletal elements to which a muscle is attached are assessed by visualizing the effects of that particular muscle as if it were acting like a recoiling rubber-band (e.g., Kutorga 1832; Bhattacharyya 1980).

(b) Handling of freshly killed animals.—The movements of skeletal elements are determined by handling anatomical material that is not hardened by fixation processes (e.g., Dubale and Rawal 1965; Cracraft 1971; Burton 1974a).

(c) Electrical stimulation of muscles in freshly killed animals.—The effects of an electrically stimulated and contracting muscle on the skeletal elements to which it attaches are observed directly (e.g., Dubale and Rawal 1965).

(d) Removal of functional units.—A particular muscle is surgically removed or anaesthesized and the functioning of the deficient apparatus is compared to the functioning of the intact apparatus (e.g., Fisher 1956; Lombard and Wake 1976).

A reductionist approach is appealing because it is simple and direct, but it has several drawbacks. It does not permit direct determination of the interactions among the different structural parts of the apparatus. For example, this approach is unlikely to depict accurately the function of one particular muscle in bringing about movements of the entire apparatus. Most movements of skeletal elements of an apparatus are generated through the coordinated actions of several muscles. Thus, a particular muscle may be responsible for only a certain aspect of the total movement of a skeletal element and its action may be modified by other, simultaneously contracting muscles.

Holistic approach.—A holistic (or system-theoretical) approach to functional analysis grows from an interest in the interactions among the functional components of an organism. These functional components can consist of the structural parts of an apparatus, the structural parts of the rest of the organism, or the environmental aspects with which the organism interacts (Dullemeijer 1974, 1980). The interactions among functional components are usually explained and described with the help of a model (e.g., Zweers 1974, 1979, 1980, 1982; Zweers et al. 1977, 1981; Lombard and Wake 1976, 1977; Dullemeijer 1974, 1980, and references therein). Models are abstractions of real situations describing only the essential aspects of a process. A model describing the real situation in all its detail would produce a replica so complex as to obscure the result in which one is interested, such as the functioning of an apparatus (Beament 1960; Gutmann 1981).

With an holistic approach, interactions among functional components can be studied on three different levels. Hence, three types of models of interactions can be distinguished:

(a) Interactions among the structures within one apparatus (e.g., "mechanical model" of Zweers 1979, 1980, 1982).

(b) Interactions of an apparatus with the environment (e.g., Homberger 1980a: interactions between surface structures of feeding apparatus and food).

(c) Interactions of the components of the apparatus associated with one particular functional complex (e.g., mechanical) with components associated with other functional complexes (e.g., nutrition, sensory and motor control) (e.g., "cybernetical model" of Zweers 1979, 1980, 1982).

Functional-anatomical studies based on an holistic approach are rare (Dullemeijer 1980). The works of Zweers (1974, 1982), Zweers et al. (1977) and Lombard and Wake (1976, 1977) may serve here as representative examples.

Synthetic approach.—In the present study of the lingual apparatus of *Psittacus* a synthetic approach to functional analysis is used. This approach focuses on the actions and interactions of the individual structural elements of an apparatus and is based on the concept that a biomechanical apparatus is constructed and works like an engineered machine (e.g., Nachtigall 1971; Mercer 1981; and references therein) that consists of a multitude of interacting components made of a variety of materials. By taking apart such a machine and reconstructing a model from its components, it should be possible to acquire an understanding of the functions, interactions, and interdependencies of the constituent parts, provided one knows the basic function of the machine, the structure and spatial arrangement of the constituent parts, the physical properties of the materials from which the parts are made, and the physical and mechanical laws governing the interactions of the parts.

In the case of the lingual apparatus of *Psittacus,* such a synthetic approach to functional analysis is possible because the necessary background knowledge of the structural components of the apparatus is available. The movements that can be performed by the lingual apparatus are known through observations of live animals (Homberger 1980a). The morphology and spatial arrangement of the various structural elements have been described in the present study. The physiological and physical properties of the various tissues have been elaborated during the past decades by various investigators. The physical and mechanical laws and biomechanical principles governing the interactions among the various tissues and structural elements either have been elaborated already by previous authors or could be synthesized from published data on physiological and physical properties of the various tissues and from empirical biomechanical data. The procedure for constructing a model that explains the functioning of a biological apparatus, through the synthesis of all these different data, is explained on pp. 96 f. The analogies between a model of an engineered machine and a model of a biomechanical apparatus are shown in Table 15.

One major difference between a man-made and a biomechanical apparatus is that the latter comprises not only tissues with predictable, relatively stable physical properties but also tissues that can vary their physical properties, i.e., muscles. For example, muscles can generate varying forces, the rigidity of the muscle tissue varies with the state of contraction of the muscle, and contracting muscles either shorten, elongate, or maintain the same length depending on the size and direction of other forces acting on the system comprising the muscles. Additional variability is introduced into a biomechanical system through the ability of muscles to generate varying forces. Because the vast majority of articulations are spanned by at least two, but usually more, muscles, it is conceivable that for any given movement a variable number of muscles may contract provided the muscles that are responsible for a particular movement generate more force than the rest of the muscles. (Such an "overactivity" of muscles has been demonstrated in children

TABLE 15

ANALOGIES BETWEEN MAN-MADE AND BIOLOGICAL MECHANICAL APPARATUS

engineered machine, e.g., car	biomechanical apparatus, e.g., avian tongue
basic function, e.g., rotation of the wheel axes	basic movements, e.g., protraction, retraction
description for purpose of assembly (blue print)	anatomical description
physical and mechanical laws governing the actions and interactions of parts, e.g., cogwheels, cylinders, transmission bars	mechanical and physical laws and biomechanical principles governing the actions and interactions of structures, e.g., muscles, articulations, bones
physical properties of materials, e.g., iron, rubber, carbon	physical and physiological properties of tissues, e.g., bone, collagen, muscle
efficiency; ratio of energy input to energy output	efficiency; ratio of metabolic cost (energy needed for maintenance of system, muscle contractions, etc.) to energy output (generation of force by muscles)

and young animals but usually disappears in adult individuals; Basmajian 1978, 1980.)

In general, muscles contract to (1) move skeletal elements, (2) counteract forces acting on skeletal elements, or (3) combine (1) and (2). Each function is variable because muscles can vary the magnitude of force they generate. Therefore, they can vary the acceleration and speed of movements and counteract varying loads on the system.

A model to account for all these variables in a biomechanical system would be extremely complex. For a simpler model, the number of variables must be minimized by establishing boundary conditions that limit the variability of some aspects of the biomechanical apparatus. The complexity of the model arising from variable loading of the system can be reduced by considering only free movements with no external loading. Muscles of a system act differently depending on whether the system is loaded or not (Molbech 1966; Carlsöö and Molbech 1966). The reason for this difference is that in a loaded system some muscles may contract in order to withstand or counteract external forces rather than just to move the apparatus.

The complexity of the model arising from variable velocities of the movements can be reduced by considering movements at only one particular speed. As will be shown below, the consideration of only slow movements further reduces the complexity of the model.

To minimize the complexity of the model arising from the fact that any combination of synchronously contracting muscles spanning a particular articulation could produce a particular movement (see above), it is necessary to distinguish the muscles that must contract to produce a particular movement from those that may contract facultatively. A criterion to that means is provided by a biomechanical principle stating that antagonistic muscles do not contract simultaneously with agonistic muscles during unloaded movements at the articulations they span, except at the end of very fast movements or to counteract gravity (Basmajian 1978:93–100; 1980:248–249). Thus, this biomechanical principle is valid only for slow, unloaded movements. This biomechanical principle of "reciprocal inhibition" (Basmajian 1978, 1980) permits the classification of each muscle as either an agonist or antagonist for each movement at any particular articulation. Therefore, a model of a biomechanical apparatus can be simplified to the degree of explaining a movement only as the result of agonistically active muscles instead of having to consider all possible combinations of synchronously contracting muscles for every particular movement.

The principle of "reciprocal inhibition" has not only been demonstrated empirically (Basmajian 1978, 1980) but can also be defended on theoretical grounds. If antagonistic muscles were to contract simultaneously with agonistic muscles, either no movement would occur at the articulation spanned by the muscles or one of two contracting muscle groups would be stretched, thus, producing movement under highly inefficient conditions. For slow, free, and unloaded movements the contraction of antagonistic muscles only costs energy without quantitatively or qualitatively modifying the movements. It is unlikely that an organism would have evolved such an inefficiently operating apparatus that required the simultaneous contraction of muscles counteracting one another during free movements.

By setting boundary conditions for the functioning of a biomechanical apparatus, a theoretical mechanical model can be formulated to explain the functioning of the apparatus and the underlying actions and interactions of the constituent structural elements. Without such boundary conditions the development of a most parsimonious, theoretical model would not be feasible on the basis of morphological, physiological and physical data alone. The use of electrophysiological data for the construction of a mechanical model would, however, hinder the testing of such models (cf. below).

Because the present model does not describe the actions of the apparatus when subjected to external forces as it is during the execution of its biological roles, it cannot be used to explain completely the adaptive value of the particular morphological construction of the lingual apparatus of *Psittacus*. These aspects of functional analysis will be dealt with in another study (cf. pp. 138 ff.). The present model is presented as a first step towards the understanding of the functioning of the lingual apparatus of *Psittacus* and is valid only within the limits defined above.

It must be stressed here that a model based on the synthetic approach is not a quantitative model nor can it be transformed into one without additional data. Because it is not based on quantitative data (e.g., forces developed by muscles, resistance at articulations, flexibility modules of bones and cartilages, and internal pressure of salivary glands), it cannot be used to make quantitative predictions about the functions and functionings of the apparatus and its components. Therefore, the theoretical model is a constructional one and will make qualitative predictions (e.g., about the synchronization of muscle contractions and the interactions between muscles and salivary glands).

The synthetic approach to functional analysis depends on the accuracy and completeness of the anatomical description. Because the present analysis focuses on the mechanical actions and interactions of the structural components of the apparatus, only mechanically active elements have to be described. The initial selection of the structural elements included in the anatomical description must be guided by the knowledge of the physical and physiological properties of the tissues and by an understanding of the biomechanical principles governing the functioning of the biological apparatus. The proper level of completeness and accuracy of the anatomical description can be reached through a process of mutual testing between the anatomical description and the mechanical model (see also

Bühler 1981). For example, one can be quite confident that all mechanically relevant structures have been described if it is possible to construct a mechanical model that can explain the observed functioning of the apparatus on the basis of the described structures and in accordance with the biomechanical and physiological principles. Should it not be possible to construct such a model, the descriptions must be checked for omissions and misinterpretations. If a review of the anatomy does not reveal any faulty descriptions or reveals a mechanically active structure that cannot be integrated into the present mechanical model, the model must be revised. The necessity for the inclusion of all mechanically active structures into the mechanical model cannot be overemphasized. All structures that can possibly affect the functioning of the apparatus must be considered. Although it is possible that upon full development of the model some of these structures may be only marginally functional, such judgments cannot be made in advance. Furthermore, a biomechanical apparatus must have evolved under the influence of selection forces to function with a certain degree of efficiency and is unlikely to contain "superfluous," non-functional components.

For the present analysis it is valid to restrict the anatomical description to the mechanically functioning structural elements because only the mechanical functioning of the apparatus is to be explained. Nevertheless, the apparatus cannot function without connections to the rest of the organism for energy supply, waste product removal, and without functional coordination through sensory receptors, nerves and blood vessels. These latter structural elements generally do not fulfill mechanical roles and can be disregarded for the construction of mechanical models. Such "connecting" structures, however, play a central role for the construction of cybernetical models in which the interactions and feed-back mechanisms between the apparatus and the rest of the organism are explored.

The synthetic approach to functional analysis permits also the testing of the internal consistency of the model. Because an apparatus generally consists of subunits and each subunit must initially be analyzed separately, the final synthesis of all the "submodels" into an integrated mechanical model of the entire system represents a test for the internal consistency, but not necessarily for the correctness of the model (pp. 138 ff.).

A mechanical model constructed on a theoretical basis as outlined here is open to testing by empirical observations through a variety of experimental techniques and approaches involving the various hypotheses and predictions contained in the model. The predictions concerning the synchronization and patterns of muscular contractions during certain movements can be tested through electromyography. The hypotheses on the changes of internal pressure in the hydraulic structures of the tongue can be tested through the use of pressure strain gauges. The biochemical nature of the fibers of individual muscles of the apparatus can be determined histochemically to see whether the physiological fiber types conform with the predicted function of particular muscles. Furthermore, the theoretical mechanical model itself can be tested when it is compared to a model describing the interactions between the apparatus and the environment because both models must be compatible and must complement each other. Comparative studies can also provide a test for a theoretical mechanical model, mainly by testing the general validity of the synthetic approach and biomechanical principles used for the construction of the model.

Physical and Physiological Properties of Tissues

A knowledge of the physical and physiological properties of the tissues of which the mechanically active structural components of the lingual apparatus are made is essential for the synthetic approach to functional analysis. Each tissue has many physical and physiological properties, but for a given function of a structural element only a few of these properties may be relevant.

The physical and physiological properties of each tissue can be compiled from a vast literature on biomechanics and physiology. In the following sections those properties have been selected and assembled which were relevant for the formulation of the present theoretical mechanical model, irrespective of whether the properties were little known or as widely known as to appear to be "truisms."

Supporting and connective tissues.—Three different types of supporting and connective tissues can be distinguished: bone, cartilage, and organized dense regular connective tissue (i.e., ligaments, fasciae, and tendons). Their mechanical properties are relatively well known and are only summarized here (for details, see Bargmann 1962; Young 1963; Alexander 1968; Bock 1974; Bloom and Fawcett 1975; Katz 1980; Kastelic and Baer 1980; Wainwright et al. 1980; Fung 1981; Vincent 1982; Weiss 1983).

A) Bone.—Bone is able to withstand compressive, tensile and shearing forces, and is relatively noncompliant. These properties make bone the premier material to transmit muscular forces for the production of movements in biomechanical systems, although it is not the only material able to fulfill this function. The lingual apparatus of *Psittacus* has several bony skeletal elements that articulate with one another and one sesamoid bone inserted in the Lig. nodulo-ceratobranchiale. All are rigid structures under the forces normally acting on them; thin flexible bones are not present in the lingual apparatus of *Psittacus.*

B) Cartilage.—The exact mechanical property of cartilage depends on the orientation and amount of collagenous (or, rarely, elastic) fibers contained in the matrix. Cartilage in general can withstand compressive forces with a reasonably high degree of noncompliance in addition to being elastic. Certain cartilages are more compliant, and deform readily. Within the lingual apparatus of *Psittacus,* three different types of cartilaginous structures can be distinguished.

1. Articular cartilages: These are found at all diarthroses of the hyoid. No cartilaginous menisci have been found.
2. Synchondroses: These are represented only by the Synchondrosis paraglossalis which forms a bridge between the anterior processes of the paraglossalia. The bridge is rather rigid, having to withstand the various forces exerted on it by the M. hypoglossus anterior (pp. 123 f.). It must also resist compressive forces when the Procc. anteriores paraglossales are pushed medially by the contraction of the M. supraglossus (pp. 124 ff.) or by the inflation of the Corpora cavernosa lateralia. Its elasticity is responsible for the readjustment of the distance between the anterior processes of the paraglossalia after the forces have receded.
3. Skeletal processes: This type of cartilaginous structure is found at the caudal end of the urohyale as Cartilago urohyalis and at the caudal end of the epibranchiale as Cartilago epibranchialis. These processes do not serve as attachment sites for muscles and are very flexible. The Cartilago urohyalis

has to be flexible to permit movements of the larynx and trachea (Homberger 1979b), and the Cartilago epibranchialis has to be flexible (i.e., compliant) to be able to adjust to different degrees of curvature during the retraction of the hyoid (p. 114).

C) Organized connective tissue.—In the lingual apparatus of *Psittacus,* organized connective tissue seems to consist mainly of collagenous fibers; therefore, the following discussion can be restricted to the mechanical properties of collagenous fibers. Collagenous fibers withstand tensile forces, but are very flexible, i.e., they cannot withstand compression and shear. They are also elastic, but only to a very small degree compliant. Within a structure, the collagenous fibers tend to be oriented along lines of tensile forces. This attribute provides some insight into the directions of the major forces acting on such a structure. The lingual apparatus of *Psittacus* includes four types of structures consisting of organized connective tissue.

1. Ligaments: Ligaments connect different skeletal elements at places within the system, where tensile stresses on these skeletal elements are resisted passively. Depending on their specific location and function, several types of ligaments can be distinguished.
2. Fasciae: Fasciae are connective tissue sheets enveloping structures or connecting different structures. In general, they function to limit or guide movements of the apparatus. Depending on their function, the constituent collagen fibers are oriented in parallel to withstand a specific force or are interwoven to withstand forces acting in various directions.
3. Tendons: Tendons connect muscles to skeletal elements or other structures. Their structure resembles that of ligaments. In the lingual apparatus of *Psittacus,* tendons are formed at the following pinnate muscles: M. ceratoglossus, M. hypoglossus obliquus, M. supraglossus, and M. hypoglossus anterior.
4. Aponeuroses: An aponeurosis is a sheet-like tendon. The lingual apparatus of *Psittacus* includes only one aponeurosis, that of the M. hypoglossus anterior.

Muscle.—The muscles responsible for the movements of the lingual apparatus of *Psittacus* are mainly striated skeletal muscles and to a much lesser extent smooth muscles associated with blood vessels. The mechanical and physiological properties of muscles are well known and need only be summarized or put in special perspective for the present mechanical analysis (for details, see Young 1963; Gans and Bock 1965; Bock 1974; Carlson and Wilkie 1974; Fung 1981; Bodine et al. 1982; Gans 1982).

One of the most important factors for functional analyses of skeleto-muscular systems based on the synthetic approach is that muscles can only actively develop tension and shorten (i.e., exert a pull) but cannot actively elongate back to their resting length (i.e., exert a push). Shortened muscle fibers have to be stretched by a contracting antagonistic muscle or by an external force, but only a small force is needed to stretch a relaxed muscle (pp. 91 ff.).

The only smooth muscles of interest to the mechanical analysis here are the presumed sphincter muscles of the cavernous veins. These control the blood flow out of the cavernous spaces and, hence, the size of the veins, and regulate the turgidity of the Corpora cavernosa. Smooth muscle contracts more slowly than skeletal muscle but can remain contracted for a longer time and at a lower metabolic cost. Because of its extended reaction time, its actions cannot be as finely synchronized with those of skeletal muscles as the actions of skeletal muscles with one another (pp. 122 ff.). Because of the simplicity of their actions and their indirect influence on the movements of the lingual apparatus of *Psittacus,* the properties of the smooth muscles need not be considered in more detail.

The remainder of this section on muscles will be devoted to skeletal muscles, considering first their physiological and mechanical properties (A) and secondly the effects of their internal architecture on their function (B).

A) Physiological and mechanical properties of muscles.—Contracting skeletal muscle fibers develop tension and, depending on the external load placed on them, react by shortening, by retaining the same length, or by elongating.

1. Isotonic contraction. Contracting muscle fibers will shorten if the load placed on them (e.g., external force, contracting antagonistic muscle) is smaller than the force (tension) they produce (i.e., they contract isotonically). Maximum tension is usually developed at rest length and decreases as the muscle shortens. As the muscle shortens it does mechanical work as well as releasing an extra amount of heat (i.e., heat of shortening). This same amount of heat is released additionally as part of the recovery heat. Therefore, isotonic muscle contractions are energetically more costly than isometric contractions in which muscles do not shorten (see below). For example, in the lingual apparatus of *Psittacus,* an isotonic contraction is executed by the M. ceratoglossus lateralis when it flexes the paraglossale during the retraction of the hyoid (Table 16).
2. Isometric contraction. Contracting muscle fibers maintain the same length if the load placed on them equals the tension they exert (i.e., isometric contraction). Because isometrically contracting muscles do not shorten, they do not generate mechanical work, the tension does not decrease during their contraction, and no shortening heat is released. Thus, isometric muscle contractions are more efficient in generating tension than isotonic muscle contractions. For example, in the lingual apparatus of *Psittacus,* a more or less isometric contraction is executed by the M. ceratoglossus lateralis when it flexes the paraglossale during the protraction of the hyoid (Table 16).
3. Elongation of contracting muscle. Contracting muscle fibers will elongate if the load placed on them exceeds the tension they produce. A contracting muscle that is stretched produces "negative work," because work is done on it while it exerts tension. Such muscular contractions occur during locomotion of terrestrial vertebrates when the load of the body weight stretches contracting limb muscles (Tucker 1975). During free unloaded movements of the lingual apparatus (i.e., the assumed condition for the present mechanical analysis), however, a contracting muscle could only be stretched by an antagonistic contracting muscle (e.g., Tables 16–19). As explained earlier, such a "negative" muscular contraction would be inefficient because energy is not only used by the contracting muscle that is stretched but also by the other contracting muscle stretching it. Therefore, it is doubtful that such a

TABLE 16

CORRELATIONS BETWEEN MOVEMENTS OF THE PARAGLOSSALE AND HYOID[a]

Movement of hyoid	Movement of paraglossale	Action of contracting two-joint muscles	
retraction (through M. serpihyoideus and M. stylohyoideus)	extension (through M. hypoglossus obliquus)	M. ho. lat.:	shortening (isotonic contraction)
	flexion (through M. ceratoglossus)	M. cg. lat. M. cg. supf.	shortening (isotonic contraction)
		M. cg. med.:	± same length (isometric contraction)
	immobilized (through M. ceratoglossus and M. hypoglossus obliquus)	M. ho. lat. M. cg. lat. M. cg. supf.	shortening (isotonic contraction)
		M. cg. med.:	elongating ("negative work")
protraction (through M. branchiomandibularis and M. ceratohyoideus)	extension (through M. ho.)	M. ho. lat.:	± same length (isometric contraction)
	flexion (through M. cg.)	M. cg. lat. M. cg. supf.	± same length (isometric contraction)
		M. cg. med.:	shortening (isotonic contraction)
	immobilized (through M. cg. and M. ho.)	M. cg. med.:	shortening (isotonic contraction)
		M. cg. lat. M. cg. supf. M. ho. lat.	elongating ("negative work")

[a] See pp. 114 ff.

contraction occurs under normal circumstances during unloaded movements of the lingual apparatus of *Psittacus*. For example, in the lingual apparatus of *Psittacus,* flattening of the dorsal surface of the tip of the tongue during flexion of the paraglossale cannot occur, unless one assumes that the contracting M. supraglossus is elongated (Table 17).

B) Muscle architecture.—The volume and external shape of a muscle body is probably influenced mainly by structural and spatial constraints set by the apparatus itself; for example, muscles must fit into the spaces available to them within the apparatus, the shape of which is dictated by structural and functional constraints. In contrast, the functional properties of muscles are dependent on three structural parameters: fiber cross-sectional area, fiber length, and angle of pinnation. These variables determine the internal architecture of a muscle, i.e., the arrangement and orientation of the muscle fibers within the muscle body. In this way, a muscle with a given shape and volume can have vastly different functional properties depending on its architecture.

The following synopsis summarizes how structural parameters can affect the function of muscles; it focuses on correlations relevant for the present mechanical analysis of the lingual apparatus of *Psittacus.*

1. Physiological cross-section. The fiber cross-sectional area (i.e., physiological cross-section) of a muscle represents the sum of the cross-sectional areas of all the muscle fibers of the muscle. The actual force exerted by a contracting

TABLE 17

CORRELATIONS BETWEEN MOVEMENTS OF THE PARAGLOSSALE AND EPITHELIAL SURFACE OF THE TONGUE'S TIP[a]

Movement of paraglossale	Simultaneously contracting muscle	Movement of central area of tip of tongue	Action of contracting muscle
flexion (through M. ceratoglossus)	M. hypoglossus ant.	from flat to depressed remaining depressed	shortening (isotonic contraction)
	M. supraglossus	from depressed to flat remaining flat	extending ("negative work")
extension (through M. hypoglossus obliquus)	M. hypoglossus ant.	from flat to depressed: remaining depressed:	± same length (isometric contraction) extending ("negative work")
	M. supraglossus	from depressed to flat remaining flat	shortening (isotonic contraction)

[a] See pp. 129 ff.

muscle is positively correlated with the number of muscle fibers contracting simultaneously. Therefore, the potential maximum force developed by a muscle is positively correlated with the number of muscle fibers contained in the muscle, hence, with the physiological cross-section of the muscle. The physiological cross-sectional area of a muscle provides the best gross morphological estimate for the maximum force a muscle is capable of generating, although under physiological conditions only up to one-third of its muscle fibers may contract simultaneously. For example, the M. hypoglossus obliquus medialis with a cross-sectional area of approximately 4×2 mm^2 at its insertion has a larger physiological cross-section and, hence, a greater potential for force production than the M. hypoglossus obliquus lateralis which has a cross-sectional area of approximately 1×2 mm^2 at its insertion.

2. Muscle fiber length. The length of muscle fibers determines the muscle properties related to shortening, namely the maximum distance of shortening and the speed of shortening. A muscle fiber, and hence a muscle, can generally shorten maximally to about 60 percent and be stretched to about 140 percent of its resting length, although most muscles in vivo operate over a much smaller range of shortening and lengthening (i.e., between 85 percent to 115 percent of their resting length). During shortening to a given fraction of its original length, a longer muscle fiber will shorten over a greater distance in a given time or will shorten a given distance in shorter time (i.e., will shorten with a greater velocity) than will a shorter one. Therefore, muscle fiber length is the best gross morphological estimate for the shortening properties of muscles. For example, the M. tracheohyoideus with a fiber length of about 100 mm can shorten with a greater maximum velocity than the M. serpihyoideus which has a fiber length of approximately 38 mm.
3. Angle of pinnation. The angle of pinnation is the angle at which muscle fibers attach to a tendon. Muscles can be characterized morphologically as either parallel-fibered or pinnate. In a pinnate muscle, the muscle fibers insert on a tendon at a certain angle to the direction of the force generated

by the muscle. In general, as the angle of pinnation increases for a given size and external shape of a muscle body, the number of muscle fibers increases and the muscle fiber length decreases. The force generated by a pinnate muscle is equal to the tension generated by the muscle fibers multiplied by the cosine of the angle of pinnation. In a parallel-fibered muscle, the angle of pinnation is 0° and consequently has a cosine of 1. For a given volume and external shape, a pinnate muscle has in general more, but shorter muscle fibers than a parallel-fibered muscle which has fewer, but longer muscle fibers. Therefore, for a comparable size and external shape of a muscle body, a parallel-fibered muscle tends to shorten more or faster, but to generate less tension than a pinnate muscle which tends to generate more force, but to shorten less. For example, the extrinsic lingual muscles are parallel-fibered and need to shorten to move the hyoid apparatus relative to the mandible, whereas the majority of the intrinsic lingual muscles are more or less pronouncedly pinnate and contract often isometrically.

Because the above parameters (B1–B3) vary with the contractile state of the muscle, artificial variables are introduced into the measurement of these parameters depending on the state of the muscles at the time of death of the specimen and on the method of preparation. In the present study of *Psittacus,* measurements of fiber lengths and cross-sectional area and statements about muscle pinnation were nevertheless included, mainly as a basis for future interspecific comparisons of particular muscles, because the various muscles within a single system are too different in these parameters to allow useful comparisons.

4. Functional grouping of muscle fibers within muscles. Even if a muscle is a single unit, morphologically different parts may contract independently and may have different functional properties (Herring et al. 1979). Within the lingual apparatus of *Psittacus,* the subdivision of some muscles into functional groups (e.g., M. ceratoglossus, M. hypoglossus obliquus) has reached an extreme. In some cases, subdivision has proceeded so far that the distinct muscle part is best recognized as a distinct muscle, e.g., the M. mesoglossus and M. supraglossus. Therefore, it will be necessary to treat the different parts of each muscle as separate units in the present analysis.

Epithelial structures.—The surface cover of an internal structure or the inner lining of an internal organ is referred to as Tunica mucosa (Mucosa) and, like the externally located skin (Cutis), consists of two layers, namely the superficial epithelium and the underlying Tunica propria (Bargmann 1962; Hodges 1974; Weiss 1983). (In the present study, the name "epithelium" (sensu lato) has generally been used instead of the less commonly used, though more accurate, name "mucosa.")

The epithelium (sensu stricto) of the avian tongue is stratified, mostly squamous, and variably keratinized (Ziswiler and Farner 1972; Hodges 1974; McLelland 1979b; Homberger and Brush, in press). It also forms various superficially projecting papillae which have important functions during food handling and swallowing (Homberger 1980a, and references therein). The thickness of the epithelium varies greatly, probably depending on the degree of wear to which it is subjected. The mechanical properties (e.g., compliance, flexibility) of the epithelium appears to be determined to a large degree by the nature of its keratinization (Wainwright et al. 1976; Brush 1980; Fraser and Macrae 1980; Vincent 1982).

The Tunica propria consists of a dense network of mainly collagenous fibers. Sometimes, it is not clearly distinguishable from the underlying submucosa (Hodges 1974; pers. obs.). The mechanical properties (e.g., distensibility, elasticity) of the mucosa probably depends largely on the organization and arrangement of the collagenous and elastic fibers within the Tunica propria (for a discussion of the mechanical properties of the analogous dermis, see Wainwright et al. 1976).

The structure and function of surface coverings of vertebrates have been studied mainly with respect to their interactions with the environment. In contrast to invertebrate skin, the vertebrate surface covering has only occasionally been investigated with respect to its mechanical interactions with the apparatus it covers (e.g., Breder 1947; Gans 1974:105ff.; Gutmann 1975; Russell 1975; Wainwright et al. 1980).

In the lingual apparatus of *Psittacus,* the mucosa fulfills a variety of mechanical functions depending on its location and structure. Accordingly, the lingual mucosa varies in the thickness of the epithelium and type of keratinization, in the arrangement of the collagenous fibers in the Tunica propria, and in the strength of the bond among the Tunica propria, submucosa, and underlying tissue. Because there are no experimental data on the mechanical properties of the different types of mucosa covering the various regions of the avian lingual apparatus, the different mucosal types are classified on the basis of criteria that are mainly inferred from morphological data, from analogous reasoning, and from the present theoretical functional model.

A) Loosely attached surface coverings.—These consist of a moderately thick, flexible mucosa which is folded in the relaxed state and only loosely attached to the submucosa and underlying structures. Their main function is to adapt to changing sizes of surface areas they cover and, therefore, to make particular extensive movements of the apparatus possible. They do not have to resist or transmit external or internal forces. This type of surface cover is represented by the mucosa covering the underside of the free portion of the tongue and by the mucosa forming the posterior lingual wings.

B) Firmly attached surface coverings.—The common feature of the mucosal types in this group is that the Tunica propria is indistinguishable from the submucosa and is firmly attached to underlying structures, such as tendons and muscles. This type of surface covering is able to resist, transmit or directly react to forces applied to it. Frequently, it also adapts to changing sizes of surface areas or to changing relief at the same time.

1. Flexible mucosa with folds. This type of epithelium has to fulfill a variety of functions at the same time, such as adapting to a changing size of surface area and interacting with forces applied directly to it. It is represented by the epithelium covering the central area of the tip of the tongue and the epithelium forming the thickened cap over the extreme tip of the tongue. Both epithelia have to adapt to changing sizes of surface areas during the inflation and deflation of the cavernous vascular tissue. The epithelium covering the central area of the tip of the tongue has to be flexible enough, but not distensible, to be able to react to and transmit the forces exerted on

it by the M. mesoglossus and M. supraglossus. The epithelium forming the thickened cap at the extreme tip of the tongue must be able to resist the hydraulic pressure of the inflated Corpus cavernosum apicale so that the turgidity of the extreme tip of tongue can increase.

2. Flexible mucosa without folds. This type of mucosa is flexible and can adapt to varying surface shapes (e.g., flat versus round), but it cannot adapt to changing sizes of surface areas it must cover. It moves with the underlying structures to which it is firmly attached, but it does not react separately to forces applied to it. It is represented by the epithelia covering the V-shaped mounds, the Dorsum linguae and the anterior lingual wings.
3. Stiff cuticle without folds. This type of epithelial covering is represented by the lingual nail. It can be bent, but not changed in size. It is firmly attached to the underlying structures with which it moves. It is naturally curved and can be straightened by forces, such as the inflation of the Corpus cavernosum laterale, applied to it. After cessation of such a force, it returns to its original shape through elastic recoil. Due to its stiffness it acts like an external skeletal element in its interactions with the M. supraglossus, M. hypoglossus anterior and M. mesoglossus (pp. 123 ff.).

Hydraulic structures.—Structurally, an hydraulic structure is characterized as a fluid enclosed by walls. It can also consist of several such compartments and is then enveloped by a separate sheet of connective tissue. The functional properties of hydraulic structures are derived from the physical properties of the fluid and of the enclosing walls. Liquids in general are practically incompressible, and a force acting on a certain point in a fluid transmits itself equally in all directions. In addition, some fluids, such as synovial fluid and mucus, are visco-elastic (Alexander 1968; Fung 1981). Visco-elastic fluids resist being squeezed out between two surfaces and tend to return to a "minimum-surface" state after deforming forces have receded. The mechanical properties of the wall (e.g., elasticity, distensibility, resistance to bending, ability to withstand tension) depend on its material composition. The shape of the hydraulic structure and the amount of its internal pressure (turgidity) can be varied by the application of external forces on the fluid container or by changes in the fluid volume (see also Alexander 1968; Chapman 1975). With increasing turgidity, hydraulic stuctures become rigid and, thus, can function as skeletal elements, so that these hydraulic structures may be referred to as hydrostatic skeletons.

The structure and function of hydrostatic skeletons have been studied extensively in invertebrates and primitive chordates (Clark 1964; Alexander 1968; Gutmann 1972, 1981; Chapman 1975; Wainwright et al. 1976; and references therein). In vertebrates, however, the existence of stuctures functioning as hydrostatic skeletal elements has only rarely been appreciated (for exceptions see, Gans 1967; Bock and Hikida 1969; Du Brul 1976; Winokur 1977, 1981, 1982; Griffiths 1978 and references therein; Russell 1981a, b; Gans and Gorniak 1982a, b; Homberger 1982; Kier and Smith 1983; Smith 1983). In addition, hydraulic properties have been alluded to for certain structures (e.g., Bargmann 1962; Ziswiler 1979, 1980).

Within the lingual apparatus of *Psittacus,* three different types of hydraulic structures, namely salivary glands, synovial bursae, and cavernous vascular tissue, were identified. Although, in principle, cartilage does act as an hydraulic structure, it is best for the present analysis to consider it as a rigid skeletal tissue. Cartilage yields very little when no muscular forces are acting on it, and the increase in rigidity of cartilage is small when compressed by surrounding muscles. Following structural and functional criteria, the various hydraulic structures can be classified into several groups (Homberger 1982).

A) Passive hydraulic structures.—These consist of fluid or an assemblage of fluid-filled cells in an envelope of tough connective tissue which is flexible but not distensible. Their hydraulic function becomes apparent only when other structures, especially muscles, exert pressure on them.

1. Cushions. These types of hydraulic structures are represented by the Gl. sublingualis and Gl. mandibularis. They serve as cushions between the M. ceratoglossus and M. mylohyoideus to prevent the constriction of a deep muscle by a superficial muscle lying across the deep one.
2. Pressure distributors. These types of hydraulic structures are repesented by the Bursae synoviales which lie between projecting bony or cartilaginous crests and overcrossing tendons. The bursae function as pressure distributors and allow the tendons to move across skeletal crests without friction. Within the lingual system of *Psittacus* there are the paired Bursae synoviales paraglossales and the Bursa synovialis urohyalis.
3. Hydroskeletons. These hydraulic structures are flaccid, but become rigid under the influence of muscle contractions so that they can function as skeletal elements (i.e., hydroskeletons). They are represented by two different salivary glands. The Gl. sublingualis serves as the insertion site for the M. mylohyoideus anterior and becomes rigid when the muscle contracts and compresses the gland (see p. 75). The Gl. lingualis is partly attached to the paraglossale and lies in a tendinous envelope onto which the M. hypoglossus obliquus attaches. It becomes rigid through a combination of forces exerted on its body by the contracting M. hypoglossus obliquus and M. supraglossus or through the flexion of the paraglossale (pp. 73 ff., 115 ff., 122 ff.).

B) Active hydraulic structures.—Active hydraulic structures consist of containers (i.e., cavernous veins) that can vary the amount of fluid they hold. The walls of the containers are flexible and more or less distensible. The containers are embedded in surrounding tissue and surrounded by an envelope of connective tissue. They function as hydraulic structures only when the containers are inflated with fluid. Depending on the nature of the surrounding tissues, these cavernous vascular tissues have two main functions.

1. Hydroskeleton. The formation of a hydroskeleton is the main function of the Corpus cavernosum apicale. The surrounding thickened epithelial cap of the extreme tip of the tongue is flexible and, at best, only slightly distensible. In this way, the extreme tip can increase its turgidity and becomes rigid, although it also increases its volume to a certain degree.
2. Movement. The movement of epithelial surface structures is the main function of the Corpus cavernosum laterale. Its surrounding envelope consists of two different parts. One part consists of non-distensible, relatively stiff, keratinized epithelium (i.e., the lingual nail) which can be pushed into certain

directions (e.g., laterally) by the inflating cavernous vascular tissue. The other part of the envelope consists of flexible, but only slightly distensible, epithelium which can be stretched to cover the increased volume of the Corpus cavernosum laterale.

Biomechanical Principles

The interactions and functioning of the various structures constituting a biomechanical apparatus are governed by biomechanical principles or rules that are derived from the shape and physical properties of the structures. These biomechanical principles must be clearly stated before one can build a mechanical model based on the anatomical structures of a biomechanical apparatus, much in the same way as engineers must know how cogwheels, pulleys, transmission bars, and pistons function and interact before they can assemble a machine (pp. 78 ff.).

In the following sections the biomechanical principles that govern the functioning of the various components in a skeletomuscular apparatus have been compiled. Many of the principles listed below are based on the properties of the biomaterials, have been described at least partially elsewhere (Nachtigall 1971; Bock 1974; Wainwright et al. 1976:Chapter 6; Gans 1982), or have been common knowledge for a long time. They have, however, never been assembled in such a context as to serve as guidelines for the construction of a theoretical mechanical model of a complete biomechanical apparatus.

Constructional basis of movements.—The movements of a biomechanical apparatus are based on the movements of skeletal elements (e.g., bony, cartilaginous, hydraulic, epithelial) that provide the structural support for all the soft tissues of the apparatus. The skeletal elements, however, do not move through an inherent mechanism but only in response to forces exerted on them, mostly by contracting muscles.

If the skeletal elements are fully rigid, as is the case for bony elements under the influence of muscular forces acting on them, they react to these forces by moving as one unit and not by deforming. Furthermore, they transmit the muscular forces to neighboring skeletal elements and surrounding tissue. A biomechanical apparatus with a rigid skeleton can, thus, be regarded as a system of levers upon which forces are acting. Therefore, the potential movements of rigid skeletal elements can be determined with a certain degree of accuracy after taking into account the various structural constraints contained within the system.

In contrast, some skeletal elements, such as epithelial and hydraulic structures, are rigid only under the influence of certain muscular or hydraulic forces. As long as these skeletal elements are flexible and compliant, they react to forces exerted upon them not only by moving but also by deforming. Furthermore, in this state, they are not efficient in transmitting forces to surrounding tissues. Therefore, the potential movements of skeletal elements with variable rigidity are much more difficult to assess accurately than the movements of permanently rigid structures (pp. 121 ff.).

Guidance and regulation of movements.—Movements of skeletal elements consist in general of a variety of translations, of angular or axial rotations, or of combinations of these motions. When skeletal muscles contract, however, they exert tension only along a straight line; this means that a muscle by itself would not be capable of pulling a skeletal element along a curved line. Furthermore, the muscle fibers of a muscle are usually not oriented in the direction of the movement of the skeletal element to which it is attached; for example, no transversely oriented muscles are responsible for the lateral excursions of the psittacine hyoid or for the lateral movements of the psittacine mandible (Homberger 1981).

The direction of the linear pull of a muscle can be modified by a variety of structures and constructions. In this way, the final directions of movements of the skeletal elements are determined not only by the directions of muscle forces but also to a major degree by modifying structures and constructions, such as the shape of the articular facets of diarthroses, the presence of ligaments, the shape of the extrinsic structural framework within which the apparatus has to move, the re-orientation of muscles over skeletal crests, and the presence of guiding fascial sheaths. Therefore, it is necessary to pay attention to the details of structures and constructions that potentially modify the action of muscles and to include these detailed observations in the description of a biomechanical apparatus if this description is to be used as the basis for a theoretical mechanical analysis.

Articulations among skeletal elements play a central role in guiding the direction of muscular forces. The shape and relief of articular facets of true diarthroses determine to a large degree the possible movements of adjoining skeletal elements because articular facets tend to remain in close contact when gliding past each other during movements of the participating skeletal elements. Therefore, it is possible to estimate the direction of the movements of skeletal elements by analyzing the surface structure and geometry of the articular facets and by experimentally manipulating the skeletal elements. The freedom of movement at an articulation may also be restricted by the presence of ligaments. Certain articular ligaments reduce the amount of displacement of the articular facets with respect to each other at particular points of the articulations. For example, the Lig. interarticulare cerato-basihyale determines the axis of angular rotation for the Art. cerato-basihyalis. The freedom of movements at articulations can also be restricted by linkage ligaments that couple movements of different skeletal elements with one another. For example, the Lig. nodulo-ceratobranchiale links the movements of the ceratobranchialia, basihyale, and Nodulus. Because of the important role of ligaments in determining the actual movements at articulations, a complete description of articulations must include the functional-anatomical aspects of the associated ligaments.

The movements of an apparatus and its skeletal components are also guided and restricted by the shape of the extrinsic framework from which the extrinsic musculature originates and within the limits of which the apparatus must move. For example, during the protraction of the hyoid, the ceratobranchialia must converge to adjust the configuration of the hyoid to the narrower space between the apically converging Rami mandibulares. Consequently, a theoretical mechanical analysis of a biomechanical apparatus must take into consideration the effects of the extrinsic framework on the movements of the skeletal elements of an apparatus.

The direction of the line of action of a muscle can also be modified by redirecting the muscle over skeletal crests. For example, the M. branchiomandibularis posterior exerts a *ventro*-apical pull on the epibranchiale, although its origin on the mandible lies about at the same level as its insertion, because it passes over the ventral edge of the mandible. Another example is the M. hypoglossus anterior which exerts a caudal pull on its tendon, but is able to pull the origin of the M. mesoglossus on its aponeurosis apically because it passes, as in a pulley system, over the anterior edge of the Synchondrosis paraglossalis. Because of the presence of such constructions, it is necessary for a complete theoretical mechanical analysis to consider the possibility that the various sections of a particular muscle may have different effects on associated structures.

Because muscles can only actively shorten but not actively elongate, a muscle can in general only pull, but not push, a skeletal element into a new position. A skeletal element can, however, be pushed into motion under special structural conditions. In the lingual apparatus of *Psittacus* several such special conditions allow the "pushing" of skeletal elements, namely "see-saw constructions," vaginal fasciae as guiding devices, and inflatable Corpora cavernosa.

In "see-saw constructions" one skeletal element articulates with the central portion, rather than with the extremity, of a distal element. In this way, one end of the distal element can be pulled down by a contracting muscle while the other end is simultaneously pushed upwards. For example, when the basihyale is rotated caudo-dorsally by the contraction of the M. stylohyoideus, the urohyale is pushed ventro-apically against the Nodulus of the Lig. nodulo-ceratobranchiale. In another example of a "see-saw construction," the posterior process of the paraglossale is pushed dorso-apically when the anterior process is pulled caudo-ventrally by the contracting M. ceratoglossus (pp. 116 f.).

Vaginal fasciae render the pushing of skeletal elements possible by forming a space into which skeletal elements can be pushed back and guided. For example, the hyoid horns are pushed back into the Fasciae vaginales hyoidei by the contracting Mm. serpihyoidei during the retraction of the hyoid apparatus.

In contrast to the two above constructions in which the pull of contracting muscles is redirected to push skeletal elements into particular directions, constructions allowing centrifugal, radial movements at the surface of an apparatus cannot be achieved through the direct action of contracting skeletal muscles. (Centripetal, radial movements, however, can be achieved by muscles pulling the surface covering towards the center of an apparatus; for example, the central area of the lingual tip can be depressed by the contracting M. mesoglossus.) One way to "push" a surface covering "away" from the center is to increase the volume of the underlying tissue which can be accomplished by active hydraulic mechanisms. For example, the inflation of the Corpora cavernosa lateralia pushes the lingual nail laterally away from the paraglossalia and, thus, broadens the lingual tip. (Another way to push a surface covering away from the center is to shorten an hydraulic structure with constant volume; such a passive hydraulic mechanism, however, does not occur in the lingual apparatus of *Psittacus*.)

In a complex system, movements need not only be guided but also be regulated in extent. Every movement about an articulation must have safeguards against excessive motions. Without such safeguards, articular capsules could be disrupted, and relaxed antagonistic muscles could be overstretched during rapid movements and during movements supported by additional external forces. Besides neurophysiological feedback mechanisms, which do not have to be considered for the construction of a purely mechanical model, passive mechanical devices, such as articular and linkage ligaments and fasciae, are the most important such safeguards. For example, the Lig. articulare paraglosso-basihyale ventrale prevents an excessive extension of the paraglossale, the Lig. nodulo-ceratobranchiale prevents an excessive flexion of the basihyale during retraction of the hyoid, and the Fascia nodulo-lingualis prevents an excessive extension of the basihyale during the protraction of the hyoid. Therefore, the correctness of an inferred direction of a movement of skeletal elements, which is arrived at through the analysis of the shape and geometry of the articular facets and of the location of articular and linkage ligaments, can be further verified by the presence of passive mechanical devices that prevent excessive movement in that particular direction.

Actions and interactions of muscles.—Muscles in a biomechanical system can be grouped in several ways: A) as extrinsic or intrinsic muscles to explain differences in their actions on the skeletal system; B) as synergistic or antagonistic muscles to explain their interactions with one another; C) in one-joint and two-joint muscles to explain differences in their actions on skeletal elements.

A) Intrinsic and extrinsic muscles.—Intrinsic muscles (Table 1) attach only to skeletal elements within the apparatus and, hence, move the skeletal elements relative to one another and change the configuration of the apparatus. They cannot, however, move the center of mass of the apparatus by themselves. Therefore, if an intrinsic muscle contracts and moves one skeletal element, other skeletal elements move too because the center of mass of the apparatus remains at the same place. In other words, the movement of the center of mass of one skeletal element is compensated by the movement of the center of mass of another skeletal element. For example, the contraction of a portion of the M. ceratoglossus can initiate a chain-reaction of skeletal movements during the side-to-side movement of the hyoid.

Extrinsic muscles (Table 1), in contrast, connect the skeletal elements of an apparatus with the structural framework and can, therefore, move the center of mass of the entire apparatus relative to the framework. In the lingual apparatus of *Psittacus,* they simultaneously move the skeletal elements of the apparatus relative to one another because the movements of the hyoid relative to the mandible are linked to movements of the basihyale relative to the ceratobranchialia. For example, the contracting extrinsic M. serpihyoideus and M. stylohyoideus retract the hyoid apparatus and simultaneously rotate the basihyale towards caudo-dorsal.

A strict distinction between extrinsic and intrinsic muscles is necessary for a theoretical analysis of a biomechanical apparatus because their functions differ so much. The functions of the extrinsic muscles may also greatly influence the functions of the intrinsic muscles. If extrinsic muscles not only move the apparatus relative to its framework but also move the skeletal elements of the apparatus relative to one another, as in the lingual apparatus of *Psittacus,* then the effects of contracting intrinsic muscles are subordinate to the effects of simultaneously contracting extrinsic muscles. For example, the functions of the various portions of the intrinsic M. ceratoglossus vary depending on whether the hyoid is simultaneously protracted or retracted by extrinsic muscles (Table 16). Therefore, if the extrinsic muscles influence the movements of the skeletal elements of an apparatus relative to one another, it is necessary to analyze the function of the extrinsic musculature before that of the intrinsic musculature. If the extrinsic musculature does not influence the movements of the skeletal elements with one another, such as in the laryngeal apparatus of *Psittacus* (Homberger 1979b), the extrinsic and intrinsic muscles can be analyzed independently.

B) Synergists and antagonists.—Every muscle or muscle group responsible for a particular movement needs to be counteracted by an antagonistic force to return the skeletal element to its initial position and, thus, to extend the agonistic muscle or muscle group. The antagonistic force need not necessarily be generated by a muscle, nor need there be a one-to-one relationship between synergists and antagonists. For example, during the retraction of the hyoid, two extrinsic muscles (M. serpihyoideus and M. stylohyoideus) retract the hyoid and the M. stylohyoideus also rotates the basihyale caudo-dorsally, whereas during the protraction one extrinsic muscle with three portions (M. branchiomandibularis) protracts the hyoid and one intrinsic muscle (M. ceratohyoideus) rotates the basihyale ventro-apically. In another example, the extension of the epibranchiale is brought about by the contraction of the M. branchiomandibularis posterior and M. branchiomandibularis anterior lateralis during the protraction of the hyoid; there is, however, no muscular antagonist present to flex the epibranchiale. The epibranchiale is forced into flexion by the curved shape of the Fascia vaginalis hyoidei during retraction of the hyoid. Yet in another example, the surface covering of the anterior end of the central area of the lingual tip is depressed by the contracting M. mesoglossus anterior whereas the surface covering is lifted and the M. mesoglossus anterior extended, at least partly, by the inflating Corpus cavernosum apicale.

Elastic recoil of compressed or stretched tissue can also provide an antagonistic force, such as the recoil of the lingual nail during the deflation of the Corpora cavernosa. However, it cannot act as the only antagonist to a movement caused by muscular contraction, because its speed, force, and duration cannot be actively controlled.

The need for an antagonistic force for every agonistic force generating a movement can be used to test the completeness of an anatomical description in a functional-anatomical study.

C) One-joint and two-joint muscles.—One-joint muscles generate movements only at the one articulation they span. In contrast, two-joint or multiple-joint muscles extend over at least two articulations. In theory, a contracting two-joint muscle affects both articulations over which it passes. In actuality, however, a contracting two-joint muscle probably induces movements at only one joint, because its function at the other articulation is influenced by movements that are generated by another, simultaneously contracting muscle or muscle group (Carlsöö and Molbech 1966; Basmajian 1978:156–164). For example, the M. ceratoglossus primarily flexes the paraglossale, but its function is influenced by the simultaneous movements at the Art. cerato-basihyalis, which are produced by extrinsic muscles during the protraction and retraction of the hyoid (Table 16).

Two-joint muscles are also known for their "paradoxical effect" (Molbech 1966, and references therein), which means that they can be active during several different movements of the apparatus to which they belong. For example, the M. ceratoglossus medialis can contract and flex the paraglossale during protraction as well as retraction of the hyoid. However, the mode of contraction differs for the two different situations; i.e., during the protraction of the hyoid the M. ceratoglossus medialis contracts isotonically, whereas it contracts more or less isometrically during the retraction of the hyoid (Table 16). Hence, two-joint muscles have the advantage of being able to move skeletal elements without shortening (i.e, by contracting isometrically), at least under certain circumstances. They are, therefore, more efficient than one-joint muscles which generally contract isotonically to move skeletal elements, although they depend on simultaneously contracting one-joint muscles that determine their functions.

In this way, the movements at joints that are spanned by a two-joint muscle are linked. This means, the function of a two-joint muscle cannot be analyzed unless the actions of the other muscles of the apparatus are known.

Synchronization of movements.—In the preceding sections those biomechanical principles that determine the actions and interactions of the various structural elements of a biomechanical apparatus were explained. But to construct a model that describes the coordinated, simultaneous movements of an entire apparatus comprising a multitude of movable skeletal elements an additional biomechanical principle is needed. This "efficiency principle" rests on the concept that every biomechanical apparatus has evolved under the influence of selection forces to function efficiently. This biomechanical principle can be applied to unloaded and slow-moving systems, which use a minimum amount of energy, as compared to loaded and fast-moving systems, which require more energy for simultaneous antagonistic muscle contractions.

The "efficiency principle" predicts that during slow, unloaded movements antagonistic muscles do not contract simultaneously with agonistic muscles. It also predicts that contracting two-joint muscles, which affect two articulations simultaneously and may act differently on each articulation depending on the synchronously contracting one-joint muscles, will not be elongated during slow, unloaded movements. The predictions allow for the determination of those movements that can be performed simultaneously by the various structural elements of an apparatus (pp. 120, 134, 138).

Construction of a Theoretical Mechanical Model

In this section, the procedures are outlined for the construction of a theoretical mechanical model described earlier.

1. The detailed and complete anatomical description of the biomechanical apparatus and its structural components is the basis for the construction of a theoretical mechanical model. The decisions about which structures to include in the description and about the degree of detail to which the structures have to be described must be guided by the knowledge of the biomechanical principles that control the mechanical functioning of such an apparatus. Only those structures mechanically interacting during movements of the apparatus must be included in the description. For example, in the lingual apparatus of *Psittacus,* the nerves and most blood vessels can be omitted from the description, but the cavernous veins must be included because they have an important mechanical function.
2. The movements performed by a biomechanical apparatus must be known before the movements of the individual skeletal elements of an apparatus

can be inferred from the anatomical description. For example, it has to be known whether *Psittacus* can move its tongue laterally in order to be sure that asymmetrical movements of the lingual apparatus are possible under physiological conditions.

3. The direction and extent of the movements at each articulation of the apparatus must be determined by analyzing the surface structure and geometry of the articular facets, the functions of ligaments and fasciae, and the influence of the extrinsic structural framework on the apparatus. For example, it is necessary to recognize that the extension of the paraglossale is linked to a simultaneous protraction to understand the function of the M. hypoglossus anterior (pp. 129 f., 132 f.).
4. The function of each muscle at every articulation over which it passes must be established. For two-joint muscles, the effects of their contractions have to be determined separately for each articulation. For example, the contracting M. ceratoglossus flexes the paraglossale, but the five heads of the muscle have different effects on the Art. cerato-basihyalis. The functional analysis of muscles must consider muscle architecture and physiological properties, as well as whether they are intrinsic or extrinsic and the number of joints over which they pass. In biomechanical systems in which extrinsic muscles not only move the apparatus with respect to extrinsic structures but also skeletal elements of the apparatus with respect to one another, the functions of the extrinsic musculature must be analyzed before the functions of the intrinsic musculature.
5. Each muscle must be matched with an antagonistic structure (muscular or hydraulic), because an integral biomechanical apparatus functions only if every structure that is moved can resume its initial position. This search for agonistic-antagonistic pairs often reveals structures that would otherwise have been overlooked. For example, the "lack" of an antagonistic muscle counteracting the extension of the epibranchiale prompted the discovery of the presence and mechanical function of the Fascia vaginalis hyoidei (p. 114).
6. In a final step, the movements of the various skeletal components of the apparatus must be "synchronized" to represent orderly and potentially functional movements in the model.

Description of the Theoretical Mechanical Model

Traditionally, the functioning of a biomechanical apparatus was explained with the help of a model built out of wood, metal, and strings (P. R. Stettenheim, pers. comm.). The construction of such analogous mechanical models has been discussed by Brown (1960). These models seem, however, to work best for relatively simple systems and are less suited for a complex apparatus that contains not only muscles and bones, but also hydraulic and epithelial structures as mechanically interacting components.

A theoretical mechanical model, such as the one presented for the lingual apparatus of *Psittacus,* explains two interrelated, but different, aspects of the functioning of an apparatus, namely (1) the actions and interactions of each structural element and (2) the effects of these actions and interactions on the entire apparatus. The latter aspect consists mainly of configurational and positional changes of the apparatus and can easily be represented through figures, whereas the former aspect is structured like a causal network of actions and interactions and is best explained verbally.

The following descriptive method for the theoretical mechanical model is adapted from the method of free-body diagrams for the analysis of muscle-bone systems elaborated by Dempster (1961) and Bock (1974). Techniques retained from this method include the consideration of a skeletal apparatus as a system of rotating lever arms, the representation of muscle forces by force lines that can be resolved into components, and the analysis of a muscle force diagram through the separate analysis of the force components. In other respects, however, the method had to be modified because the present mechanical model describes dynamic events and does not include quantitative measurements (relative or absolute) of muscle forces. Therefore, no mathematical equations could be formulated to describe the muscle force correlations or the movements within the apparatus.

The figures explaining the configurational and positional changes of the apparatus show, in general, the extreme stages of particular movements (e.g., flexion, protraction, and inflation) and include the muscle forces that will transform the initial configuration into the opposite configuration (e.g., flexor muscle forces are shown acting on skeletal elements in an extended configuration). The forces exerted by contracting muscles are indicated by force lines with arrows pointing in the direction in which the movable skeletal element is pulled towards the stationary element (the contracting muscle itself, of course, exerts tension on both skeletal elements to which it attaches). Separate, free-standing arrows indicate motions of skeletal elements as well as external forces acting on skeletal elements which are, in general, reaction forces from the extrinsic framework. For example, free-standing arrows in Figure 48A indicate the retraction of the hyoid, the caudo-dorsal angular rotation of the basihyale, the apico-ventral angular rotation of the urohyale, the apical translation of the Nodulus, and the axial rotation of the ceratobranchiale. In another example in which external forces are acting (Fig. 51A), a free-standing arrow indicates the potential dorsal rotation of the caudal end of the ceratobranchiale and the ventrally directed reaction force from the palate, in which case no dorsal movement occurs at the caudal end of the ceratobranchiale. In order to demonstrate the three-dimensionality of certain movements, the same configuration is usually shown in dorsal as well as lateral view (e.g., Fig. 48). If a single movement requires many different muscle forces, it is shown in two separate, but congruent, figures each illustrating a separate set of muscles in order to keep the figures clear and simple (e.g., Figs. 50A, B). The muscle forces are not resolved into components to avoid overcrowding the figures.

In the verbal descriptions, the actions and interactions of the various structural components of the apparatus are explained as if they were resulting from a linear chain reaction. In reality, these actions and interactions probably occur more or less simultaneously, but such a cluster of events cannot easily be described verbally. The directions of exerted forces and movements of skeletal elements are described according to a system of perpendicularly arranged axes, e.g., transverse (medial, lateral, dextral, sinistral), vertical (dorsal, ventral), and sagittal (caudal, apical). This system coincides with the system of axes of the mandible which represents the framework within which the hyoid apparatus moves. For example, the ventro-latero-caudal force exerted by the M. serpihyoideus on the ceratobranchiale can be resolved into a ventral, lateral and caudal force component, and the effects of each component can be discussed separately. Although the force components are not shown in the figures, it is not difficult to visualize the components by reading the verbal descriptions in connection with the figures.

Movements of the Hyoid Relative to the Mandible

The mandible forms the framework within which the hyoid apparatus moves because all extrinsic lingual muscles, except the M. tracheohyoideus, originate from the mandible. Within this framework, the hyoid can move along longitudinal, transversal and vertical axes. For descriptive purposes, the movements along these three axes will be analyzed first separately before the correlations between them are discussed. However, most actual movements of the hyoid relative to the mandible are based on some combination of those basic movements.

Back-and-Forth Movement of the Hyoid

The back-and-forth movement of the hyoid apparatus is based on coordinated actions of three extrinsic lingual muscles (M. serpihyoideus, M. stylohyoideus, M. branchiomandibularis) and one intrinsic lingual muscle (M. ceratohyoideus). For the analysis of this movement, it was assumed that the muscles act symmetrically and that the ceratobranchialia maintain the same orientation relative to the longitudinal axis of the mandible during the excursions. Although the back-and-forth movement is tied to a synchronous up-and-down displacement of the hyoid apparatus, this latter movement will be disregarded until later. The back-and-forth movement of the hyoid is also tied to a configurational change of the hyoid itself, due to the restrictions placed on the hyoid system by the specific structures of the mandible and of the Art. cerato-basihyalis.

Retraction of the hyoid (Fig. 46).—Movements of the skeletal elements of the hyoid will be described from the initial protracted position of the hyoid (Fig. 46) to the final retracted position (Fig. 47). When the hyoid is protracted, the ceratobranchialia lie relatively close together (Fig. 46B) and the basihyale points slightly upwards, forming an angle of about 150° with the dorsal surface of the ceratobranchialia (Fig. 46A). The Nodulus with its surrounding connective tissue is located near the caudal end of the urohyale, and the Lig. nodulo-ceratobranchiale is taut. The extrinsic muscles acting as retractors of the hyoid are the M. serpihyoideus (M. sh.) and the M. stylohyoideus (M. st.).

When the paired Mm. serpihyoidei contract, the ceratobranchialia are drawn diagonally in a caudo-lateral direction. The lateral components of these tensile forces keep the ceratobranchialia gliding along the medial surfaces of the jaw muscles covering the medial sides of the Rami mandibulares (Fig. 15B). Because the Rami mandibulares diverge caudally, the ceratobranchialia spread apart when gliding backwards. Due to the particular structure of the articular facets of the Art. cerato-basihyalis, a lateral angular rotation of the ceratobranchiale about this articulation also results in either a dorso-apical angular rotation of the ceratobranchiale or a dorso-caudal angular rotation of the basihyale, or a combination of both, so that the angle enclosed between the dorsal surfaces of the ceratobranchiale and basihyale is reduced. Because the caudal end of the ceratobranchiale abuts dorsally against the pterygoid musculature of the palate, it cannot rotate dorsally about the Art. cerato-basihyalis. Therefore, the lateral angular rotation of the ceratobranchiale forces the basihyale to rotate caudo-dorsally about the Art. cerato-basihyalis into a more vertical orientation (Fig. 47A).

This caudo-dorsal angular rotation of the basihyale automatically results in a ventro-apical rotation of the urohyale and in an increase of the angle enclosed by this element and the ceratobranchialia (compare Figs. 46A, 47A and Figs. 46B, 47B). The rotating urohyale is pushed against the Nodulus which is connected to the ceratobranchialia through the paired segments of the noncompliant Lig. nodulo-ceratobranchiale. As the urohyale moves away from the ceratobranchialia, the Nodulus must glide apically along the urohyale so that the non-stretchable Lig. nodulo-ceratobranchiale can still "span" the "increasing" linear distance between the urohyale and ceratobranchialia. However, a forward movement of the Nodulus is opposed by the contracting M. sh. exerting a caudal pull on the Lig. nodulo-ceratobranchiale, on the Nodulus, and on the midsagittal raphe extending caudally from the Nodulus. In contrast, the simultaneous caudo-ventral pull exerted by the M. st. on the Proc. parahyalis of the basihyale reinforces the ventro-apical angular rotation of the urohyale during the retraction of the hyoid. As a net result of the various forces acting on the Nodulus and the urohyale, the Nodulus is pushed forward along the urohyale during the retraction of the hyoid. This means that the position of the Nodulus relative to the urohyale is determined by the configuration of the hyoid, i.e., by the position of the hyoid relative to the mandible. Therefore, the force of the contracting M. sh. on the Nodulus and Lig. nodulo-ceratobranchiale is transmitted to the ceratobranchialia and urohyale and serves to pull back the entire hyoid. In this way, the caudal component of the pull exerted by the M. st. on the basihyale retracts the hyoid (cf. pp. 105 f.).

Simultaneously with its lateral angular rotation during the retraction of the hyoid, the ceratobranchiale undergoes an axial rotation due to the particular structure of the articular facets of the Art. cerato-basihyalis. Through this axial rotation, the Tuberositas ceratobranchialis, which serves as attachment site for the Lig. nodulo-ceratobranchiale, turns slightly ventro-medially so that it lies closer to the urohyale. In this way, the increase of the linear distance between the ceratobranchiale and urohyale brought about by the lateral angular rotation of the ceratobranchiale is, to a small extent, counteracted by the axial rotation of the ceratobranchiale. Therefore, the linear distance between the ceratobranchiale and urohyale spanned by the Lig. nodulo-ceratobranchiale does not increase in the same proportion during the retraction of the hyoid as does the angle enclosed by the ceratobranchiale and urohyale. Hence, the apical displacement of the Nodulus along the urohyale is somewhat less extensive than if no axial rotation of the ceratobranchiale occurred during the retraction of the hyoid.

During the retraction of the hyoid, the M. ceratoglossus lateralis and M. ceratoglossus superficialis may contract synchronously with the M. sh. and M. st. to support the movements of the ceratobranchialia and basihyale at the Art. cerato-basihyalis (pp. 106 ff.).

The Lig. nodulo-ceratobranchiale plays an important role in preventing an excessive flexion, i.e., caudo-dorsal rotation, of the basihyale during the retraction of the hyoid. Namely, the basihyale can rotate caudo-dorsally only until the two limbs of the Lig. nodulo-ceratobranchiale lie perpendicular to the urohyale. This position of the ligament indicates the apicalmost position of the Nodulus on the urohyale. Any further caudo-dorsal angular rotation of the basihyale, i.e., ventro-apical angular rotation of the urohyale, and simultaneous additional spreading of the ceratobranchialia would increase the linear distance between the urohyale and ceratobranchialia. This is prevented by the resistance of the non-stretchable Lig. nodulo-ceratobranchiale bridging the gap between the urohyale and ceratobranchialia.

Protraction of the hyoid (Fig. 47).—The movements of the skeletal elements of the hyoid will be described from the initial retracted position of the hyoid (Fig. 49) to the final protracted position (Fig. 48). When the hyoid is retracted, the ceratobranchialia are spread apart (Fig. 49B) and the basihyale points obliquely upwards, forming an angle of about 120° with the dorsal sides of the ceratobranchialia (Fig. 49A). The Nodulus with its surrounding connective tissue is located near the apical end of the urohyale, and the Lig. nodulo-ceratobranchiale is taut. The muscles acting as protractors of the hyoid are the three portions of the extrinsic M. branchiomandibularis (M. bm.) and the intrinsic M. ceratohyoideus (M. ch.).

When the paired Mm. branchiomandibulares contract, the ceratobranchialia and epibranchialia are pulled forward. (In the following paragraphs, the epibranchiale is treated as a part of the ceratobranchiale since there is little movement at the Art. epi-ceratobranchialis.) Whereas the pull of the contracting M. bm. anterior lateralis is directed almost exactly apically (Fig. 49B), the forces exerted by the M. bm. posterior and M. bm. ant. medialis are directed slightly diagonally towards apico-lateral. (For a discussion of the dorsal component of the M. bm. ant. in Fig. 49A, see pp. 103 f., 113 f.) The lateral component of these forces keep the ceratobranchialia gliding along the medial surfaces of the jaw muscles covering the medial surfaces of the Rami mandibulares. Because, however, the Rami mandibulares converge apically, the ceratobranchialia are pushed closer together as they glide forward. Due to the particular structure of the articular facets of the Artt. cerato-basihyales, a medial angular rotation of the ceratobranchiale about this articulation also results in a ventro-caudal angular rotation of the ceratobranchiale, or a ventro-apical angular rotation of the basihyale, or a combination of both, so that the angle formed between the dorsal surfaces of the basihyale and ceratobranchiale is increased. The ceratobranchiale, however, is prevented from performing a ventro-caudal angular rotation about the Art. cerato-basihyalis by the dorsal component of the apico-dorsal pull of the contracting M. bm. ant. which keeps the caudal end of the ceratobranchiale gliding along the pterygoid musculature of the palate (Fig. 47A). (For more details on the dorsal component of the force exerted by the M. bm. ant., see pp. 103 f.) Therefore, the medial angular rotation of the ceratobranchialia forces the basihyale to rotate ventro-apically about the Art. cerato-basihyalis to a more horizontal orientation (Fig. 46A).

This ventro-apical angular rotation of the basihyale automatically results in a dorso-caudal angular rotation of the urohyale and in a reduction of the angle enclosed by this bone and the ceratobranchialia (compare Figs. 47A and 46A). The angular rotation of the urohyale is supported and guided by the simultaneously contracting M. ch. which exerts a caudo-dorsal pull on the Nodulus and an apico-medio-ventral pull on the ceratobranchialia (Fig. 47). The caudal component of the caudo-dorsal force of the M. ch. pulls the Nodulus caudally along the urohyale so that the paired limbs of the Lig. nodulo-ceratobranchiale remain taut during the protraction of the hyoid. When the Lig. nodulo-ceratobranchiale is taut, the caudo-dorsal pull exerted by the M. ch. on the Nodulus is transmitted to the urohyale and supports the dorso-caudal angular rotation of the urohyale (Fig. 47A). At the same time, the medial components of the apico-medio-ventral forces on the ceratobranchialia by the M. ch. pull the ceratobranchialia closer together during the protraction of the hyoid, so that they fit into the anteriorly narrowing space between the Rami mandibulares (Fig. 47B). The ventral component of the apico-medio-ventral pull on the ceratobranchiale by the M. ch. does not make the ceratobranchiale rotate caudo-ventrally about the Art. cerato-basihyalis because it is opposed by the dorsal component of the apico-dorsal force exerted on the caudal end of the ceratobranchiale by the simultaneously contracting M. bm. ant. (Fig. 47A).

Simultaneously with its medial angular rotation during the protraction of the hyoid, the ceratobranchiale undergoes an axial rotation due to the particular structure of the articular facets of the Art. cerato-basihyalis. Through this axial rotation, the Tuberositas ceratobranchialis, which serves as attachment site for the Lig. nodulo-ceratobranchiale, turns slightly dorso-laterally so that its distance to the urohyale is slightly increased. In this way, the decrease of the linear distance between the ceratobranchiale and urohyale brought about by the medial angular rotation of the ceratobranchiale is partly counteracted by the axial rotation of the ceratobranchiale. Therefore, the linear distance between the ceratobranchiale and urohyale spanned by the Lig. nodulo-ceratobranchiale does not decrease in the same proportion during the protraction of the hyoid as does the angle enclosed by the ceratobranchiale and urohyale. Hence, the caudal displacement of the Nodulus, which ensures that the Lig. nodulo-ceratobranchiale is kept taut over the gap between the ceratobranchiale and urohyale, is somewhat less extensive than if no axial rotation of the ceratobranchiale occurred during the protraction of the hyoid.

During the protraction of the hyoid, the M. ceratoglossus medialis may contract synchronously to support the movements of the ceratobranchiale and urohyale at the Art. cerato-basihyalis (pp. 106 ff.).

The Lig. articulare cerato-basihyale plays an important role in preventing an overextension, i.e., an excessive apico-ventral rotation, of the basihyale during the protraction of the hyoid by limiting the extension of the Art. cerato-basihyalis. An overextension of the basihyale is also prevented by the Fascia nodulo-lingualis which anchors the Nodulus and the Lig. nodulo-ceratobranchiale apically so that their caudal excursions are limited. This limitation prevents any further caudo-dorsal angular rotation of the urohyale, which would be responsible for the extension of the basihyale.

Up-and-Down Movement of the Hyoid

The up-and-down movement of the hyoid is based mainly on the coordinated, symmetrical action of three different extrinsic lingual muscles, namely the M. genioglossus, the M. tracheohyoideus, and the M. mylohyoideus. The up-and-down movement, however, cannot be performed independently. It is superimposed on the back-and-forth movement of the hyoid because some muscles acting primarily as elevators or depressors of the hyoid simultaneously also protract or retract the hyoid, and vice versa. In this way, raising of the hyoid is tied to a protraction of the tongue (the reverse, however, is not necessarily the case; see pp. 104 f.) and lowering of the hyoid is tied to a retraction of the tongue (the reverse is the case only to a limited extent; see pp. 105 f.). Therefore, the muscles responsible for the up-and-down movement have to coordinate their actions with those of the muscles responsible for the back-and-forth movement.

Lifting of the retracted, lowered hyoid (Figs. 48A, B).—The movements of the skeletal elements will be described from the initial, retracted and lowered position of the hyoid (Figs. 48A, B) to the final, protracted and lifted position of the hyoid (Figs. 49A, B). In the retracted position of the hyoid the ceratobranchialia are maximally spread apart (Fig. 47B), and in the lowered position of the hyoid the tip of the tongue touches the bottom of the lower mandible (Fig. 48A; note that the apical tip of the tongue is not identical with the tip of the paraglossale, but lies beyond it).

During raising of the hyoid, one intrinsic and two extrinsic lingual muscles, the M. ceratohyoideus (M. ch.), the M. mylohyoideus (M. mh.), and the M. branchiomandibularis (M. bm.), contract.

When the M. mh. ant. contracts, it exerts a dorso-apical pull on the body of the Gl. sublingualis which is attached through connective tissue to the basihyale and the musculature covering its ventral surface (Fig. 48A). This pull is transmitted to the basihyale which, thus, is simultaneously lifted and protracted. In addition, the Fascia facialis, to which the caudal borders of the M. mh. ant. and the Gl. sublingualis are attached, is pulled along with the hyoid. Due to the particular structure of the articular facets of the Art. cerato-basihyalis, the apical displacement of the basihyale and, hence, of the entire hyoid, is tied to the simultaneous ventro-apical angular rotation of the basihyale and caudo-medial angular rotation of the ceratobranchialia about the Artt. cerato-basihyales. These two angular rotations are supported by the simultaneously contracting M. ch. which exerts a caudo-dorsal pull on the Nodulus (and, thus, on the urohyale) and an apico-ventro-medial pull on the ceratobranchialia (Figs. 47A, B, 48B). At the same time, the contracting M. mh. post. exerts a dorso-apical pull on the Nodulus (Fig. 48A). The apical component of this pull is opposed by the caudal component of the caudo-dorsal pull on the Nodulus exerted by the M. ch. (Fig. 48B). Hence, the dorso-apical force exerted by the M. mh. post. on the Nodulus is transmitted to the urohyale and, consequently, to the basihyale, and supports the dorso-apical force exerted by the M. mh. ant. on the basihyale. The dorso-apical force of the M. mh. post. does not produce a rotational moment for the urohyale because its force line always passes ventro-laterally over the Art. cerato-basihyalis which is the rotational axis of the urohyale (Figs. 48A, C).

The ventro-apical angular rotation of the basihyale, which is tied to an apical displacement of the basihyale, is opposed by the dorsal component of the dorso-apical force exerted by the M. mh. ant. on the basihyale. Therefore, the ventro-apical angular rotation of the basihyale transforms itself into a dorso-caudal angular rotation of the urohyale and region of the Art. cerato-basihyalis, using the point of support of the basihyale on the M. mh. ant. as fulcrum. Lifting of the region of the Art. cerato-basihyalis leads to a lifting of the anterior end of the ceratobranchiale which corresponds to an angular rotation of the anterior end of the ceratobranchiale about its center of inertia towards apico-dorsal and tends to have the posterior end of the ceratobranchiale perform an angular rotation towards caudo-ventral (Fig. 48A). This caudo-ventral angular rotation of the caudal end of the ceratobranchiale about its center of inertia is opposed and prevented by the dorsal component of the apico-dorsal pull exerted simultaneously by the M. bm. ant. on the caudal end of the ceratobranchiale (Fig. 48B). This apico-dorsal pull of M. bm. ant. holds the posterior end of the ceratobranchiale against the ventral surface of the palatal pterygoid musculature during the protraction of the hyoid. (For details on the functions of the dorsal and ventral components of the M. bm. on the epibranchiale, see pp. 113 f.) Therefore, the apico-dorsal angular rotation of the anterior end of the ceratobranchiale is performed about the posterior end of the ceratobranchiale and its contact point with the pterygoid musculature of the palate as a fulcrum. Lifting of the region of the Art. cerato-basihyalis can be characterized as occurring through a caudo-dorsal angular rotation at the apical end of the ceratobranchiale about its caudal end.

The apical components of the apico-dorsal, respectively apico-ventral, forces exerted on the epibranchiale and ceratobranchiale by the M. bm. support the apical components of the dorso-apical forces exerted on the hyoid by the M. mh. in pulling the hyoid forward (Fig. 48B). Because no muscular forces that could counteract the apical displacement during the raising of the hyoid can be generated within the lingual apparatus, the lifting motion of the hyoid is always tied to a simultaneous protraction. The reverse, however, is not necessarily the case (see below).

Protraction of the hyoid without simultaneous lifting (Fig. 48C).—Although lifting of the hyoid apparatus is tied to a synchronous protraction, protraction of the hyoid is not necessarily tied to a simultaneous lifting because the dorsal motion of the hyoid can be modified by the action of the M. genioglossus (M. gg.). For descriptive purposes, the starting point of the movement is assumed to be at an intermediate stage of the combined protraction and lifting of the hyoid as described in the previous section (Fig. 48C). It is also assumed that during protraction of the hyoid, the paraglossale remains immobilized with respect to the basihyale. The movements of the skeletal elements will be described from the initial, halfway protracted and lifted position of the hyoid (Fig. 48C) to the final, fully protracted, but still only halfway lifted position of the hyoid (Fig. 46A).

During protraction of the hyoid without simultaneous lifting, the muscle actions of the M. mylohyoideus (M. mh.), M. branchiomandibularis (M. bm.), and M. ceratohyoideus (M. ch.) are the same as those described for lifting of the retracted and lowered hyoid. The simultaneous contraction of the M. gg., however, exerts a ventro-apical force on the paraglossale. With the paraglossale assumed to be immobilized, this ventro-apical pull is transmitted to the apical end of the basihyale and supports a ventro-apical rotation of the basihyale which is associated with a protraction of the hyoid. If the M. mh. is contracting at the same time, the ventral component of this ventro-apical force generated by the M. gg. counteracts the dorsal component of the dorso-apical force exerted by the M. mh. ant. on the basihyale (Fig. 48C). In this way, the basihyale is not lifted during the protraction. Consequently, the region of the Art. cerato-basihyalis is raised to a lesser extent because it is lifted only through a dorso-caudal angular rotation of the urohyale about the anterior end of the basihyale, where it is supported by the M. mh. ant., and through a simultaneous apico-dorsal angular rotation of the anterior end of the ceratobranchiale about its posterior end. Through the dorso-caudal angular rotation of the urohyale without simultaneous lifting of the hyoid, the force line of the apico-dorsal pull of the M. mh. post., which is exerted on the Nodulus, but transmitted to the urohyale, is shifted from a position latero-ventral to the Art. cerato-basihyalis to one more caudal to it. In this case, the dorsal component of the apico-dorsal pull of the M. mh. post. supports the dorso-caudal angular rotation of the urohyale about the Art. cerato-basihyalis.

If the M. mh. does not contract at the same time as the M. gg., the ventral component of the ventro-apical force of the M. gg. depresses the basihyale and, with it, the entire hyoid. Therefore, as a general rule, the hyoid is raised, lowered or kept at the same level during its protraction depending on whether only the M. mh., only the M. gg., or both muscles together, contract.

The apical component of the ventro-apical pull exerted by the M. gg. on the paraglossale supports the apical components of the forces exerted on the hyoid by the M. bm. and M. mh. until the insertion of the M. gg. lies directly dorsally to its origin on the mandible, i.e., until the force exerted by the M. gg. consists only of a ventral component. Any further protraction of the hyoid increases the distance between origin and insertion of the M. gg. so that the M. gg. is elongated. If the M. gg. were still contracting, it would produce "negative work," i.e., the movement would be inefficient (p. 84).

Contraction of the M. gg. does not affect the orientation of the paraglossale with respect to the basihyale because the insertion of the M. gg. on the paraglossale extends about equally apically and caudally beyond the central part which forms the Art. paraglosso-basihyalis; thus, the resultant force of the M. gg. passes through the Art. paraglosso-basihyalis which represents the rotational axis of the paraglossale. Orientation of the paraglossale towards the basihyale is determined by the M. ceratoglossus and M. hypoglossus obliquus (pp. 114 ff.).

Lowering of the hyoid (Figs. 49A, B).—The movements of the skeletal elements will be described from the initial, protracted and lifted position of the hyoid (Figs. 49A, B) to the final, retracted and lowered position of the hyoid (Figs. 48A, B). In the protracted position of the hyoid, the ceratobranchialia lie close together (Fig. 46B), and in the lifted position of the hyoid, the M. mh. is maximally contracted.

During lowering of the hyoid, three pairs of extrinsic lingual muscles, namely the M. serpihyoideus (M. sh.), M. stylohyoideus (M. st.), and M. tracheohyoideus (M. th.), contract in a coordinated way. When the M. th. contracts, it exerts a caudo-ventral pull on the Proc. parahyalis of the basihyale (Fig. 49A). The caudal component of this force results in a caudal displacement of the basihyale, and consequently, in a retraction of the entire hyoid. Simultaneously, it supports the caudo-dorsal angular rotation of the basihyale, which is tied to a retraction of the hyoid. The retraction of the hyoid and the caudo-dorsal angular rotation of the basihyale are supported by the caudal component of the caudo-ventral pull exerted by the simultaneously contracting M. st. on the Proc. parahyalis of the basihyale (Fig. 49B). During retraction of the hyoid, the ceratobranchialia glide along the medial surfaces of the jaw musculature covering the Rami mandibulares and, therefore, spread apart. These movements of the ceratobranchialia and the retraction of the hyoid are supported by the synchronously contracting M. sh. (For details on the interactions among the basihyale, ceratobranchialia, Nodulus, M. sh., and M. st. during the caudal displacement of the hyoid, see pp. 99 ff.)

The caudo-ventral component of the caudo-latero-ventral pull of the M. sh.

rotates the anterior end of the ceratobranchiale towards caudo-ventral. This caudo-ventral angular rotation of the anterior end of the ceratobranchiale about its center of inertia forces the caudal end of the ceratobranchiale to rotate dorso-apically. In doing this, however, the caudal end of the ceratobranchiale abuts against the pterygoid musculature of the palate (Fig. 49B). Therefore, the anterior end of the ceratobranchiale performs its caudo-ventral angular rotation about the caudal end of the ceratobranchiale, using its point of contact with the pterygoid musculature as a fulcrum. Through this action, the ceratobranchiale, the basihyale, and, thus, the entire hyoid, are lowered. Ventral displacement of the basihyale induced by the M. sh. is supported by the ventral components of the M. st. and M. th.

In the course of the retraction of the hyoid, the direction of the force exerted by the M. sh. on the hyoid also changes. At the beginning of the movement, all muscle fibers of the M. sh. exert caudo-ventral forces on the hyoid. During the retraction, which is linked to a simultaneous lowering of the hyoid, however, the ventral components of the caudo-ventral forces are more and more reduced, so that some muscle fibers will eventually exert purely caudal forces and some others even dorso-caudal forces (Figs. 49B, C).

Although the M. th. and M. st. are similar because they both exert caudo-ventral forces on the Proc. parahyalis of the basihyale (Figs. 49A, B), they differ in the way the proportions of the ventral and caudal components of their forces change in the course of the retraction of the hyoid. The ventral component of the caudo-ventral force of the M. st. decreases during a retraction of the hyoid until it reaches an equilibrium with the dorsal component of the caudo-dorsal force exerted on the hyoid by the simultaneously contracting M. sh. (cf. Figs. 49B, C). When this equilibrium is reached, the hyoid can be retracted without simultaneous lowering, provided the M. th. does not contract at the same time. In contrast, the ventral component of the caudo-ventral force of the M. th., which originates from the sternum and attaches to the trachea, increases in the course of the retraction of the hyoid. Thus, the pull of the M. th. becomes more efficient in pulling the hyoid down as the tongue is retracted (cf. Figs. 49A, C). In this way, the contracting M. th. is capable of pulling the hyoid farther down at a late stage of the retraction of the hyoid, when the ventral component of the caudo-ventral pull exerted on the hyoid by the M. st. is counterbalanced by the dorsal component of the caudo-dorsal force exerted on the hyoid by the M. sh.

Because no muscular forces that could counteract the caudal displacement during the lowering of the hyoid are generated within the lingual apparatus, the depression of the hyoid is tied to its retraction or protraction (pp. 104 f.).

Side-to-Side Movement of the Hyoid

The side-to-side movement of the hyoid is based on asymmetrical actions of several pairs of extrinsic and intrinsic lingual muscles and can best be characterized as a unilateral protraction and simultaneous contralateral retraction of the hyoid. For the analysis of the side-to-side movement as opposed to the up-and-down or back-and-forth movement, it is assumed that the hyoid as a whole does not shift along the longitudinal or vertical axes of the mandible during the sideways movement, although particular elements may do so. It is also assumed that the ceratobranchialia remain aligned with the medial surfaces of the jaw musculature covering the rami of the mandible since they must glide within the Fasciae vaginales hyoidei. Furthermore, any movement of the paraglossale is disregarded, and it is postulated that any force applied to the paraglossale is transferred to the basihyale (for details on the coordination of movements of the paraglossale with those of the hyoid, see pp. 117–120).

For descriptive purposes, the starting position of the hyoid is assumed to be at an intermediate stage between the protracted and retracted position (Figs. 50–52). The configuration of the hyoid at the start of the movement is symmetrical; i.e., the midsagittal axis of the basihyale coincides with that of the mandible (Figs. 50A, B; the final lateral position of the hyoid with its asymmetrically arranged skeletal elements is not illustrated). Because sideways movements of the hyoid are based on an asymmetrical combination of a protraction and retraction, the movements at the two Artt. cerato-basihyales are different from each other. To visualize these movements and their effects on the configuration of the entire hyoid, it is best to describe them first as if they resulted directly from contractions of intrinsic muscles. For descriptive purposes, the M. ceratoglossus was chosen as an example of such a muscle because it affects and is affected by movements at the Art. cerato-basihyalis and because it consists of several parts that function differently at this articulation. Also for descriptive purposes, only those portions of the M. cg. are mentioned that actually shorten during the described movements at the Artt. cerato-basihyales, although one has to keep in mind that the other portions of the M. cg. may contract simultaneously. Nevertheless, the actual sideways movements of the hyoid are initiated and executed primarily by the extrinsic lingual muscles because intrinsic muscles change only the configuration of the hyoid, but not its position within the framework of the mandible. Whether the intrinsic muscles, such as the M. ceratoglossus, M. hypoglossus obliquus, and M. supraglossus, actually contract simultaneously or not probably depends on whether movements of the paraglossale and of the epithelial surface of the tip of the tongue are performed at the same time.

In the following description, a distinction will be made between movements to the left (sinistral) and movements to the right (dextral) to avoid confusion when discussing lateral and medial movements. In the following explanatory example, a movement of the tip of the hyoid toward the left side of the head will be described. Return of the hyoid to the original midsagittal position is accomplished by the equivalent action of the contralateral muscles pulling the tip of the hyoid toward the right side of the head.

Functions of intrinsic lingual muscles (Figs. 50A, 51A, 52A).—When the left M. ceratoglossus lateralis (M. cg. lat.) contracts, it reduces the angle enclosed between the dorsal surfaces of the left ceratobranchiale and the basihyale by pulling both skeletal elements closer together (Fig. 51A). If both skeletal elements were unrestrained, this movement would be achieved by a synchronous angular rotation about the left Art. cerato-basihyalis of the basihyale toward sinistro-dorso-caudal and of the left ceratobranchiale toward sinistro-dorso-apical (Figs. 50A, 51A). At the same time, the left Art. cerato-basihyalis would be pushed down (Fig. 51A). However, whereas the basihyale is free to perform its angular rotation, the caudal end of the left ceratobranchiale is prevented from rotating because it abuts dorsally against the pterygoid musculature of the palate and laterally against the medial surface of the pterygoid musculature covering the mandibular ramus (Figs. 50A,

51A). Therefore, as a reaction, the anterior end of the left ceratobranchiale rotates toward dextro-ventro-caudal about the caudal end of the left ceratobranchiale using the point of contact of the latter with the pterygoid musculature of the palate as a fulcrum (Figs. 50A, 51A). This motion pushes the left region of the Art. cerato-basihyalis towards dextro-ventro-caudal (Figs. 50A, 51A). As a result, the right region of the Art. cerato-basihyalis and, consequently, the anterior end of the right ceratobranchiale perform an angular rotation towards sinistro-dorso-apical about the center of inertia of the basihyale between the two Artt. cerato-branchiales. This movement of the apical end of the right ceratobranchiale should cause its caudal end to perform an angular rotation toward dextro-ventro-apical about the center of inertia of the right ceratobranchiale. However, this movement of the caudal end of the right ceratobranchiale cannot be performed because the right ceratobranchiale abuts laterally (i.e., dextrally) against the medial surface of the pterygoid musculature covering the medial surface of the right Ramus mandibularis (Fig. 50A). The movement is also ventrally opposed by the dorsal component of the dextro-dorso-apical pull exerted by the synchronously contracting extrinsic M. branchiomandibularis anterior on the caudal end of the right ceratobrachiale (Fig. 52B). Therefore, the sinistro-dorso-apical movement of the right region of the Art. cerato-basihyalis is more pronounced than it would be if the caudal end of the ceratobranchiale moved.

The left M. ceratoglossus superficialis (M. cg. supf.) contracts synchronously with the left M. cg. lat. It exerts an apico-dorsal pull on the Lig. nodulo-cerato-branchiale, which is transmitted to the Nodulus (Figs. 50A, 51A). This apico-dorsal pull supports the actions of the M. cg. lat., but during the sideways movement of the hyoid, one of its main functions is to oppose with its apical component that caudal component of the caudo-lateral force of the left M. serpihyoideus which is exerted on the Nodulus. Another function of the apico-dorsal force of the M. cg. supf. is to counterbalance the caudal component of the caudo-dextro-dorsal force of the left M. ceratohyoideus and the apical component of the apico-dextro-dorsal force of the left M. mylohyoideus posterior. However, the force of the M. cg. supf. appears not to be indispensable for the proper execution of sideways movements of the hyoid because the muscle is missing in certain individuals.

Simultaneous contraction of the right M. ceratoglossus medialis (M. cg. med.) increases the angle enclosed by the dorsal surfaces of the right ceratobranchiale and the basihyale by pulling both skeletal elements farther apart (Fig. 52A). If both elements were unrestrained, this movement would be achieved through a synchronous angular rotation about the right Art. cerato-basihyalis of the basihyale toward sinistro-ventro-apical and of the right ceratobranchiale toward sinistro-ventro-caudal. At the same time, the right Art. cerato-basihyalis would be lifted (Fig. 52A). Whereas the basihyale is free to rotate about the Art. cerato-basihyalis, the caudal end of the right ceratobranchiale is not. It is prevented from rotating sinistro-ventro-caudally by the dextral and dorsal components of the dextro-dorso-apical pull exerted by the simultaneously contracting right M. branchio-mandibularis anterior medialis on the caudal end of the right ceratobranchiale (Figs. 50B, 52B). Thus, the anterior end of the right ceratobranchiale and, consequently, the region of the right Art. cerato-basihyalis perform an angular rotation towards dextro-dorso-apically about the caudal end of the ceratobranchiale using the point of contact of the latter with the M. branchiomandibularis ant. med. as a fulcrum (Figs. 50A, 52A). A dextro-dorso-apical movement of the right region of the Art. cerato-basihyalis results in a simultaneous, sinistro-ventro-caudal angular rotation of the left region of the Art. cerato-basihyalis about the center of inertia of the basihyale between the two joints. The movement of the left region of the Art. cerato-basihyalis pushes the apical end of the left ceratobranchiale towards dextro-ventro-caudal (Figs. 50A, 51A). This movement of the anterior end of the left ceratobranchiale, in turn, induces its caudal end to perform an angular rotation about the center of inertia of the left ceratobranchiale towards sinistro-dorso-apical (Figs. 50A, 51A). An angular rotation of the caudal end of the left ceratobranchiale, however, cannot be performed because the left ceratobranchiale abuts dorsally against the pterygoid musculature of the palate and laterally against the pterygoid musculature covering the medial surface of the left mandibular ramus (Figs. 50A, 51A, B). Therefore, the position of the hyoid within the framework of the mandible has to be readjusted through contraction of the extrinsic left M. serpihyoideus and M. stylohyoideus to provide for the space requirements of the new configuration of the hyoid.

To summarize the previous paragraphs, synchronous actions of the left M. cg. lat. and of the right M. cg. med. complement each other exactly. Action of the left M. cg. lat. on the left ceratobranchiale starts a chain reaction that affects the basihyale and the right ceratobranchiale, whereas action of the right M. cg. med. on the right ceratobranchiale starts a chain reaction that affects the basihyale and the left ceratobranchiale in the opposite way. The action of the left M. cg. lat. on the left half of the hyoid mirrors the action of the M. cg. med. on the right half of the hyoid and, therefore, the action of the left M. cg. lat. on the left ceratobranchiale is the same as the effect of the right M. cg. med. on the left ceratobranchiale, and vice versa.

Simultaneously with the angular rotations about the Artt. cerato-basihyales, the basihyale and both ceratobranchialia perform slight axial rotations about their respective longitudinal axes due to the structure of the articular facets of the Art. cerato-basihyalis. When the left M. cg. lat. contracts, reducing the angle enclosed by the dorsal surfaces of the basihyale and left ceratobranchiale and, consequently, increasing the linear distance between the urohyale and the left ceratobranchiale (Figs. 50A, 51A), the basihyale and urohyale actually perform an axial rotation in such a way that their left sides are turned slightly toward dextro-dorsal. In this way, the ventral surface of the urohyale, against which the Nodulus lies, is turned towards the left ceratobranchiale. At the same time, the left ceratobranchiale performs an axial rotation from the left to medio-ventral, so that the Tuberositas ceratobranchialis, which serves as the attachment site for the Lig. nodulo-ceratobranchiale, comes to lie closer to the urohyale. In this way, the increased linear distance between the left ceratobranchiale and the urohyale brought about by the angular rotations of the basihyale and left ceratobranchiale about the Art. cerato-basihyalis is somewhat offset by the axial rotations of the urohyale and left ceratobranchiale. Synchronously with the contraction of the left M. cg. lat., the right M. cg. med. contracts, so that the angle enclosed between the dorsal surfaces of the basihyale and right ceratobranchiale is increased and, consequently, the linear distance between the urohyale and right ceratobranchiale is decreased (Figs. 50A, 52A). At the same time, the basihyale and urohyale actually perform an axial rotation so that their right sides are turned sinistro-ventrally and the ventral surface of the urohyale is turned away from the right ceratobranchiale and towards the left ceratobranchiale. The axial rotation of the basihyale and urohyale corresponds to a dextro-dorsal axial rotation of the left sides of the basihyale and urohyale resulting from the contraction of the left M. cg. lat. At the same time, the right ceratobranchiale performs an axial rotation from the right to sinistro-dorsal so that the Tuberositas ceratobranchialis turns slightly away from the urohyale. In this way, decrease of the linear distance between the urohyale and ceratobranchiale brought about by the angular rotations of the basihyale and the right ceratobranchiale about the Art. cerato-basihyalis is slightly offset by axial rotations of the urohyale and right ceratobranchiale.

Due to the dextro-ventro-caudal angular rotation of the left region of the Art. cerato-basihyalis and the simultaneous sinistro-dorso-apical angular rotation of the right region of the Art. cerato-basihyalis about the center of inertia of the basihyale between the two joints, the basihyale and urohyale perform an axial rotation from right towards dorso-sinistral, so that the dorsal surface of the basihyale faces sinistro-dorso-caudally at the end of the sideways movement of the hyoid. This axial rotation overrides the axial rotation of the basihyale as described relative to the axial rotations of the ceratobranchialia.

The Nodulus must adjust its position during the sideways movement of the hyoid so that the limbs of the non-stretchable Lig. nodulo-ceratobranchiale can adapt to changing linear distances between the urohyale and ceratobranchialia, which they have to bridge. Because of the asymmetry of the changes in the linear distances between the urohyale and the left and right ceratobranchialia, the Lig. nodulo-ceratobranchiale cannot adjust to these changes by simply moving the Nodulus forward or backward along the urohyale as it did during the back-and-forth movement of the hyoid. Rather, the Nodulus has to glide from side to side over the Bursa synovialis urohyalis during sideways movements of the hyoid. If the hyoid moves towards the left, the Nodulus also moves towards the left, thus allowing the left limb of the Lig. nodulo-ceratobranchiale to bridge the increasing gap between the urohyale and left ceratobranchiale while keeping the right limb of the Lig. nodulo-ceratobranchiale taut over the decreasing gap between the urohyale and the right ceratobranchiale (Fig. 50B). Because of the axial rotations of the ceratobranchialia and urohyale, which somewhat counteract the changes in the linear distances between the urohyale and the ceratobranchialia, the lateral displacement of the Nodulus is less extensive than if no such axial rotations took place.

Functions of extrinsic lingual muscles (Figs. 50B, 51B, 52B).—The muscles supporting the possible actions of the M. cg. lat. and M. cg. med. during sideways movements of the hyoid include the extrinsic M. serpihyoideus (M. sh.), M. stylohyoideus (M. st.), M. mylohyoideus (M. mh.) and M. branchiomandibularis (M. bm.), and the intrinsic M. ceratohyoideus (M. ch.). These muscles regulate and adjust the position of the hyoid within the framework of the mandible and adjust the position of the Nodulus with respect to the urohyale. Regulation and adjustment of the position of the hyoid by the extrinsic muscles are necessary because the contractions of intrinsic muscles only change the configuration of the hyoid. However, the hyoid can change its configuration only if it is properly positioned within the mandible so that its new space requirements are met. Also, adjustment of the position of the Nodulus along the urohyale is necessary because of the asymmetrical changes of the linear distances between the urohyale and the left and right ceratobranchialia when the hyoid moves sideways, as described above. This adjustment of the Nodulus is guided and supported by the M. sh. and M. ch., both of which insert on the Nodulus.

The synchronous actions and functions of the various muscles can be summarized as follows:

1. Left M. serpihyoideus (Figs. 50B, 51B).—The lateral (sinistral) component of the sinistro-caudo-ventral pull exerted by the left M. sh. on the left ceratobranchiale counteracts the medial (dextral) component of the dextro-ventro-caudal displacement of the anterior end of the left ceratobranchiale and of the left region of the Art. cerato-basihyalis which are brought about by the contraction of the left M. cg. lat. Thus, the left ceratobranchiale remains aligned along the medial surface of the pterygoid musculature covering the left mandibular ramus (cf. Figs. 50A and B). The ventral component of the sinistro-caudo-ventral pull exerted by the M. sh. on the left ceratobranchiale supports a ventral displacement of the anterior end of the left ceratobranchiale and of the left region of the Art. cerato-basihyalis which are achieved by the contraction of the left M. cg. lat. (Figs. 51A, B). The caudal component of the pull exerted by the left M. sh. moves the left ceratobranchiale caudally (Figs. 50B, 51B) and, thus, supports the caudal displacement of the left region of the Art. cerato-basihyalis described as a result of the contraction of the left M. cg. lat. (Fig. 50A). The sinistral component of the sinistro-caudal force exerted by the M. sh. on the Nodulus and Lig. nodulo-ceratobranchiale pulls the Nodulus sinistrally so that the Lig. nodulo-ceratobranchiale can bridge the increased gap between the urohyale and left ceratobranchiale. The caudal component of this force may be opposed by the apical force of the M. ceratoglossus supf. and is transmitted to the urohyale and to the left ceratobranchiale and region of the Art. cerato-basihyalis. The caudal force exerted on the Nodulus by the combined resultant forces of the left M. sh., which, in combination with the force exerted by the M. st. on the basihyale, would push the Nodulus apically along the urohyale, is counterbalanced by the caudal component of the force exerted by the right M. ch. on the Nodulus.

2. Left M. stylohyoideus (Figs. 50B, 51B).—The sinistro-ventro-caudal force exerted by the left M. st. on the Proc. parahyalis of the basihyale supports and guides the angular rotation of the basihyale about the Art. cerato-basihyalis towards sinistro-dorso-caudal as well as the dextro-ventro-caudal displacements of the left region of the Art. cerato-basihyalis and of the apical end of the left ceratobranchiale, which are brought about by the contraction of the left M. cg. lat. (Figs. 50, 51). In addition, the caudo-sinistral component of the M. st. pulls the left Proc. parahyalis and, thus, the basihyale and left ceratobranchiale, caudo-sinistrally and, therefore, supports the action of the left M. sh. (Figs. 50B, 51B).

3. Right M. branchiomandibularis anterior medialis (Figs. 50B, 52B).—The dorsal and dextral components of the dextro-dorso-apical pull exerted by the right M. bm. ant. med. on the caudal end of the right ceratobranchiale counteract the sinistro-ventro-caudal angular rotation of the right ceratobranchiale about the Art. cerato-basihyalis brought about by the contraction of the right M. cg. med. As a result, the right ceratobranchiale is held against the pterygoid musculature covering the palate and the medial surface of the right mandibular ramus (Figs.

50, 52). The apical component of the force exerted by the right M. bm. ant. med. protracts the right ceratobranchiale (Figs. 50B, 52B) and, thus, supports the apical displacement of the right region of the Art. cerato-basihyalis during contraction of the right M. cg. med. (Fig. 50A). The combined forces exerted by the simultaneously contracting right M. bm. ant. lat. and M. bm. post. on the right epibranchiale support the action of the M. bm. ant. med. on the right ceratobranchiale (for details, see pp. 113 f.).

4. Right M. ceratohyoideus (Figs. 50B, 52B).—The sinistro-ventro-apical pull exerted by the right M. ch. on the right ceratobranchiale should support the angular rotation of the caudal end of the right ceratobranchiale about the Art. cerato-basihyalis toward sinistro-ventro-caudal, which was brought about by the contraction of the right M. cg. med. (Figs. 50A, 52A). However, this movement is prevented by the dextro-dorso-apical pull exerted by the right M. bm. ant. med. on the ceratobranchiale (Figs. 50B, 52B). At the same time, the right M. ch. exerts a dextro-dorso-caudal pull on the Nodulus. The caudal component of this pull counterbalances the resultant force of the left M. sh. and M. st., which would push the Nodulus apically along the urohyale. The dextral component of this dextro-dorso-caudal force exerted by the M. ch. is overriden by the summation of the forces exerted by the left M. sh., M. st. and Lig. nodulo-ceratobranchiale, all of which pull the Nodulus sinistrally. In this way, the Nodulus is not displaced forward or backward during the sideways movements of the hyoid but, rather, is pulled laterally (in the present example, to the left) so that both limbs of the Lig. nodulo-ceratobranchiale can span the different gaps between the urohyale and the left and right ceratobranchialia during the sideways movement of the hyoid.

5. Right M. mylohyoideus posterior (Fig. 50B).—The dextro-dorso-apical force exerted by the M. mh. post. on the Nodulus is transmitted directly to the urohyale because the Nodulus is immobilized with respect to the urohyale by the combined actions of the left M. sh. and M. st. and the right M. ch. Since the right M. mh. post. passes over the right Art. cerato-basihyalis, its dextro-dorso-apical force supports the dextro-dorso-apical rotation of the right region of the Art. cerato-basihyalis and the simultaneous sinistro-ventro-caudal rotation of the left region of the Art. cerato-basihyalis such as results from the contraction of the right M. cg. med. The dextral component of the dextro-dorso-apical force of the right M. mh. post. is, however, opposed by the sinistral components of the caudo-sinistral forces of the left M. sh. and M. st.

6. Right M. mylohyoideus anterior (not illustrated).—The dextro-dorso-apical force exerted by the M. mh. ant. on the Gl. sublingualis is transmitted directly to the basihyale since the Gl. sublingualis is firmly attached to the basihyale through connective tissue. The dextro-dorso-apical force of the M. mh. ant. supports the action of the M. mh. post. so that the dorsal surface of the basihyale performs an axial rotation towards sinistro-dorsal. The dextral component of the dextro-dorso-apical force is counteracted by the sinistral component of the caudo-sinistro-ventral pull exerted by the left M. st. on the basihyale and by the sinistral pull exerted by the left M. sh. on the left ceratobranchiale and transmitted to the basihyale. In this way, the basihyale still points to the left at the end of the sideways movement.

Discussion.—Despite the apparent complexity of the actions of the various muscles, the side-to-side movement of the hyoid consists of nothing more than a unilateral retraction and simultaneous contralateral protraction of the hyoid. The caudal ends of both ceratobranchialia are pushed laterally against the jaw muscles and the points of contact serve as fulcra for the rotation of the hyoid. The lateral rotation of the hyoid can be summarized as resulting from an approximate force couple formed by the right M. bm. and M. mh. and the left M. st. and M. sh. This force couple rotates the hyoid about a central point lying on the midsagittal axis of the basihyale on the borderline towards the urohyale. Whether the basihyale moves laterally by remaining on the same level or by being raised or lowered depends largely on the net balance of the dorso-ventral components of the simultaneously acting muscular forces.

The most extensive lateral excursion of the hyoid occurs when one Art. cerato-basihyalis is maximally flexed and the other maximally extended. In this position, the linear distance between the caudal ends of the ceratobranchialia is about equal to the corresponding distance when the hyoid is in the maximally retracted position (e.g., Fig. 47B). In the extremely protracted position (e.g., Fig. 46B), however, no lateral excursion is possible because both Artt. cerato-basihyales are fully extended. This prevents any further extension of one of the articulations which would be necessary for any sideways movement. This means that the most extensive lateral excursions can be performed when the hyoid is retracted and that the range of the lateral excursions decreases as the hyoid is protracted.

Movements of the Epibranchiale Relative to the Ceratobranchiale

In the previous sections, the epibranchiale has not been treated as a separate skeletal element for the motion analysis of the hyoid, but simply as an extension of the ceratobranchiale. This approximation is valid for the protraction of the hyoid (see below) and has been useful in simplifying the mechanical analysis. However, the epibranchiale is connected to the ceratobranchiale through a diarthrosis and can perform angular rotations about the longitudinal axis of the articulation from caudo-medio-ventral to apico-latero-dorsal. Furthermore, the epibranchiale serves as an insertion site for two muscles, namely for the M. branchiomandibularis posterior (M. bm. post.) and M. branchiomandibularis anterior lateralis (M. bm. ant. lat.) and, hence, can be moved by these muscles. Therefore, it is necessary to analyze the movements of the epibranchiale separately. Movements of the epibranchiale, however, are linked to movements of the hyoid because both the M. bm. post. and M. bm. lat. also protract the hyoid.

Movements of the Epibranchiale During Protraction of the Hyoid

The movements of the epibranchiale will be described from an initial stage with retracted hyoid and latero-dorsally flexed epibranchiale (Fig. 47) to a final stage with protracted hyoid and straight epibranchiale (Fig. 46).

During protraction of the hyoid, both the M. bm. post. and M. bm. ant. lat. contract simultaneously. The M. bm. ant. lat. inserts on the apical part of the epibranchiale, forming a sleeve around the Art. epi-ceratobranchialis and exerting an apico-dorsal pull on the epibranchiale. When the sleeve-like M. bm. ant. lat. contracts, it compresses the Art. epi-ceratobranchialis equally from all sides, so that the articulation is immobilized in a straight position. Thus, the epibranchiale is oriented in a straight line with the ceratobranchiale, which was achieved through an angular rotation of the epibranchiale towards ventral. This ventral rotation, however, is linked to a simultaneous angular rotation towards caudo-medial. This slight medial movement keeps the epibranchiale in line with the ceratobranchiale which is pulled medially during protraction of the hyoid. The dorsal component of the apico-dorsal pull exerted by the M. bm. ant. lat. on the epibranchiale is counterbalanced by the ventral component of the apico-ventro-lateral pull exerted by the M. bm. post. which inserts on the dorsal side of the posterior part of the epibranchiale (Fig. 47A). The lateral component exerted by the pull of the M. bm. post. on the epibranchiale ensures that the epibranchiale glides along the medial surface of the jaw musculature covering the medial surface of the mandibular ramus during protraction of the hyoid (Fig. 47B). The apical components of the forces exerted on the epibranchiale by the M. bm. post. and M. bm. ant. lat. pull the epibranchiale forward and, thus, protract the hyoid.

Movements of the Epibranchiale During Retraction of the Hyoid

The movements of the epibranchiale will be described from an initial stage with protracted hyoid and straight epibranchiale (Figs. 46A, B, 49A) to a final stage with retracted hyoid and dorsally flexed epibranchiale (Figs. 47A, B, 49C).

During retraction of the hyoid, the hyoid horn, which consists of the ceratobranchiale and epibranchiale, is pushed caudally. No muscles actually pull the hyoid horn backwards, but the proper direction of the movement is assured by the sleeve-like Fascia vaginalis hyoidei within which the hyoid horn glides. This Fascia vaginalis curves dorsally at the caudal end of the mandible to attach along the neck musculature. When the hyoid horn glides caudally, it has to adapt its configuration to this curvature. The flexible cartilaginous process of the epibranchiale easily adapts to the curvature, and the epibranchiale itself rotates dorsally to do so. This dorsal angular rotation, however, is linked to a simultaneous angular rotation towards apico-lateral. This slight lateral movement brings the epibranchiale in line with the ceratobranchiale which is pulled laterally and assumes a more oblique orientation during the retraction of the hyoid.

Movements of the Paraglossale Relative to the Basihyale

The movements of the paraglossale relative to the basihyale include basically an up-and-down and a slight side-to-side rotation about the Art. paraglosso-basihyalis. They are based on the coordinated actions of a set of intrinsic lingual muscles, namely the M. ceratoglossus, M. supraglossus, and M. hypoglossus obliquus. (The actions of the M. supraglossus, which acts synergistically with the M. hypoglossus obliquus on the Art. paraglosso-basihyalis and Art. cerato-basihyalis and which acts additionally on the epithelial surface of the tip of the tongue, will be discussed separately.) The action of the M. genioglossus on the paraglossalia was discussed elsewhere (pp. 104 f.).

Up-and-Down Movement of the Paraglossale

The up-and-down movement of the paired paraglossalia (in the following called "paraglossale" for convenience because the collateral parts move as a unit) consists of an angular rotation about the Art. paraglosso-basihyalis, which is linked to a back-and-forth translation along the longitudinal axis of the articulation. This movement of the paraglossale is based on the coordinated, symmetrical action of the M. ceratoglossus (M. cg.) as the depressor of the anterior tip of the paraglossale (i.e., flexor of the paraglossale) and of the M. hypoglossus obliquus (M. ho.) as the elevator of the anterior tip of the paraglossale (i.e., extensor of the paraglossale). Because the majority of the muscle portions acting on the Art. paraglosso-basihyalis are two-joint muscles acting also on the Art. cerato-basihyalis, movements of the hyoid at the Art. cerato-basihyalis influence the function of most of the muscles involved in the up-and-down movement of the paraglossale. Therefore, movements of the paraglossale have to be analyzed first independently, and then in connection with the movements of the hyoid at the Art. cerato-basihyalis.

Extension of the paraglossale with stationary hyoid (Fig. 53).—The movements of the paraglossale will be described from an initial, flexed and retracted position (Fig. 53A) to a final, extended and protracted position (Fig. 53C). For descriptive purposes, it is assumed that the Art. cerato-basihyalis is immobilized during the movement of the paraglossale, i.e., that the hyoid is stationary. The paraglossale is extended through a contraction of the M. hypoglossus obliquus (M. ho.) which consists of two parts, the M. ho. lateralis and M. ho. medialis. These two muscle parts have different attachments and directions of pull; thus, their actions on the paraglossale will be described separately.

The M. ho. med. is a one-joint muscle. Its apical portion, which inserts on the paraglossale (Figs. 53A, B; muscle P_4–B_2), applies an apico-ventral force to the Proc. posterior paraglossalis and, therefore, raises the anterior tip of the paraglossale in an angular rotation about the Art. paraglosso-basihyalis. At the same time, the entire paraglossale is displaced towards apico-ventral in a linear translation along the midsagittal crest of the saddle-shaped articular facet of the basihyale (Fig. 53A). In this way, the apical portion of the M. ho. med. is a direct antagonist of the M. cg. prof. (pp. 116 f.).

The caudal portion of the M. ho. med. inserts on the tendinous envelope of the lingual salivary gland (Figs. 55A, B; muscle G_3–B_1). When it contracts, the ventral component of its force directed medio-caudo-ventrally supports the ventral component of the pull exerted by the apical portion of the M. ho. med. on the Proc. posterior paraglossalis to which the envelope of the lingual salivary gland is attached. The caudal component of the force generated by the caudal portion of the M. ho. med. keeps the floor of the glandular envelope taut by pulling it caudally (Fig. 53A). This force is also transmitted to the paraglossale and opposes the apical component of the apico-ventral force of the apical portion of the M. ho. med. The caudal pull on the paraglossale is, however, overridden during the extension of the paraglossale because this extension is tied to an apical translation of the paraglossale along the midsagittal crest of the articular facet of the basihyale. The medial component of the force generated by the caudal portion of the M. ho. med. pulls the tendinous envelope of the Gl. lingualis medially and is counterbalanced by the lateral component of the latero-caudo-ventral pull exerted by the M. ho. lat. (see below; Fig. 53B). The Gl. lingualis acts as an hydroskeletal extension of the Proc. posterior paraglossalis when it is compressed and turgid through the combined forces exerted on it by the M. ho. and the synergistic M. supraglossus (for the actions of the M. supraglossus, see pp. 124 ff., 130 ff., 133 f.).

Therefore, the Gl. lingualis is not bent down relative to the paraglossale by the ventral components of the forces generated by the contracting M. ho. lat. (see below) and posterior portion of the M. ho. med.

The M. ho. lat. is a two-joint muscle, extending from the ceratobranchiale to the tendinous floor of the envelope of the Gl. lingualis (Figs. 53A, B; muscle G_4–C_8). It probably contracts synchronously with the M. ho. med. because both muscles insert on the floor of the tendinous envelope of the Gl. lingualis and cooperate in keeping the tendinous floor taut. This function could probably not be achieved with the contraction of only one of the muscles. When the M. ho. lat. contracts, it exerts a latero-caudo-ventral force on the tendinous floor of the envelope of the Gl. lingualis. The ventral component of this force supports the ventral components of the forces generated by the M. ho. med. exerted on the Proc. posterior paraglossalis and Gl. lingualis. The caudal component of the pull exerted by the M. ho. lat. keeps the glandular envelope taut and prevents any downward bending of the glandular body relative to the paraglossale (Fig. 53A). The lateral component of the force exerted by the M. ho. lat. on the glandular envelope counteracts the medial component of the pull exerted by the caudal portion of the M. ho. med. on the glandular envelope (Fig. 53B). At the same time, the contraction of the M. ho. lat. affects also the Art. cerato-basihyalis by exerting a medio-apico-dorsal pull on the ceratobranchiale (pp. 117 ff.).

The Lig. articulare paraglosso-basihyale ventrale is important in preventing an excessive protraction of the paraglossale. At the same time, it prevents a lateral luxation of the anterior end of the Art. paraglosso-basihyalis when the paraglossale is extended and the ligament taut.

Flexion of the paraglossale with stationary hyoid (Fig. 53).—The movements of the paraglossale will be described from an initial, extended and protracted position (Fig. 53C) to a final, flexed and retracted position (Fig. 53A). For descriptive purposes, it is assumed that the Art. cerato-basihyalis remains immobilized during the movement of the paraglossale, i.e., that the hyoid is stationary.

Contraction of the entire M. ceratoglossus (M. cg.; or of any portion of it, but see below) generates a caudo-ventral pull acting on the Proc. anterior paraglossalis. This pull depresses the anterior end and raises the posterior end of the paraglossale so that the paraglossale performs an angular rotation about the Art. paraglosso-basihyalis. At the same time, the paraglossale is displaced towards caudo-dorsal along the midsagittal crest of the saddle-shaped articular facet of the basihyale (Fig. 53C).

With the exception of the M. cg. prof. (see below), all parts of the M. cg. are functional two-joint muscles and act also on the Art. cerato-basihyalis by exerting a pull either on the ceratobranchiale or on the Lig. nodulo-ceratobranchiale (Fig. 53C). The Art. cerato-basihyalis is immobilized when all parts of the M. cg. contract simultaneously. In such a case, the M. cg. med., a contraction of which tends to increase the angle enclosed by the dorsal surfaces of the basihyale and ceratobranchiale by rotating the basihyale ventro-apically (Figs. 52A, 55C), is counteracted by the M. cg. lat. A contraction of this muscle tends to decrease the angle enclosed by the dorsal surface of the basihyale and ceratobranchiale by rotating the basihyale dorso-caudally (Figs. 51A, 54B). The M. cg. supf. supports the action of the M. cg. lat. by exerting a dorso-apical pull on the Lig. nodulo-ceratobranchiale and, thus, by supporting an apical displacement of the Nodulus along the urohyale. This movement of the Nodulus is tied to a dorso-caudal angular rotation of the basihyale about the Art. cerato-basihyalis, which is brought about by a contraction of the M. cg. lat. (see pp. 99 ff. for details on the forward movement of the Nodulus during the dorso-caudal rotation of the basihyale).

Unlike the other parts of the M. cg., the M. cg. prof. has only a negligible effect on the Art. cerato-basihyalis because it originates so close to the rotational axis of the ceratobranchiale and basihyale. This means that, although the M. cg. prof. originates from the ceratobranchiale and is, therefore, morphologically a two-joint muscle, it functions as a one-joint muscle that can flex the paraglossale without simultaneously inducing movements at the Art. cerato-basihyalis or without being affected by simultaneous movements at this articulation (Fig. 53C).

An excessive flexion of the paraglossale is prevented by the Lig. articulare paraglosso-basihyale dorsale.

Movements of the paraglossale during the retraction of the hyoid (Fig. 54).— When the hyoid is retracted, the angle enclosed by the dorsal surfaces of the ceratobranchiale and basihyale is reduced, i.e., the basihyale performs an angular rotation about the Art. cerato-basihyalis towards caudo-dorsal. At the same time, the ceratobranchialia spread apart, i.e., perform angular rotations about the Artt. cerato-basihyales towards apico-lateral (Figs. 46A, B). Regardless of the movements at the Art. paraglosso-basihyalis, this rotation of the basihyale results in a reduction of the linear distance between origin and insertion of the majority of the functional two-joint muscles responsible for the up-and-down movement of the paraglossale (i.e., M. ho. lat., M. cg. lat., M. cg. supf.). It results simultaneously in an increase of the distance between origin and insertion of the M. cg. med. (cf. Figs. 54C and 55C; consult Fig. 56A).

The extension and protraction of the paraglossale during retraction of the hyoid will be described from an initial position with protracted hyoid and flexed, retracted paraglossale (Fig. 54A) to a final position with retracted hyoid and extended, protracted paraglossale (Fig. 55B). The paraglossale is extended through a contraction of the M. ho., of which the M. ho. med. is a one-joint muscle spanning only the Art. paraglosso-basihyalis, whereas the M. ho. lat. is a two-joint muscle spanning also the Art. cerato-basihyalis (Fig. 54A). Therefore, only the M. ho. lat. affects and is affected by movements at the Art. cerato-basihyalis. The contracting M. ho. lat. exerts an apico-dorso-medial force on the ceratobranchiale and a caudo-ventro-lateral force on the lingual salivary gland, which is transmitted to the paraglossale (Figs. 53B, 54A). The caudal component of this force on the paraglossale is transmitted to the basihyale and supports the reduction of the angle between the dorsal surfaces of the basihyale and ceratobranchiale, which occurs during any retraction of the hyoid. Because the caudal end of the ceratobranchiale abuts dorsally against the pterygoid musculature of the palate and, therefore, cannot follow the apico-dorsal pull exerted by the M. ho. lat., a reduction of the angle at the Art. cerato-basihyalis can be achieved only through a caudo-dorsal angular rotation of the basihyale (Fig. 54A). As a result of the combined movements of the paraglossale and hyoid, the M. ho. lat. shortens during its contraction; i.e., it contracts isotonically (see Table 16).

During the retraction of the hyoid and caudo-dorsal rotation of the basihyale, the ceratobranchialia simultaneously spread apart, i.e., perform latero-apical angular rotations. Therefore, the lateral component of the force exerted by the M.

ho. lat. on the Gl. lingualis increases during the retraction of the hyoid (consult Fig. 53B). This increasing lateral tension on the Gl. lingualis is counteracted by the medial tension exerted by the non-elongating tendinous fibers of the portion A of the M. supraglossus on the caudo-dorsal tip of the Gl. lingualis (Figs. 21A, 22, 58A). In this way, the lingual salivary gland is compressed and its turgidity increases so that it can function as an hydrostatic skeleton (pp. 73 ff., 135 f.).

The flexion and retraction of the paraglossale during retraction of the hyoid will be described from an initial position with protracted hyoid and extended, protracted paraglossale (Fig. 54B) to a final position with retracted hyoid and flexed, retracted paraglossale (Fig. 55A). The paraglossale is flexed through a contraction of the M. cg. of which only the deep part (M. cg. prof.) functions as a one-joint muscle affecting only the Art. paraglosso-basihyalis. All the other parts of the M. cg. are functional two-joint muscles spanning also the Art. cerato-basihyalis. They, therefore, affect and are affected by movements at this articulation. The various two-joint muscle parts are, however, affected differently by a reduction of the angle enclosed by the dorsal surfaces of the ceratobranchiale and basihyale during a retraction of the hyoid because they pass over the articulation at different places.

Contractions of the M. cg. lat. and M. cg. supf. support a reduction of the angle between the dorsal surfaces of the ceratobranchiale and basihyale and, thus, a caudo-dorsal angular rotation of the basihyale (Fig. 51A; cf. Figs. 54B and 55B). This means that the M. cg. lat. and M. cg. supf. shorten (i.e., contract isotonically) during a flexion of the paraglossale if the hyoid is simultaneously retracted because the linear distances between their origins and insertions decrease due to movements at both the Art. paraglosso-basihyalis and Art. cerato-basihyalis (cf. Figs. 54B and 55A; see Table 16).

Because the M. cg. med. passes ventrally over the Art. cerato-basihyalis (Fig. 54B), a caudo-dorsal angular rotation of the basihyale during retraction of the hyoid increases the distance between origin and insertion of the M. cg. med. and, therefore, tends to elongate the muscle (cf. Figs. 54B and 55B). At the same time, however, flexion of the paraglossale brings the insertion of the M. cg. med. closer to its origin. In this way, the total distance between origin and insertion of the M. cg. med. remains more or less unchanged during a flexion of the paraglossale if the hyoid is retracted because the effects on the length of the M. cg. med. by the simultaneous movements at the Art. cerato-basihyalis and Art. paraglosso-basihyalis cancel each other. Therefore, if the M. cg. med. contracts synchronously with the other portions of the M. cg., it does not shorten, i.e., its contraction is more or less isometric (Table 16).

During retraction of the hyoid, the paraglossale can also be immobilized, if the M. ho. and M. cg. contract simultaneously. In such a case, the rotational effects of the simultaneously applied forces on the paraglossale cancel each other (Fig. 54C). Retraction of the hyoid with fixed paraglossale will be described from an intitial, protracted position of the hyoid (Fig. 54C) to a final, retracted position (Fig. 55C). The contracting M. ho. lat. and M. cg. supf. should support a caudo-dorsal angular rotation of the basihyale during a retraction of the hyoid (Fig. 54C; the M. cg. prof. and M. ho. med. function as one-joint muscles and are not shown). The movement at the Art. cerato-basihyalis, however, increases the distance between origin and insertion of the M. cg. med. so that the M. cg. med. is elongated.

If this muscle contracted simultaneously with the other parts of the M. cg., it would produce "negative work" (Table 16).

Movements of the paraglossale during protraction of the hyoid (Fig. 55).—When the hyoid is protracted, the angle enclosed by the dorsal surfaces of the ceratobranchiale and basihyale is increased, i.e., the basihyale performs an angular rotation about the Art. cerato-basihyalis towards apico-ventral. At the same time, the ceratobranchialia are drawn together, performing angular rotations about the Artt. cerato-basihyales towards caudo-medial (Figs. 47A, B). Regardless of the movements at the Art. paraglosso-basihyalis, this rotation of the basihyale results in an increased distance between origins and insertions of most of the two-joint muscle parts responsible for the movements of the paraglossale (i.e., M. ho. lat., M. cg. lat., M. cg. supf.). At the same time, the rotation of the basihyale results in a decrease of the distance between origin and insertion of the M. cg. med. (cf. Figs. 55C and 54C; cf. Fig. 56A).

The extension and protraction of the paraglossale during protraction of the hyoid will be described from an initial position with retracted hyoid and flexed, retracted paraglossale (Fig. 55A) to a final position with protracted hyoid and extended, protracted paraglossale (Fig. 54B). The paraglossale is extended through a contraction of the M. ho. of which only the lateral part (M. ho. lat.) is a two-joint muscle and, therefore, affects and is affected by movements at the Art. cerato-basihyalis. On the one hand, an apico-ventral angular rotation of the basihyale during protraction of the hyoid increases the distance between origin and insertion of the M. ho. lat. and, thus, tends to elongate the muscle (cf. Figs. 55A and 54A). On the other hand, an extension of the paraglossale brings the insertion of the M. ho. lat. closer to its origin (cf. Figs. 54A and C) and, thus, offsets the increase of distance due to the movement at the Art. cerato-basihyalis. In this way, the contracting M. ho. lat. does not change its length when it extends the paraglossale during protraction of the hyoid (cf. Figs. 55A and 54A), i.e., it contracts more or less isometrically (Table 16).

The flexion and retraction of the paraglossale during protraction of the hyoid will be described from an initial position with retracted hyoid and extended, protracted paraglossale (Fig. 55B) to a final position with protracted hyoid and flexed, retracted paraglossale (Fig. 54A). The paraglossale is flexed through a contraction of the M. cg. of which only the deep portion (M. cg. prof.) functions as a one-joint muscle, and, thus, functions independently from movements at the Art. cerato-basihyalis. All other parts of the M. cg. are two-joint muscles spanning also the Art. cerato-basihyalis and, therefore, affecting and being affected by movements at this articulation.

The apico-ventral angular rotation of the basihyale during protraction of the hyoid increases the distance between origin and insertion of the M. cg. lat. and, hence, tends to elongate this muscle (cf. Figs. 55B and 54C). The distance between origin and insertion of the M. cg. supf. is also increased because the Lig. nodulo-ceratobranchiale is pulled caudally (Figs. 47A, B). At the same time, however, a flexion of the paraglossale brings the origins and insertions of the M. cg. lat. and M. cg. supf. closer together and, therefore, offsets the increased distance caused by the movements at the Art. cerato-basihyalis. In this way, the contracting M. cg. lat. and M. cg. supf. do not change their length when flexing the paraglossale during protraction of the hyoid, i.e., when both muscles contract more or less isometrically (Table 16).

The increase of the angle enclosed by the dorsal surfaces of the ceratobranchiale and basihyale through an apico-ventral angular rotation of the basihyale during protraction of the hyoid reduces the distance between origin and insertion of the M. cg. med. (Figs. 52A, 55B). Because flexion of the paraglossale reduces the distance between origin and insertion of the M. cg. med. even more, the contracting M. cg. med. shortens during a flexion of the paraglossale and simultaneous protraction of the hyoid, i.e., the M. cg. med. contracts isotonically (Table 16).

During protraction of the hyoid, the paraglossale may also be held immobilized if both the M. ho. and M. cg. contract simultaneously. In such a case, the rotational effects of the muscle forces applied to the paraglossale cancel each other (Fig. 55C). Protraction of the hyoid with immobilized paraglossale will be described from an initial, retracted position of the hyoid (Fig. 55C) to a final, protracted position of the hyoid (Fig. 56C). The contracting M. cg. med. shortens, i.e., contracts isotonically (Table 16), and supports the apico-ventral angular rotation of the basihyale during the protraction of the hyoid (Fig. 55C; the M. cg. prof. and M. ho. med. function as one-joint muscles and are not shown). The movement at the Art. cerato-basihyalis increases, however, the distances between the origins and insertions of the M. cg. lat., M. cg. supf. and M. ho. lat. so that these muscles are elongated. If these muscles contract at the same time at which they are elongated, they produce "negative work" (Table 16) (pp. 84 ff.).

Discussion.—As long as the movements of the paraglossale are synchronized with back-and-forth movements of the hyoid, the lingual muscles responsible for up-and-down movements of the paraglossale contract either isotonically or more or less isometrically (Table 16). If, however, the paraglossale is kept immobilized through simultaneous contractions of the antagonistic M. ho. and M. cg. during back-and-forth movements of the hyoid, one group of muscles will contract isotonically whereas the other group of contracting muscles is elongated. These latter muscles produce "negative work" (Table 16). Because elongation of contracting muscles uses additional energy that is not transformed into mechanical energy, movements during which contracting muscles are elongated are inefficient. It is, therefore, doubtful that the paraglossale is actually kept immobilized through a contraction of all the muscles responsible for its movements during a protraction or retraction of the hyoid. Conceivably the paraglossale could be immobilized during back-and-forth movements of the hyoid through a contraction of only those muscles that would not have to elongate. For example, during retraction of the hyoid, the paraglossale may be immobilized through a simultaneous contraction of the M. ho., M. cg. lat. and M. cg. supf., but not of the M. cg. med. During protraction of the hyoid the paraglossale may be immobilized only through a simultaneous contraction of the M. ho. med. and M. cg. med. (Table 16). Whether this is actually the case or whether movements of the paraglossale are firmly linked to movements of the hyoid during free, unloaded, and slow tongue movements cannot be determined through a theoretical mechanical analysis alone (p. 77).

Side-to-Side Movement of the Paraglossale

The side-to-side movements of the paraglossale are produced by asymmetrical contractions of the M. ceratoglossus.

Movements of the paraglossale will be described from an initial, straight position, in which the midsagittal axis of the paraglossale is in line with that of the basihyale (Fig. 56A), to a final position, in which the midsagittal axis of the paraglossale points apico-laterally (Fig. 56B), and then back to the initial position.

For descriptive purposes, these movements are assumed to start with an extended and straight paraglossale. When the right M. cg. contracts, it exerts a caudo-ventral pull on the Proc. anterior paraglossalis (Fig. 56A). The caudo-ventral component of this pull flexes the paraglossale. At the same time, this component compels the paraglossale to perform an axial rotation towards dextro-ventral about the midsagittal crest of the articular facet of the basihyale because the force generated by the contracting right M. cg. is not counterbalanced by the relaxed left M. cg. (Fig. 14B). Due to the special structure of the articular facets of the Art. paraglosso-basihyalis, the right anterior tip of the paraglossale is simultaneously pulled to the right (cf. Figs. 56A and B), i.e., the anterior tip of the paraglossale performs an angular rotation to the right. The return of the paraglossale to the straight position (i.e., its rotation to the left side) is under the control of the left M. cg., a contraction of which has the same, but contralateral, effect as the one described for the right M. cg., with the difference that it affects the left side of the paraglossale (cf. Figs. 56A and B).

Lateral movements of the paraglossale can be performed irrespective of whether the hyoid is simultaneously protracted or retracted because the various parts of the M. cg. can adapt to any movement at the Art. cerato-basihyalis. They contract either isotonically or isometrically as long as the paraglossale moves. (For interactions between the two-joint muscle parts of the M. cg. and movements at the Art. cerato-basihyalis, see pp. 116–120 and Table 16.)

For several reasons, the paraglossale cannot be moved from side-to-side by asymmetrical contractions of the M. hypoglossus obliquus. If the M. ho. were to contract unilaterally, the paraglossale would have to move laterally in an extended and protracted position. This movement, however, is not possible without luxation of the anterior part of the articulation due to the structure of the articular facets of the Art. paraglosso-basihyalis. In addition, contraction of the M. ho. alone could not be used to bend the paraglossale sideways (e.g., to the left) because the medial component of the caudo-ventro-medial pull exerted by the M. ho. med. on the left posterior end of the paraglossale and on the left Gl. lingualis would be counteracted by the lateral component of the simultaneous caudo-ventro-lateral pull exerted by the left M. ho. lat. on the left Gl. lingualis (Figs. 53B, 56A).

Movements at the Tip of the Tongue

The surface relief and shape of the tip of the tongue, the skeletal support of which is provided mostly by the paraglossale, can be modified by a complex system of interacting structural elements. These include vascular cavernous tissue, epithelial structures such as the lingual nail, and muscles such as the M. supraglossus (M. sg.), M. hypoglossus anterior (M. hg. ant.), M. hypoglossus transversus (M. ho. trans.), and M. mesoglossus (M. mg.). The last two of these muscles are restricted to the tip of the tongue. The M. sg. and M. hg. ant., however, attach to the basihyale and to the ceratobranchiale, respectively, and, therefore, affect or are affected by movements at the Art. paraglosso-basihyalis and Art. cerato-basihyalis.

Movements at the tip of the tongue are always symmetrical due to the structure of the responsible muscles and Corpora cavernosa, which prevents asymmetrical movements. Both the M. sg. and M. hg. ant. converge into a single central tendon or aponeurosis, and the collateral limbs of the Corpus cavernosum laterale communicate with each other through the Corpus cavernosum apicale.

Movements of the Epithelial Surface of the Tip of the Tongue

Movements at the surface of the tip of the tongue are produced by muscles and cavernous vascular bodies and do not result from actions of a pair of antagonistic muscles as is the case for movements about articulations of skeletal elements. Because the reaction time for filling or emptying a cavernous vascular body is longer than for the contraction or relaxation of a skeletal muscle, actions of these two different organs cannot be synchronized as accurately as actions of skeletal muscles alone. This means that in the tip of the tongue, the cavernous vascular tissue may inflate or deflate independently from contractions of skeletal muscles, and vice versa, and may influence the function of the muscles. This means, the function of a muscle may vary depending on whether the cavernous vascular tissue is inflated or not.

In the following synopsis, the basic movements that are possible at the tip of the tongue are listed together with the respective structures responsible for these movements. This summary shows that certain movements are brought about by different combinations of co-operating structures and that some structures are responsible for several basic movements at the same time.

1. Medio-lateral movements.
   Narrowing of the tip of the tongue: M. supraglossus, M. hypoglossus anterior, lingual nail.
   Broadening of the tip of the tongue: M. hypoglossus transversus, Corpus cavernosum laterale, Corpus cavernosum apicale.
2. Caudo-apical movements.
   Backward pull of the dorsal epithelial surface: M. supraglossus.
   Forward pull of the dorsal epithelial surface: M. mesoglossus, M. hypoglossus anterior.
3. Dorso-ventral movements.
   Raising of the central area of the dorsal epithelial surface: M. supraglossus, Corpus cavernosum laterale, Corpus cavernosum apicale, elastic recoil of underlying tissue.
   Depression of the central area of the dorsal epithelial surface: M. mesoglossus, M. hypoglossus anterior, M. hypoglossus transversus.

For descriptive purposes, the actions of the muscles and of the cavernous vascular tissue are analyzed separately in the next sections. The actions of the two dynamic systems, namely the muscular and the hydraulic one, can overlap, resulting in a variety of movements that are difficult to describe simultaneously, but can easily be understood separately. For descriptive purposes also, it is assumed that no movement occurs about the Art. paraglosso-basihyalis. (Interactions between the muscles of the tip of the tongue and the movements of the hyoid skeleton are analyzed in later sections.)

Depression of the central area of the dorsal epithelial surface (Figs. 57A–59B).— A spoon-shaped dorsal surface at the tip of the tongue is formed through coordinated contractions of the M. mesoglossus (M. mg.) and M. hypoglossus anterior (M. hg. ant.) which counteract the M. supraglossus (M. sg.). This movement will be described from an initial stage with a flattened dorsal surface of the tip of the tongue (Figs. 57A, 59B) to a final stage with a depressed central area at the tip of the tongue (Figs. 57B, 59A). For descriptive purposes, possible simultaneous actions of the Corpora cavernosa are disregarded here.

When the M. hg. ant. contracts, it exerts a caudo-dorsal pull on its tendon which lies along the ventral surface of the Synchondrosis paraglossalis (Fig. 57A). The tendon is attached to the inner surface of the ventral side of the lingual nail; therefore, the caudo-dorsal pull of the M. hg. ant. tilts the anterior part of the lingual nail ventrally, using the Synchondrosis paraglossalis as a fulcrum (Fig. 57A). The lingual nail is tilted, and not simply pulled caudally, because its ventral side is tied to the apico-lateral corners of the paraglossale by connective tissue fibers. The force generated by the contracting M. hg. ant. is transmitted from its tendon to its aponeurosis which curves around the apical edge of the Synchondrosis paraglossalis to the dorsal side of the paraglossale and finally passes through the Foramen interparaglossale to attach to the ventro-apical tip of the basihyale (Figs. 43–44B, 57A, B). Therefore, the force generated by the contraction of the M. hg. ant. makes its aponeurosis "roll" from dorsal to ventral over the Bursae synoviales paraglossales (Figs. 57A, B). Consequently, the aponeurosis is pulled taut, and the origins of the M. mg. posterior and the insertion of the tendinous portion E of the M. sg., both located on the dorsal surface of the aponeurosis of the M. hg. ant., are pulled ventro-apically (Figs. 57A, B). At the same time, the Lig. interparaglossale apicale is pushed ventrally by the tightening aponeurosis of the M. hg. ant., and the connective tissue of the Fascia paraglossalis dorsalis is compressed between the aponeurosis and the synchondrosis (Fig. 57A). Simultaneously, the tendinous fibers, which branch off the caudo-lateral border of the aponeurosis of the M. hg. ant. to insert within the connective tissue underlying the lateral part of the V-shaped mound and the dorsal border of the lingual nail, are pulled ventro-apico-medially. Therefore, they exert a ventro-apico-medial pull on the V-shaped mound and on the dorsal edge of the lingual nail (Figs. 58A, 59B). This pull moves the tendinous portions B and C of the M. sg. forward because they attach to the dorsal part of the lingual nail before inserting on the dorsal margin of the paraglossale (pp. 55 ff.). The same pull also exerts pressure on the Corpus cavernosum laterale (Fig. 58A). The ventro-apico-medial pull on the dorsal edge of the posterior part of the lingual nail supports the ventral tilt of the anterior end of the lingual nail about the Synchondrosis paraglossalis described above (Fig. 57A) and also pulls the lingual nail and the limbs of the V-shaped mound medially, narrowing the tip of the tongue and its central area (Fig. 58A).

The M. mg. contracts synchronously with the M. hg. ant. The contracting M. mg. ant., which originates from the connective tissue underlying the lingual nail, pulls the epithelial surface of the tongue perpendicularly down, depressing the apical part of the central area of the tip of the tongue and, thus, making the lingual tip spoon-shaped (Figs. 57A, B). The contracting M. mg. ant. also influences the Corpus cavernosum apicale by exerting pressure on it through its bulging muscle fibers (Figs. 57A, B). The M. mg. post. contracts synchronously with the M. mg.

ant. and exerts a ventro-apical pull on the epithelial surface of the posterior part of the central area of the tip of the tongue. When the more steeply oriented apical muscle bundles of the M. mg. post. contract, they mainly depress the epithelial surface and only slightly pull it forward. The more obliquely oriented caudal muscle bundles, however, pull the epithelial surface and the underlying tissue more forward than downward when they contract. In this way, the depression of the central area of the tip of the tongue is more pronounced in its apical than in its caudal part. (Contraction of the M. mg. post. does not result in a lifting of the aponeurosis of the M. hg. ant. because the pull exerted by the M. mg. post. is counteracted here by the simultaneous contraction of the M. hg. ant.) By exerting an apico-ventral pull on the epithelial surface of the central area, the contracting M. mg. post. also pulls forward the tendinous portions C and E of the M. sg. which insert on the connective tissue underlying the central area (Fig. 57A). The contracting peripheral muscle bundles of the M. mg. post. also pull the V-shaped mound apically and, with it, the tendinous portion C and the main tendon of portion D of the M. sg. (Fig. 58A; see also Fig. 31A). During contraction of the M. mg. post. and the bulging of its muscle fibers, the mass of connective tissue and muscle fibers underlying the central area of the tip of the tongue increases its turgidity and exerts pressure laterally against the Corpus cavernosum laterale.

The sum of the forces generated by the contracting M. hg. ant. and M. mg. exerts a forward pull on the various tendinous portions of the M. sg. and, thus, on the M. sg. itself; this means, the M. hg. ant. and the M. mg. act together as muscular antagonists of the M. sg. When the central tendon of the M. sg. is pulled forward, the origin of the tendinous portion A of the M. sg., which inserts on the caudo-dorsal end of the Gl. lingualis, is also pulled apically (compare Figs. 64B and 61A). In this way, the position of the tendinous portion A is readjusted after a contraction of the M. sg.

If a M. hypoglossus transversus (M. hg. trans.) is present, it contracts simultaneously with the M. hg. ant. and M. mg. The contracting M. hg. trans. exerts a latero-ventral pull on both halves of the paraglossale (Fig. 31B). In this way, the two halves of the paraglossale perform a slight axial rotation towards latero-ventrally. A pull exerted by the M. hg. trans. counteracts the axial rotation towards medio-dorsal of the paraglossalia which is caused by the downward pressure exerted on the Synchondrosis paraglossalis by the aponeurosis of the contracting M. hg. ant. It also counteracts the medio-dorsal axial rotation of the paraglossalia which is caused by the medio-caudal pull exerted on the Margo dorsalis paraglossalis by portion C of the contracting M. sg. The function of the M. hg. trans., however, seems not to be indispensable for the proper functioning of the tip of the tongue since it is only rarely present.

Flattening of the dorsal epithelial surface (Figs. 57A–59B).—A flat surface of the tip of the tongue is formed through contraction of the M. supraglossus (M. sg.) which counteracts the M. hypoglossus anterior (M. hg. ant.) and M. mesoglossus (M. mg.). The movement will be described from an initial stage with a depressed central area of the tip of the tongue (Figs. 57B, 59A) to a final stage with a flattened dorsal surface of the tip of the tongue (Figs. 57A, 59B). For descriptive purposes, possible simultaneous actions of the Corpora cavernosa are disregarded here.

When the M. sg. contracts, it exerts a caudo-dorsal pull on its tendinous portion

E, which inserts on the aponeurosis of the M. hg. ant., so that, through elastic recoil, the Lig. interparaglossale apicale returns to its normal position and the tissue of the Fascia paraglossalis dorsalis expands to its original volume. At the same time, the force exerted on the aponeurosis by portion E of the M. sg. is transmitted to the tendon of the M. hg. ant., which "rolls" from ventral to dorsal over the Bursae synoviales paraglossales on the apical edge of the Synchondrosis paraglossalis (Fig. 57B). The lingual nail, the ventral side of which is tightly bound to the tendon of the M. hg. ant., is pulled forward in such a way that it tilts its anterior end upwards, using the Synchondrosis paraglossalis as a fulcrum (Fig. 57B), thus counteracting the ventral tilt of the apical end of the lingual nail which had occurred during contraction of the M. hg. ant. (Fig. 57A; pp. 123 ff.).

When the M. sg. contracts, it also exerts a caudal pull on all the rest of its tendinous fibers (Figs. 57B, 58B). In this way, the tendinous fibers of portions C and D of the M. sg., which insert on the connective tissue underlying the central area of the tip of the tongue, are pulled caudally. Thus, the dorsal epithelial surface is pulled backwards. Hence, the muscle bundles of the M. mg. post., which insert on the epithelium of the central area of the tip of the tongue, are pulled dorso-caudally and are extended (cf. Figs. 57A and B). Simultaneously, those tendinous fibers of portion E of the M. sg. that insert on the epithelial surface of the posterior half of the central area of the tip of the tongue pull the epithelial surface caudo-ventrally. This prevents bulging of the lingual surface when the mass of connective tissue and muscle fibers of the M. mg. post. underlying the central area is pulled caudo-dorsally by the other portions of the M. sg. (Fig. 57B). In this way, contraction of the M. sg. mainly raises the anterior part of the central area of the tip of the tongue and only slightly raises the posterior part. Therefore, the effect of the contraction of the M. sg. on the central area is rather an overall flattening than an overall raising of the surface.

Upon contraction of the M. sg., the tendinous portion C of the M. sg., which attaches tangentially to the inner surface of the dorsal border of the lingual nail before inserting on the Margo dorsalis paraglossalis, pulls the dorsal border of the lingual nail and the limbs of the V-shaped mound caudo-dorso-medially, thus, narrowing the tip of the tongue and its central area (Fig. 58B) and exerting pressure on the Corpus cavernosum laterale. This caudo-dorso-medial force of portion C of the M. sg. exerts a backward pull on the tendinous fibers that originate caudo-laterally from the aponeurosis of the M. hg. ant. and insert along the lateral part of the V-shaped mound and along the dorsal border of the lingual nail. This force is transmitted to the aponeurosis of the M. hg. ant. and aids portion E of the M. sg. in lifting and pulling back the aponeurosis. Simultaneously, the caudo-dorso-medial force on the portions C and B of the M. sg. on the dorsal border of the lingual nail supports the upward tilt of the anterior end of the lingual nail about the Synchondrosis paraglossalis described above. The caudal pull on the main tendons of portion D of the M. sg. pulls the peripheral muscle bundles of the M. mg. post. backwards and also contributes to the caudo-medial movement of the limbs of the V-shaped mound (Fig. 58B).

The sum of forces generated by the contracting M. sg. exerts a caudo-dorsal pull on the M. mg. and M. hg. ant.; this means, the M. sg. acts as a muscular antagonist to both the M. mg. and M. hg. ant. If the contracting M. sg. shortens, the central tendon of the M. sg. is pulled backward and so are the origins of the tendinous portions of the M. sg. When portions B and C of the M. sg. are put under tension by a contraction of the M. sg., they exert pressure on the anterior part of the body of the Gl. lingualis over which they pass (Figs. 41A–42B). In this way, turgidity of the Gl. lingualis increases every time the M. sg. contracts.

Inflation of the Corpora cavernosa (Figs. 59A, C).—Inflation of the Corpora cavernosa results in broadening the tip of the tongue or in stiffening the V-shaped mound and extreme apical tip of the tongue depending on whether or not the muscles of the lingual tip (M. supraglossus, M. hypoglossus anterior, M. mesoglossus) are contracting. The movement will be described from an initial deflated stage of the Corpora cavernosa (Figs. 59A, C) to a final inflated stage (Figs. 59B, D). The effects of the inflation of the Corpora cavernosa will be described first by assuming that the lingual muscles do not contract, and then by assuming a simultaneous contraction of various lingual muscles. For descriptive purposes also, inflation of the Corpus cavernosum laterale (C. c. lat.) and of the Corpus cavernosum apicale (C. c. ap.) will be analyzed separately, although they actually inflate synchronously (p. 71).

When the C. c. lat. inflates, its ventral portion, which lies laterally between the paraglossale and the lingual nail, pushes the lingual nail laterally (Fig. 59A). Because the lingual nail is a stiff cuticle, it does not bulge, and its dorsal part follows the lateral movement of its ventral part. Lateral bending of the dorsal part of the lingual nail is supported by the simultaneous inflation of the underlying dorsal portion of the C. c. lat. (Fig. 59A). When the dorsal part of the C. c. lat. inflates, it pushes apart the tendinous fibers of the M. hg. ant. and those fibers of portion C of the M. sg. among which the cavernous veins are embedded and, therefore, it expands medio-laterally. Inflation of the C. c. lat. causes a lateral expansion of the tip of the tongue rather than a medial compression of the tissues situated in the center of the tongue, because the dorsal part of the C. c. lat. is attached to the lingual nail which is pushed laterally by the inflation of the ventral part of the C. c. lat. (Fig. 59A). Furthermore, the inflating C. c. lat. expands laterally, and not dorsally, because the tendinous fibers of the M. hg. ant. and M. sg. can be pushed apart, but not elongated. Also, because the tendinous fibers of portion C of the M. sg. attach tangentially to the inside of the dorsal border of the lingual nail and insert on the Margo dorsalis paraglossalis, the height to which the C. c. lat. can inflate is limited.

When the lingual nail is pushed laterally by the inflating C. c. lat., the V-shaped mound is also pulled laterally and the central area of the tip of the tongue is consequently broadened (Figs. 59B, D). At the same time, the epithelial surface of the central area of the tip of the tongue is pulled laterally, tightened, and, therefore, lifted (Figs. 59A, B). Lateral expansion of the C. c. lat. exerts a lateral pull on the tendinous fibers of portions B and C of the M. sg., which are attached to the lingual nail, and on the main tendon of portion D of the M. sg., which is closely associated with the V-shaped mound (Fig. 58B). The symmetrical bilateral pull exerted on the tendinous fibers by portions C and D of the M. sg., which attach to the central area of the tip of the tongue, supports the lifting of the epithelial surface of the central area by putting these fibers under tension (Fig. 58B). Lifting of the epithelial surface of the central area of the tip of the tongue during the inflation of the C. c. lat. exerts a dorsal pull on the muscular bundles of the M. mg. post. (Figs. 37, 43, 59A). At the same time, inflation of the C. c.

lat. exerts a lateral pull on the tendinous fibers which arise from the caudo-lateral border of the aponeurosis of the M. hg. ant. to insert along the dorsal border of the lingual nail (Fig. 58A).

Depending on whether or not the lingual muscles contract simultaneously, inflation of the C. c. lat. has a different effect on the tip of the tongue. If the M. sg. does not contract, the lateral pull exerted by the inflating C. c. lat. on the tendinous fibers of the various portions of the M. sg. is transmitted as an apico-lateral force to the muscular body of the M. sg. The lateral components of these symmetrically applied forces cancel each other, whereas the apical components extend the muscle by pulling it apically. If the M. hg. ant. does not contract simultaneously, the lateral pull exerted by the inflating C. c. lat. on the tendinous fibers of the aponeurosis of the M. hg. ant. is transmitted as a caudo-dorso-lateral force to the aponeurosis. Again, the lateral components of these symmetrically applied forces cancel each other. The dorsal components of these forces lift the aponeurosis, and the ensuing enlarged space between the aponeurosis and the paraglossale is filled by the recoiling Lig. interparaglossale apicale and the connective tissue of the Fascia paraglossalis dorsalis. The caudal components of these forces exert a caudal pull on the aponeurosis and, thus, extend the muscular body of the M. hg. ant. If the M. mg. does not contract synchronously, the dorsal pull exerted by the inflating C. c. lat. on the epithelial surface of the central area is transmitted to the M. mg. post. and extends its muscle fibers. Summarizing the effects of the inflating C. c. lat. on the non-contracting lingual muscles, the C. c. lat. acts as an antagonist to the M. sg., M. mg., and M. hg. ant.

If, however, the M. sg. or both the M. hg. ant. and M. mg. contract when the C. c. lat. inflates or maintains its inflated stage, the generated forces compress the C. c. lat. so that its internal pressure, or turgidity, increases.

If the M. hg. trans. is present and contracts simultaneously with the inflation of the C. c. lat., it exerts a latero-ventral pull on both halves of the paraglossale (Fig. 31B). This pull counteracts the dorso-medial push exerted against the paraglossalia by the inflating ventral part of the C. c. lat. (Fig. 59A). It seems, however, that the action of the M. hg. trans. is not indispensable for proper functioning of the tip of the tongue because the majority of the specimens do not possess this muscle.

The Corpus cavernosum apicale (C. c. ap.) inflates synchronously with the C. c. lat. When it does so, at first it pushes apart the concentrically arranged ligamentous fibers among which the cavernous veins are located (Figs. 43, 59C). Therefore, the extreme apical tip of the tongue, in which the C. c. ap. is located, expands radially so that the cap of thickened epithelium covering the extreme apical tip of the tongue (Figs. 43–45B) is pushed ventro-apically and laterally. Hence, the inflating C. c. ap. pushes the apical end of the lingual nail ventrally and supports the C. c. lat. in broadening the tip of the tongue (Fig. 59C). Together with the anterior end of the C. c. lat., the inflating C. c. ap. tightens and lifts the epithelial surface of the anterior part of the central area of the tip of the tongue. Due to the cap of thickened, only slightly distensible epithelium covering the extreme tip of the tongue, the inflation of the C. c. ap. leads to an increase of turgidity of the extreme tip of the tongue in addition to the increase in size once the non-elongating ligamentous fibers are maximally stretched by the inflating cavernous veins.

If the M. sg. does not contract synchronously, inflation of the C. c. ap. exerts an apical pull on the main tendon of portion D of the M. sg. (Fig. 58A). This pull is transmitted to the muscular part of the M. sg. which is then extended. If the M. mg. ant. does not contract, the dorsal pull exerted by the inflating C. c. ap. on the epithelial surface of the anterior part of the central area of the tip of the tongue is transmitted to the muscle fibers of the M. mg. ant. which are then extended.

If, however, the M. sg. or the M. mg. contract as the C. c. ap. inflates or maintains its inflated stage, the generated muscular forces compress the C. c. ap. so that its internal pressure, or its turgidity, increases.

Deflation of the Corpora cavernosa (Figs. 59B, D).—Deflation of the Corpora cavernosa results in a narrowing of the tip of the tongue or in a softening and relaxation of the V-shaped mound and of the extreme tip of the tongue depending on whether or not the muscles of the tip of the tongue (M. supraglossus, M. mesoglossus, M. hypoglossus anterior) contract at the same time. The movement will be described from an initial inflated stage of the Corpora cavernosa (Figs. 59B, D) to a final deflated stage (Figs. 59A, C). The effects of deflating the Corpora cavernosa will be described first for the case in which no lingual muscle contracts at the same time, and then for the case in which lingual muscles contract simultaneously. For descriptive purposes, deflation of the C. c. lat. and of the C. c. ap. will be analyzed separately although they actually deflate simultaneously.

When the C. c. lat. deflates and collapses, the tendinous fibers originating from the aponeurosis of the M. hg. ant. and those of portion C of the M. sg. attaching to the dorsal border of the lingual nail, among which the cavernous veins are embedded, relax and return to a more densely packed arrangement. Therefore, the lateral part of the lingual nail returns through elastic recoil to a position closer to the paraglossale (Figs. 59B, D). Recoil of the lingual nail also pushes the V-shaped mounds medially so that the epithelial surface of the central area of the tip of the tongue is compressed and collapses ventrally (cf. Figs. 59B and 59A).

If the lingual muscles do not contract simultaneously, deflation of the C. c. lat. leads to a relaxation of the tendinous and muscular fibers attaching to the epithelial structures of the tip of the tongue and to a softening of the V-shaped mound and extreme apical tip of the tongue. When the C. c. lat. deflates, the tendinous fibers of portions B and C of the M. sg. attaching to the lingual nail are relaxed due to the elastic recoil of the lingual nail (Fig. 58B). The tendinous fibers of portions C and D of the M. sg., which underlie the central area of the tip of the tongue, are relaxed because of the medial motion of the V-shaped mound (Fig. 58B) so that the epithelial surface of the central area of the tip of the tongue relaxes and collapses ventrally (Fig. 59B). The ventral collapse of the epithelial surface of the central area also results in a relaxation of the muscle fibers of the M. mg. (Figs. 37, 43, 59B). The medial motion of the lingual nail and V-shaped mound also relaxes the tendinous fibers originating from the aponeurosis of the M. hg. ant. so that the aponeurosis of the M. hg. ant. collapses ventrally, compressing the tissue of the Fascia paraglossalis dorsalis (Fig. 59B).

If, however, the M. sg. or both the M. hg. and M. mg. contract during the deflation of the C. c. lat., the generated muscular forces compress the C. c. lat. and support a narrowing of the tip of the tongue caused by the deflation of the C. c. lat. In this way, contractions of the muscles help to maintain turgidity of the C. c. lat. during its deflation. If the M. sg. contracts during the deflation of the C. c. lat., the tendinous fibers of its portions B and C, which attach on the inside of the lingual nail, exert a caudo-medial pull on the lingual nail which is pushed medially against the C. c. lat. (Fig. 58B). If the M. hg. ant. contracts during the deflation of the C. c. lat., the tendinous fibers originating from the aponeurosis of the M. hg. ant. are pulled ventro-medio-apically so that they exert a ventro-medio-apical pull on the dorsal border of the lingual nail which is pushed ventro-medially against the C. c. lat. If the M. mg. contracts during the deflation of the C. c. lat., its muscular mass bulges laterally and exerts pressure laterally against the C. c. lat.

The Corpus cavernosum apicale (C. c. ap.) deflates synchronously with the C. c. lat. When the C. c. ap. deflates and collapses, the concentrically arranged connective tissue fibers, among which the cavernous veins are embedded, are relaxed and become more densely packed (Fig. 43). In this way, the extreme tip of the tongue reduces its volume and size (Fig. 59B).

If the M. sg. does not contract during the deflation of the C. c. ap., the main tendon of portion D of the M. sg. is relaxed (Fig. 58B). If the M. mg. does not contract during the deflation of the C. c. ap., the muscle fibers of the M. mg. ant. are relaxed (Figs. 57A, B).

If, however, the M. sg. or M. mg. contract during the deflation of the C. c. ap., the generated muscular forces compress the C. c. ap. and support its deflation. In this way, turgidity of the C. c. ap. can be maintained during its deflation. If the M. sg. contracts during the deflation of the C. c. ap., the main tendon of portion D exerts a caudal pull, so that the C. c. ap. is pushed against the mass of muscle bundles of the anterior portion of the M. mg. If the M. mg. ant. contracts during deflation of the C. c. ap., its muscle mass bulges and pushes radially against the C. c. ap.

Movements of the Epithelial Surface of the Tip of the Tongue During Movements of the Paraglossale

Because the M. supraglossus (M. sg.) and the M. hypoglossus anterior (M. hg. ant.) attach not only to the epithelial structures of the tip of the tongue and to the paraglossale, but also pass over the Art. paraglosso-basihyalis, interactions between movements of the dorsal surface of the tip of the tongue and movements at the Art. paraglosso-basihyalis must be analyzed. For descriptive purposes, it is presently assumed that the Art. cerato-basihyalis does not move (pp. 134 ff.).

Depression of the dorsal epithelial surface during flexion of the paraglossale (Fig. 60; Table 17).—When the paraglossale is flexed, its anterior end performs an angular rotation towards caudo-ventral about the Art. paraglosso-basihyalis (Figs. 60A, B). This rotation alone would not change the distance between origin and insertion of the M. hg. ant. However, due to the structure of the articular facets of the Art. paraglosso-basihyalis, this rotation is linked to a simultaneous caudo-dorsal translation along the midsagittal crest of the articular facet of the basihyale. Due to this translational movement, the distance between origin and insertion of the M. hg. ant. decreases because the Synchondrosis paraglossalis, over which the M. hg. ant. passes, is pulled closer to the insertion of the M. hg. ant. on the basihyale (Figs. 60A, B).

Depression of the central area of the tip of the tongue during the flexion of the paraglossale (Table 17) will be described from an initial stage with flattened central area of the tip of the tongue, relaxed aponeurosis of the M. hg. ant., and extended paraglossale (Fig. 60A) to a final stage with depressed central area of the tip of the tongue, taut aponeurosis of the M. hg. ant., and flexed paraglossale (Fig. 60B). For descriptive purposes, only the action of the M. hg. ant. is considered here, although the depression of the central area of the tip of the tongue is dependent on the coordinated contractions of both the M. hg. ant. and M. mesoglossus. The M. mg., however, does not pass over the Art. paraglosso-basihyalis and is, therefore, not directly affected by movements occurring at this articulation. When the aponeurosis of the M. hg. ant. is pulled taut by its contracting muscle during the depression of the central area of the tip of the tongue, the M. hg. ant. shortens. Through the simultaneous flexion of the paraglossale (due to the contraction of the M. ceratoglossus), the distance between origin and insertion of the M. hg. ant. is reduced, so that a contraction of the M. hg. ant. not only pulls its aponeurosis taut but also supports the flexion of the paraglossale. In this way, the M. hg. ant. contracts isotonically.

The epithelial surface of the central area of the tip of the tongue can also remain depressed during the flexion of the paraglossale (Table 17). Flexion of the paraglossale with maintained depression of the epithelial surface will be described from an initial stage with already depressed central area of the tip of the tongue, taut aponeurosis of the M. hg. ant., and extended paraglosssale (Fig. 60D) to a final stage with flexed paraglossale (Fig. 60B). At the beginning of this movement, the M. hg. ant. is already contracted and, therefore, its aponeurosis is taut. Continuing contraction of the M. hg. ant. supports flexion of the paraglossale, thus reducing the distance between origin and insertion of the M. hg. ant. In this way, the M. hg. ant. also contracts isotonically, although it shortens less (in absolute terms) during the actual flexion of the paraglossale than if the epithelial surface were depressed synchronously with the flexion of the paraglossale (see above).

If the Corpora cavernosa inflate during the combined flexion of the paraglossale and depression of the central area of the tip of the tongue, the contracting M. hg. ant. shortens less than it would if the Corpora cavernosa were inactive, because the Corpora cavernosa act as antagonists of the M. hg. ant. In contrast, if the Corpora cavernosa deflate during these movements, the contracting M. hg. ant. shortens even more than it would if the Corpora cavernosa maintained the same volume.

Flattening of the dorsal epithelial surface during flexion of the paraglossale (Figs. 61A–64B; Table 17).—With the flexion of the paraglossale, not only is the anterior end of the paraglossale rotated caudo-ventrally, but also the Proc. posterior paraglossalis and the attached Gl. lingualis perform an angular rotation towards apico-dorsal about the Art. paraglosso-basihyalis (Figs. 61A, 63A). This rotation increases the distances between the origins and insertions of the various portions of the M. sg. This increase of the distance between origin and insertion is more pronounced for the portions of the M. sg. lying farther away from the Art. paraglosso-basihyalis (representing the rotational axis) than for those portions lying closer to the articulation. Thus, the increase of the distance between origin and insertion is smallest for portion E of the M. sg. and largest for portion A, and is smaller for the M. sg. basihyalis (M. sg. bas.) than for the M. sg. ceratobranchialis (M. sg. cer.). A simultaneous caudo-dorsal translation of the paraglossale, which is linked to its flexion, only minimally offsets the increase of the distance between origins and insertions of the M. sg. This is due to the angular rotation of the paraglossale, by moving the insertions of the various portions of the M. sg. slightly more caudally, i.e., closer to their origins. As a net result, flexion of the paraglossale leads to an increase of distances between origins and insertions of the M. sg.

In addition, flexion of the paraglossale affects the Gl. lingualis. Flexion of the paraglossale not only increases the distance between origin and insertion of portion A of the M. sg. cer. but also increases the ventral component of the medio-apico-ventral pull exerted by the tendinous fibers of portion A on the dorso-caudal tip of the Gl. lingualis (cf. Figs. 61A and 63B or Figs. 63A and 61B). This increasing ventral pull compresses the glandular body which, in turn, increases its internal pressure after its duct system has been compressed. With increased internal pressure, turgidity and stiffness of the glandular body increase so that the gland assumes the function of an hydrostatic skeletal element (pp. 72 f.).

Flattening of the central area of the tongue's tip during the flexion of the paraglossale (Table 17) will be described from an initial stage with depressed central area of the tongue's tip, taut aponeurosis of the M. hg. ant., extended paraglossale, and uncompressed Gl. lingualis (Figs. 61A, 63A) to a final stage with flattened central area of the tongue's tip, relaxed aponeurosis of the M. hg. ant., flexed paraglossale, and compressed Gl. lingualis (Figs. 63B, 61B). When the dorsal surface of the central area is flattened, the tendinous fibers of portions B to E of the M. sg. are pulled medio-caudally by the contracting M. sg. In addition, the tendinous fibers of portion A of the M. sg. exert a medio-apico-ventral pull on the caudo-dorsal tip of the Gl. lingualis. If no movement occurred at the articulations of the hyoid, the M. sg. would shorten and the tendinous fibers of portion A of the M. sg. would relax because their origin on the central tendon of the M. sg. would have glided caudally (Fig. 58B; cf. Figs. 61A and 64B or Figs. 63A and 62B). Through the simultaneous flexion of the paraglossale, however, the distances between the origins and insertions of the various parts of the M. sg. are increased. Therefore, the contracting muscle fibers cannot shorten but have to elongate (the tendinous fibers cannot increase their length). Hence, when the central area of the tongue's tip is flattened during the flexion of the paraglossale, the contracting M. sg. is elongated and produces "negative work." In addition, the Gl. lingualis is forcefully compressed by the ventral pull exerted by the M. sg. cer. on its caudo-dorsal tip and by the pressure exerted by the tendinous portions B and C of the M. sg. on its anterior part.

Maintaining a flattened central area of the tongue's tip during the flexion of the paraglossale (Table 17) will be described from an initial stage with the already flattened central area of the tongue's tip, relaxed aponeurosis of the M. hg. ant., slightly compressed Gl. lingualis, and extended paraglossale (Figs. 62B, 64B) to a final stage with compressed Gl. lingualis and flexed paraglossale (Figs. 61B, 63B). The M. sg. is already contracted at the beginning of this movement and, thus, the central area of the tip of the tongue is already flattened. Because flexion of the paraglossale increases the distances between origins and insertions of the various portions of the M. sg., the paraglossale can now be flexed only by elongating the contracting M. sg. In this way, a flat central area of the tongue's tip can be maintained during flexion of the paraglossale only if the M. sg. contracts and elongates, producing "negative work."

If the Corpora cavernosa inflate during the combined flexion of the paraglossale and flattening of the central area of the tip of the tongue, the contracting M. sg. is even more elongated than it would be if the Corpora cavernosa were inactive. In contrast, if the Corpora cavernosa deflate during these movements, the contracting M. sg. must lengthen less extensively than it would have to if the Corpora cavernosa maintained the same volume.

Depression of the dorsal epithelial surface during extension of the paraglossale (Fig. 60; Table 17).—When the paraglossale is extended, its anterior end performs an angular rotation towards apico-dorsal about the Art. paraglosso-basihyalis (Fig. 60C). This rotation alone would not alter the distance between the origin and insertion of the M. hg. ant. However, due to the structure of the articular facets of the Art. paraglosso-basihyalis, this rotation is linked to a simultaneous apico-ventral translation of the paraglossale along the midsagittal crest of the articular facet of the basihyale. Due to this translational movement, the distance between origin and insertion of the M. hg. ant. increases during extension of the paraglossale because the Synchondrosis paraglossalis, over which the M. hg. ant. passes, is pushed away from the origin of the M. hg. ant. on the basihyale (Figs. 60C, D).

Depression of the central area of the tongue's tip during extension of the paraglossale (Table 17) will be described from an initial stage with flattened central area of the tip of the tongue, relaxed aponeurosis of the M. hg. ant., and flexed paraglossale (Fig. 60C) to a final stage with depressed central area of the tongue's tip, taut aponeurosis of the M. hg. ant., and extended paraglossale (Fig. 60D). For descriptive purposes, only the action of the M. hg. ant. is considered here, although the synchronous contraction of the M. mesoglossus is necessary for depressing the central area of the tongue's tip. The M. mg., however, is not affected by movements of the paraglossale because it does not pass over the Art. paraglosso-basihyalis. When the aponeurosis of the M. hg. ant. is pulled taut by the contraction of its muscle during the depression of the central area of the tongue's tip, the contracting M. hg. ant. tends to shorten. Simultaneous extension of the paraglossale, however, increases the distance between origin and insertion of the M. hg. ant. As a result, the contracting M. hg. ant. does not shorten and contracts more or less isometrically during extension of the paraglossale.

Maintaining a depressed central area of the tongue's tip during extension of the paraglossale (Table 17) will be described from an initial stage with already depressed central area of the tip of the tongue, taut aponeurosis of the M. hg. ant., and flexed paraglossale (Fig. 60B) to a final stage with extended paraglossale (Fig. 60D). At the beginning of the movement, the M. hg. ant. is already contracted, and its aponeurosis is taut. Because an extension of the paraglossale increases the distance between origin and insertion of the M. hg. ant., the paraglossale can now be extended only by elongating the contracting M. hg. ant. In this way, depression of the central area of the tongue's tip can be maintained during extension of the paraglossale only if the contracting M. hg. ant. is elongated and produces "negative work."

If the Corpora cavernosa inflate during the combined extension of the paraglossale and depression of the central area of the tongue's tip, the contracting M.

hg. ant. is even more elongated than it would be if the Corpora cavernosa were inactive, because the Corpora cavernosa act as antagonists of the M. hg. ant. In contrast, if the Corpora cavernosa deflate during these movements, the contracting M. hg. ant. must elongate less extensively than it would have to if the Corpora cavernosa maintained the same volume.

Flattening of the dorsal epithelial surface during extension of the paraglossale (Figs. 61A–64B; Table 17).—With the extension of the paraglossale, not only is the Proc. anterior paraglossalis rotated apico-dorsally, but the Proc. posterior paraglossalis and the attached Gl. lingualis also perform an angular rotation towards caudo-ventral about the Art. paraglosso-basihyalis (Figs. 62A, 64A). This rotation reduces the distances between origins and insertions of the various portions of the M. sg. Such a reduction of the distances is more pronounced for those portions of the M. sg. lying farther away from the rotational axis, i.e., the Art. paraglosso-basihyalis, than for those portions lying closer to the articulation. Thus, a reduction of the distance between origin and insertion is smallest for portion E of the M. sg. and largest for portion A, and is smaller for the M. sg. basihyalis (M. sg. bas.) than for the M. sg. ceratobranchialis (M. sg. cer.). A simultaneous apico-ventral translation of the paraglossale, which is linked to its rotation, only minimally offsets a reduction of the distance between origins and insertions of the various portions of the M. sg., which is due to an angular rotation of the paraglossale, by moving the insertions of the M. sg. on the paraglossale slightly more apically and farther away from the origin. As a net result, extension of the paraglossale leads to a reduction of the distances between origins and insertions of the various portions of the M. sg.

In addition, extension of the paraglossale affects the Gl. lingualis. An extension of the paraglossale not only reduces the distance between origin and insertion of portion A of the M. sg. cer. but also reduces the ventral component of the medio-apico-ventral pull exerted by the tendinous fibers of portion A of the dorso-caudal tip of the Gl. lingualis (cf. Figs. 62A and 64B or Figs. 64A and 62B). As the ventral pull on the glandular body decreases, the internal pressure of the gland decreases, and the Gl. lingualis returns to its uncompressed stage.

Flattening of the central area of the tongue's tip during the extension of the paraglossale (Table 17) will be described from an initial stage with depressed central area of the tongue's tip, taut aponeurosis of the M. hg. ant., flexed paraglossale, and compressed Gl. lingualis (Figs. 62A, 64A) to a final stage with flattened central area of the tongue's tip, relaxed aponeurosis, extended paraglossale, and less compressed Gl. lingualis (Figs. 64B, 62B). When the central area of the tongue's tip is flattened, the tendinous fibers of portions A to E of the M. sg. are pulled caudally by the contracting muscle. In addition, the tendinous fibers of portion A of the M. sg. cer. exert a medio-apico-ventral pull on the caudo-dorsal tip of the Gl. lingualis. If no movement occurred at the articulations of the hyoid, the entire M. sg. would shorten and the tendinous fibers of portion A would relax because their origin on the central tendon of the M. sg. would have glided caudally (Fig. 58B; cf. Figs. 61A and 64B, or Figs. 63A and 62B). The simultaneous extension of the paraglossale, however, reduces the distances between the origins and insertions of the various portions of the M. sg., so that contraction of this muscle not only flattens the central area of the tongue's tip, but also supports the extension of the paraglossale. In this way, the M. sg. contracts more or less isotonically during the entire movement. With regard to the effect on the Gl. lingualis, reduction of the distance between origin and insertion of portion A of the M. sg. reduces the tension placed upon the Gl. lingualis by the tendinous fibers of portion A of the M. sg. so that the turgidity of the Gl. lingualis is reduced although the tendinous portions B and C of the M. sg. maintain some pressure on the anterior end of the glandular body.

Maintaining a flattened central area of the tongue's tip during extension of the paraglossale (Table 17) will be described from an initial stage with already flattened central area of the tongue's tip, relaxed aponeurosis of the M. hg. ant., compressed Gl. lingualis, and flexed paraglossale (Figs. 61B, 63B) to a final stage with uncompressed Gl. lingualis and extended paraglossale (Figs. 62B, 64B). Because the M. sg. is already contracted at the beginning of this movement and, thus, the central area of the tongue's tip is already flattened, a continuing contraction of the M. sg. only supports a further reduction of the distance between origins and insertions of the M. sg., which is brought about by an extension of the paraglossale. In this way, the M. sg. contracts more or less isotonically.

If the Corpora cavernosa inflate during the combined extension of the paraglossale and depression of the central area of the tongue's tip, the contracting M. sg. does not shorten as much as it would if the Corpora cavernosa were inactive because the Corpora cavernosa act as antagonists of the M. sg. In contrast, if the Corpora cavernosa deflate during these movements, the contracting M. sg. has to shorten even more than it would if the Corpora cavernosa maintained the same volume.

Discussion (Table 17).—Depending on the particular combination of movements of the paraglossale and of the epithelial surface of the tongue's tip, the M. sg. and M. hg. ant. contract either isotonically or more or less isometrically, or are elongated during their contraction (Table 17). It seems doubtful, however, that the movements requiring elongation of a contracting muscle are performed under normal, unloaded conditions given their higher energy costs.

Movements of the Epithelial Surface of the Tongue's Tip During Movements of the Hyoid

Of the muscles that control the movements of the epithelial surface of the tongue's tip, only one part of the M. supraglossus, namely the M. sg. ceratobranchialis (M. sg. cer.), passes over the Art. cerato-basihyalis in addition to the Art. paraglosso-basihyalis. At least some parts of the tendinous fibers of portions B and C originate from the M. sg. cer., whereas all the tendinous fibers of portion A originate from it. In this way, a large part of the movements of the epithelial surface of the tongue's tip is controlled by the M. sg. cer. and is, therefore, influenced by movements at the Art. cerato-basihyalis.

Probably both parts of the M. sg., i.e., the M. sg. cer. and the M. sg. basihyalis, always contract more or less simultaneously, because both muscular heads are so closely associated with each other at their insertion on the common central tendon. Because both muscular parts of the M. sg. contribute fibers to all tendinous portions (except portion A to which only the M. sg. cer. contributes), contraction of only one of the muscular heads of the M. sg. would not produce different movements at the epithelial surface of the lingual tip. The two muscular parts have only different functions with respect to the Art. cerato-basihyalis.

TABLE 18

CORRELATIONS AMONG THE MOVEMENTS OF THE HYOID, PARAGLOSSALE, AND EPITHELIAL SURFACE OF THE TONGUE'S TIP[a]

Movement of hyoid	Movement of paraglossale	Action of contracting M. supraglossus during flattening of central area of tip of tongue
retraction (through M. serpihyoideus and M. stylohyoideus)	flexion (through M. ceratoglossus)	extending ("negative work")
	extension (through M. hypoglossus obliquus)	shortening (isotonic contraction)
protraction (through M. ceratohyoideus and M. branchiomandibularis)	flexion (through M. ceratoglossus)	extending ("negative work")
	extension (through M. hypoglossus obliquus)	M. sg. bas.: shortening (isotonic contraction) M. sg. cer.: ± same length (isometric contraction)

[a] See pp. 134 ff.

Because the M. sg. cer. interacts simultaneously with the Art. cerato-basihyalis, the Art. paraglosso-basihyalis and the epithelial surface of the tongue's tip, the interdependencies of the movements of these three structures must be analyzed (Table 18).

Movements of the epithelial surface and paraglossale during retraction of the hyoid (Figs. 61A–64B; Table 18).—When the hyoid is retracted by the contracting M. serpihyoideus and M. stylohyoideus, the angle enclosed between the dorsal surfaces of the ceratobranchiale and basihyale is reduced, i.e., the basihyale performs an angular rotation towards caudo-dorsal about the Art. cerato-basihyalis (Fig. 46A). Independent of movements at the Art. paraglosso-basihyalis and epithelial surface of the tongue's tip, this rotation of the basihyale alone results in a reduction of the distance between origin and insertion of the M. sg. cer. (cf. Figs. 61A and 63A or Figs. 62A and 64A).

At the same time as the caudo-dorsal angular rotation of the basihyale, the ceratobranchialia perform angular rotations about the Artt. cerato-basihyales towards apico-lateral (Fig. 46B). These rotations, however, do not affect the distance between origin and insertion of the M. sg. cer. which only changes the direction of its force line. With the spreading apart of the ceratobranchialia, the origins of the collateral parts of the M. sg. cer. move laterally so that the collateral muscles have an increasingly oblique orientation. This change in orientation does not affect the direction of the force exerted on the tendinous fibers by the M. sg. cer. because all muscle fibers converge towards the common central tendon and because the lateral components of the caudo-lateral forces of the collateral muscles cancel each other. However, the caudal component of the force exerted by the M. sg. cer. decreases during the retraction of the hyoid so that the caudal pull generated by the contracting M. sg. cer. becomes increasingly inefficient during retraction of the hyoid.

Contraction of the entire M. sg. during the flexion and retraction of the paraglossale caused by the contracting M. ceratoglossus and the simultaneous retraction of the hyoid (Table 18) will be described from an initial stage with protracted hyoid, extended and protracted paraglossale, uncompressed Gl. lingualis, taut aponeurosis of the M. hg. ant., and depressed central area of the tip of the tongue (Fig. 61A) to a final stage with retracted hyoid, flexed and retracted paraglossale, compressed Gl. lingualis, relaxed aponeurosis of the M. hg. ant., and flattened central area of the tongue's tip (Fig. 61B). When the central area of the tongue's tip is flattened by the contracting M. sg. and the paraglossale is simultaneously flexed by the contracting M. ceratoglossus, the distance between origin and insertion of the M. sg., including the M. sg. cer., is increased irrespective of movements at the Art. cerato-basihyalis. In addition, the Gl. lingualis is compressed. During a simultaneous retraction of the hyoid, the distance between the origin and insertion of the M. sg. cer. is decreased. This decrease, however, does not offset an increase of the distance that is caused by a flexion of the paraglossale. As a result of all movements, the entire contracting M. sg. is elongated and, therefore, produces "negative work" (Table 18). However, due to the caudo-dorsal rotation of the basihyale about the Art. cerato-basihyalis, which reduces the distance between origin and insertion of the M. sg. cer., the increase of the distance between origin and insertion of the M. sg. caused by the flexion of the paraglossale is less extensive than if the hyoid remained stationary. In this way, the contracting M. sg. cer. is less elongated and, thus, produces less "negative work" (cf. Figs. 61A, B with Figs. 61A, 63B; cf. pp. 73ff., 117ff., 130ff.).

The contraction of the M. sg. during extension and protraction of the paraglossale caused by a contraction of the M. hypoglossus obliquus and the simultaneous retraction of the hyoid (Table 18) will be described from an initial stage with a protracted hyoid, flexed and retracted paraglossale, somewhat compressed Gl. lingualis, taut aponeurosis of the M. hg. ant., and depressed central area of the tongue's tip (Fig. 62A) to a final stage with retracted hyoid, extended and protracted paraglossale, compressed Gl. lingualis, relaxed aponeurosis of the M. hg. ant., and flattened central area of the tongue's tip (Fig. 62B). When the central area of the tongue's tip is flattened by the contracting M. sg. and the paraglossale is simultaneously extended by the contracting M. hypoglossus obliquus, the distance between origin and insertion of the M. sg., including the M. sg. cer., is reduced, irrespective of movements at the Art. cerato-basihyalis. In addition, the ventral component of the compressive force exerted by the tendinous fibers of portion A of the M. sg. on the caudo-dorsal tip of the Gl. lingualis is reduced. A simultaneous caudo-dorsal rotation of the retracted basihyale also reduces the distance between origin and insertion of the M. sg. cer., so that a contraction of the M. sg. cer. supports not only the extension of the paraglossale but also the caudo-dorsal angular rotation of the basihyale. As a result of all movements, the entire contracting M. sg. shortens, i.e., contracts isotonically. However, due to the caudo-dorsal rotation of the basihyale about the Art. cerato-basihyalis, the contracting M. sg. cer. shortens more than it would if the hyoid remained stationary (cf. Figs. 62A, B with Figs. 62A, 64B).

Movements of the epithelial surface and paraglossale during protraction of the hyoid (Figs. 61A–64B; Table 18).—When the hyoid is protracted by the contracting M. branchiomandibularis and M. ceratohyoideus, the angle enclosed between the dorsal surfaces of the ceratobranchiale and basihyale is increased, i.e., the basihyale performs an angular rotation towards apico-ventral about the Art. cerato-basihyalis (Fig. 47A). Independent of movements of the paraglossale and epithelial surface of the tongue's tip, this rotation of the basihyale alone results in an increase of the distance between origin and insertion of the M. sg. cer. (cf. Figs. 63A and 62B or Figs. 64A and 63B).

Simultaneously with the apico-ventral rotation of the basihyale, the ceratobranchialia perform angular rotations about the Artt. cerato-basihyales towards caudo-medial, i.e., they are drawn together (Fig. 47B). These rotations, however, do not affect the distance between origin and insertion of the M. sg. cer. When the ceratobranchialia converge, the origins of the collateral parts of the M. sg. cer. move medially so that the collateral muscles assume a less oblique orientation. In this way, the caudal component of the force exerted by the M. sg. cer. increases during the protraction of the hyoid at the expense of the lateral component. Therefore, if the M. sg. contracts during the protraction of the hyoid, the caudal pull exerted by the M. sg. cer. on the epithelial structures becomes increasingly efficient during protraction of the hyoid.

Contraction of the M. sg. during flexion and retraction of the paraglossale caused by the contracting M. ceratoglossus and the simultaneous protraction of the hyoid (Table 18) will be described from an initial stage with retracted hyoid, extended and protracted paraglossale, uncompressed Gl. lingualis, taut aponeurosis of the M. hg. ant., and depressed central area of the tongue's tip (Fig. 63A) to a final stage with protracted hyoid, flexed and retracted paraglossale, compressed Gl. lingualis, relaxed aponeurosis of the M. hg. ant., and flattened central area of the tongue's tip (Fig. 63B). When the central area of the tongue's tip is flattened by the contracting M. sg. and the paraglossale is simultaneously flexed by the contracting M. ceratoglossus, the distance between origin and insertion of the M. sg., including the M. sg. cer., is increased irrespective of movements at the Art. cerato-basihyalis. At the same time, the Gl. lingualis is compressed. During a simultaneous apico-ventral rotation of the basihyale, the distance between origin and insertion of the M. sg. cer. is further increased. In this way, the entire contracting M. sg. is elongated during protraction of the hyoid and simultaneous flexion of the paraglossale, i.e., the M. sg. produces "negative work" (Table 18). The M. sg. cer. is elongated to an even greater degree during this movement than it would be if the hyoid remained stationary (cf. Figs. 63A, B with Figs. 63A, 61B).

Contraction of the M. sg. during extension and protraction of the paraglossale caused by the contracting M. hypoglossus obliquus and the simultaneous protraction of the hyoid (Table 18) will be described from an initial stage with retracted hyoid, flexed and retracted paraglossale, somewhat compressed Gl. lingualis, taut aponeurosis of the M. hg. ant., and depressed central area of the tongue's tip (Fig. 64A) to a final stage with protracted hyoid, extended and protracted paraglossale, less compressed Gl. lingualis, relaxed aponeurosis of the M. hg. ant., and flattened central area of the tongue's tip (Fig. 64B). When the central area of the tongue's tip is flattened by the contracting M. sg. and the paraglossale is simultaneously extended by the contracting M. hypoglossus obliquus, the distance between origin and insertion of the M. sg., including the M. sg. cer., is reduced irrespective of movements at the Art. cerato-basihyalis. In addition, the compressive force exerted by the tendinous fibers of portion A of the M. sg. on the Gl. lingualis is decreased, so that the internal pressure of the salivary gland is reduced. However, the simultaneous apico-ventral rotation of the basihyale increases the distance between origin and insertion of the M. sg. cer. and, therefore, offsets the reduction of this distance brought about by the extension of the paraglossale. As a result of all movements, the M. sg. cer. contracts more or less isometrically, whereas the M. sg. bas. shortens and, hence, contracts isotonically. In this way, contraction of the M. sg. cer. is able to produce force more economically than it would if the hyoid were stationary, in which case the M. sg. cer. would contract isotonically.

Discussion.—Depending on the particular combination of movements of the hyoid, paraglossale, and epithelial surface of the tongue's tip, the M. sg. cer. contracts either isotonically or more or less isometrically, or is elongated during its contraction (Table 18). It seems doubtful, however, that movements during which the contracting M. sg., including the M. sg. cer., is elongated are actually performed under normal, unloaded conditions given their high energy costs.

With respect to its actions on the Art. cerato-basihyalis and Art. paraglosso-basihyalis, the M. sg. cer. is a synergist of the M. hypoglossus obliquus lateralis from which it differs only by its additional actions on the epithelial surface of the tongue and on the Gl. lingualis (Table 18; pp. 115 f., 117 ff.).

Conclusions

Before the functioning of the lingual apparatus of *Psittacus* can be compared to that of other avian lingual apparatus, the various submodels that explain the functioning of the subunits of the lingual apparatus (described above) must be synthesized to an integrated model that describes the functioning of the psittacine lingual apparatus as a whole. Such a synthesis is achieved through the synchronization of the various submodels. The procedure for constructing a synthetic model follows the guidelines that were formulated on the basis of various biomechanical principles. The same guidelines were already applied, for example, to the synchronization of the movements of the paraglossale, basihyale and dorsal surface of the lingual tip (Tables 17, 18), or to the synchronization of the lifting and lowering the hyoid with the protraction and retraction of the lingual apparatus. The result of the synthesis of all submodels is summarized in Table 19.

The synthetic model illustrates that the lingual apparatus consists of a set of hierarchically arranged, interdependent subsystems. The extrinsic lingual muscles and the intrinsic M. ceratohyoideus move the hyoid skeleton with respect to the mandible and, at the same time, determine the movements of the basihyale and urohyale with respect to the ceratobranchialia as well as the movements of the epibranchialia with respect to the ceratobranchialia. In contrast, the intrinsic muscles control the movements of the paraglossale relative to the basihyale and the movements of the surface of the lingual tip with respect to the paraglossale. However, most of these muscles (i.e., M. ceratoglossus, M. hypoglossus obliquus, and M. supraglossus) possess at least one part that spans the Art. cerato-basihyalis and that is, therefore, affected by the movements of the basihyale and ceratobranchialia. Hence, the movements of the paraglossale and surface of the lingual tip are not independent from the movements of the hyoid and basihyale.

The synthetic model also shows that, under the assumed boundary conditions of slow, unloaded movements, the movements of the various subunits of the hyoid apparatus can combine in almost any way without extending contracting muscles (Table 19). However, exceptions are known. For example, the surface of the tongue's tip cannot be flattened simultaneously with a flexion of the paraglossale (Table 17, 18, 19), or the hyoid cannot be lifted at the same time as it is retracted

TABLE 19

SYNCHRONIZATION OF MUSCLE CONTRACTIONS DURING VARIOUS MOVEMENTS OF THE HYOID APPARATUS, WITHOUT EXTENSION OF CONTRACTING MUSCLES

Movement of hyoid and basihyale		Movement of paraglossale	Movement of surface of lingual tip
protraction of hyoid and flexion of basihyale (through M. branchiomandibularis and M. ceratohyoideus)	lifting of hyoid (through M. mylohyoideus)	flexion (through M. ceratoglossus) extension (through M. hypoglossus obliquus)	depression (through M. mesoglossus, M. hypoglossus anterior, and M. hypoglossus transversus)
		extension (through M. hypoglossus obliquus)	flattening (through M. supraglossus)
	lowering of hyoid (through M. genioglossus)	flexion (through M. ceratoglossus) extension (through M. hypoglossus obliquus)	depression (through M. mesoglossus, M. hypoglossus anterior, and M. hypoglossus transversus)
		extension (through M. hypoglossus obliquus)	flattening (through M. supraglossus)
retraction of hyoid and extension of basihyale (through M. serpihyoideus and M. stylohyoideus)	lowering of hyoid (through M. tracheohyoideus)	flexion (through M. ceratoglossus) extension (through M. hypoglossus obliquus)	depression (through M. mesoglossus, M. hypoglossus anterior, and M. hypoglossus transversus)
		extension (through M. hypoglossus obliquus)	flattening (through M. supraglossus)

(Table 19). These restrictions notwithstanding, the basic versatility of the lingual apparatus is maintained mainly because the intrinsic two-joint muscles can contract and function during different movements of the hyoid skeletal elements, although in different ways, i.e., by contracting either isotonically or isometrically (Tables 17, 18, 19).

The psittacine tongue has a much more complex structure than other avian tongues (cf. Boullion and Homberger 1982; Cummins and Homberger, unpub. data; W. J. Bock, pers. comm.; pers. obs.). The increased structural complexity of the psittacine tongue is superimposed on a generalized avian construction and consists of the same basic inventory of skeletal elements and muscles as that found in other birds. The psittacine-specific modifications of the hyoid skeleton involve mainly an increased massiveness of the bony elements and a greater complexity of the articular facets of the diarthroses. The most distinctive psittacine-specific modifications affect the lingual muscles. Muscles that are generally simple in birds are split into several parts with different attachments in parrots. In extreme cases, this splitting of muscles gave rise to "new" muscles (cf. Bock and Morony 1978b). For example, the M. supraglossus originated from a part that was split off the M. hypoglossus obliquus, and the M. mesoglossus is derived from a part of the M. hypoglossus anterior. Added to the "primary" complexity of the lingual muscles is the "secondary" complexity provided by hydraulic structures (salivary glands, cavernous vascular tissue). The total structural complexity of the psittacine tongue can be interpreted as having evolved in adaptation to the specialized diet of shelled seeds.

Parrots have evolved from an ancestral seed-eating species that shelled the seeds before swallowing them (Homberger 1980a, b). With a single exception, all parrots always shell the seeds they eat and use the same method to do so. In shelling a seed, a parrot places a seed between the transversal, cutting edge of the lower mandible and the projecting ridge on the rhamphothecal palate. The seed is prevented from slipping by the corrugated inner surface of the tip of the upper mandible and is held in place with the tongue's tip (Homberger and Ziswiler 1972; Homberger 1980a, b). In doing so, the tongue has to be protracted and lifted and the anterior tip of paraglossale has to be rotated dorso-apically to push the seed against the projecting palatal ridge and the inner surface of the tip of the upper rhamphotheca. After the cutting edge of the lower mandible has penetrated the seed shell and removed the half of the shell that faces the inner surface of the tip of the upper rhamphoteca, the seed is rotated by the tongue. When rotating the seed "downside up," the tongue is protracted and lifted and the tongue's tip is raised. Simultaneously, the relief of the dorsal surface of the tongue is continuously modified to allow for adjustments of the seed position. The remaining part of the seed shell now faces the inner surface of the tip of the upper rhamphotheca and is removed by the lower mandible in the same way as was used for the first part of the shell. Finally, the shelled seed is placed on the tongue's tip and transported into the pharynx by the retracting tongue.

During drinking with the method characteristic of the Psittacinae, of which *Psittacus* is a member, water is ladled with the spoon-shaped tip of the tongue and swallowed by pushing the tongue against the palate. During the simultaneous gradual flattening of the surface of the lingual tip, the water is pushed along the lingual surface into the pharynx in a similar fashion as humans use to swallow liquid (Homberger 1980a, b). Thus, during drinking the tongue is first protracted, but not necessarily lifted, and the tongue's tip is flexed by flexing the paraglossale. As the paraglossale is successively extended and the hyoid lifted, the dorsal surface of the lingual tip becomes spoon-shaped. When the dorsal surface of the lingual tip has reached the palate, it flattens gradually. Finally, the hyoid is retracted and lowered.

These movements of the tongue during seed shelling and drinking, by being used to counteract external forces, differ from the unloaded, slow movements that were assumed as boundary conditions for the theoretical mechanical analysis. Thus, they are loaded movements. During seed shelling, the lingual apparatus has to counteract mostly ventro-caudally directed forces derived from the seed's tendency to fall into the mouth cavity. Also during drinking, the lingual apparatus must overcome ventrally and, to a lesser degree, caudally directed forces when its dorsal surface is pushed against the palate (cf. Homberger 1983). It can be concluded that the specialized, parrot-specific activities of the psittacine tongue require a maximized ability to resist or counteract ventro-caudally directed external forces by increasing the musculature generating forces that move the hyoid apparatus apico-dorsally. At the same time, the specialized functioning of the tongue's tip during seed shelling and drinking also requires an ability to modify the relief of the surface of the lingual tip. A comparison of the lingual apparatus of *Psittacus* to that of non-psittacine birds shows that the psittacine apparatus is actually distinguished by an increase in complexity and quantity of precisely those muscles that control the dorsal and apical movements of the hyoid and the movements of the dorsal surface of the lingual tip. For example, the M. branchiomandibularis, which protracts and lifts the hyoid, consists of three instead of the usual two parts: the M. mylohyoideus, which lifts and protracts the hyoid, consists of two parts instead of being undivided as in most other birds; the usually undivided M. ceratoglossus consists in parrots of at least four parts, two of which support a dorsal rotation of the basihyale; and both the usually undivided M. hypoglossus obliquus and the psittacine-specific M. supraglossus rotate the anterior tip of the paraglossale dorsally and possess each one part that supports the dorsal rotation of the basihyale. Of the three muscles controlling the movements of the surface of the lingual tip, two (M. supraglossus and M. mesoglossus) are unique to parrots. Hence, especially those psittacine structures responsible for the psittacine-specific, specialized movements have a greater degree of complexity than the homologous structures in non-psittacine birds. This primary complexity engenders a secondary complexity of the supporting and antagonistic structures, so that the final result is what appears to be a disproportionately complex system.

Given the extraordinary number of muscles present in the psittacine tongue, the question arises whether the lingual apparatus contains redundant structural elements, i.e., elements that perform functions that are also executed by other elements. Redundant elements are, therefore, not vital for the proper functioning of an apparatus and might be identified as such if their removal did not affect the function of the other elements of the apparatus. Such a "natural" experiment is provided by the individual variations in the configuration and presence of muscles, as described in the present study of *Psittacus*. Seemingly redundant muscles are represented by individual parts of several lingual muscles, e.g., the M. ceratoglossus superficialis, the medial portion of the M. hypoglossus anterior, and the basihyal and ceratobranchial slips of the M. tracheohyoideus. In these cases, the presence of the variable muscle parts may provide an advantage by increasing the number of muscle fibers without changing the functions of the particular muscles to which they belong. This advantage may, however, not be important enough to influence markedly the overall fitness of the individual animal and may, therefore, not have become fixed as an expression of genetic information. The occurrence of variable presence in only some few muscles of the lingual apparatus might also be interpreted differently. Variability due to varying developmental processes is likely to occur with a similar frequency in all structures, but those variations that affect the proper functioning of the lingual apparatus are not encountered in a "normal" adult specimen because of selective pressures against an animal with a non-functional tongue. This hypothesis is supported by the fact that individual variations in the occurrence of muscles and their parts are observed only for few selected muscles, but never for others, e.g., the M. serpihyoideus, M. ceratohyoideus, or M. branchiomandibularis. The latter muscles are responsible for particular movements that cannot be executed by any other structure or muscle. The variable presence of the M. hypoglossus transversus, however, presents a different problem. Its function cannot exactly be taken over by another muscle, yet it is only rarely present. This muscle could be either on its way to reduction and final loss in *Psittacus* or could be a new mutational variation. The final answer to this problem will have to be postponed until more information is available from comparative studies on other parrot species.

Individual variation not only encompasses the presence or absence, but also the size of the lingual muscles and the location of their attachments on particular

TABLE 20

COMPARISONS OF MUSCLE ACTIONS IN *PSITTACUS, ANAS,* AND *GALLUS*

	Psittacus (Homberger)	*Anas* (Zweers et al. 1977)	*Gallus* (Suzuki et al. 1975)
M. serpihyoideus	retraction of hyoid	retraction of hyoid	protraction of hyoid
M. stylohyoideus	retraction of hyoid	retraction of hyoid	protraction of hyoid
M. branchiomandibularis	protraction of hyoid	protraction of hyoid, extension of basihyale	retraction of hyoid
M. mylohyoideus	lifting of hyoid	lifting of hyoid	lifting of hyoid
M. genioglossus	lowering of hyoid	—	—
M. tracheohyoideus	lowering of hyoid	—	retraction of hyoid
M. ceratohyoideus	extension of basihyale (during protraction of hyoid)	extension of basihyale (during protraction of hyoid)	protraction of hyoid
M. ceratoglossus	flexion of paraglossale	flexion of paraglossale	extension of paraglossale
M. hypoglossus obliquus	extension of paraglossale	extension of paraglossale	extension of paraglossale
M. supraglossus	flattening of surface of lingual tip	—	—
M. hypoglossus anterior	depression of surface of lingual tip	depression of lingual tip	—
M. mesoglossus	depression of surface of lingual tip	—	—

skeletal elements. This type of variation has the potential to affect the overall efficiency of an apparatus, e.g., by varying the efficiency with which a force can be generated by a lever system of muscles and bones. The relative efficiency of an apparatus may, in turn, have an effect on the fitness of the animal. Individual variability of structures, however, has customarily been ignored in anatomical studies, although it has the same role in providing the raw material for evolutionary change as does the variability of external or behavioral characters, the role of which has been recognized since Darwin's time.

The only other models of avian lingual movements are provided by Suzuki and Nomura (1975) for the domestic chicken (*Gallus gallus*) and by Zweers et al. (1977) for the domestic duck (*Anas platyrhynchos*). A direct comparison of these models with the present model for *Psittacus* is difficult for several reasons. First, the structures of the lingual apparatus of *Gallus* and *Anas* are very different from one another and from those of *Psittacus*. Second, the level of detail of the anatomical descriptions varies considerably among the three studies. Third, the models provide different kinds of explanations for the functioning of the lingual apparatus. The models of Suzuki and Nomura (1975) and of Zweers et al. (1977) correlate muscle activities directly with movements of the whole apparatus, whereas my model correlates muscle activities with movements of particular skeletal elements. Fourth, in contrast to the other models, the model of Zweers et al. (1977) is concerned with movements during which the lingual apparatus interacts with external forces during straining of food and water. Thus, in this case, it is not always possible to decide whether a particular muscle force is exerted to move the hyoid apparatus or to resist or counteract particular external forces. To permit a comparison of the three models, the information given in the papers of Suzuki and Nomura (1975) and Zweers et al. (1977) was transformed to correlate muscle activity with movements of skeletal elements (Table 20).

In a comparison of the three models (Table 20), it is remarkable that the actions of the particular muscles appear to be the same in *Anas* and *Psittacus* (provided that the data of *Anas* have been transformed correctly), despite the vast morphological and functional-adaptional differences between the overall structure of the two lingual systems. In contrast, according to the model of Suzuki and Nomura (1975), the actions of the M. serpihyoideus, M. stylohyoideus, M. branchiomandibularis, and M. ceratoglossus in *Gallus,* seem to differ diametrically from those in both *Psittacus* and *Anas.* My preliminary study of the lingual apparatus of *Gallus* indicates, however, that these muscles function similar to those in *Psittacus* and *Anas* (Boullion and Homberger 1982). These comparisons invite speculations and conjectures about the mode of evolutionary changes of the avian lingual apparatus in general. Such changes may have proceeded mainly through differentiation of the mechanical elements, whereas the nervous control patterns and nervous pathways may have remained conservative and retained their primitive conditions. However, more data on a variety of birds are needed before this hypothesis can be evaluated.

The last and perhaps most important aspect of the present study of the lingual apparatus of *Psittacus* to be discussed is its potential significance and value as a basis for future research.

In one way, the synthetic method of functional analysis developed and formulated in this study can be applied to other complex morphological systems, the functioning of which may otherwise prove difficult to analyze (e.g., Homberger 1979b, 1981, 1982; Boullion and Homberger 1982; Cummins and Homberger, unpub. data).

In another way, the study of *Psittacus* may serve as a reference for future comparative studies of the avian lingual apparatus. Such studies could be based on the identification of structural differences in comparison to the reference work. Because the functional properties are known for the particular structures forming the apparatus that is used as the reference, structural differences can be evaluated in terms of their effect on the functioning of the integral apparatus. In this way, the adaptive value of structural changes and, thus, the polarity of evolutionary change in a morphocline may be determined. On the basis of such comparative studies within the order Psittaciformes, it will be possible to trace the process of adaptive radiation of the biomechanical parts of the lingual apparatus and to test existing hypotheses on the systematics and phylogeny of the Psittaciformes. If this comparative method is applied to a variety of avian taxa of higher categories, it may eventually shed some light on the phylogeny of avian families and orders, a topic that represents one of the least understood aspects of avian systematics.

Finally, one has to keep in mind that the lingual apparatus is only a part of the integral feeding apparatus. Functional-anatomical studies of the jaw and laryngeal apparatus must follow those of the lingual apparatus. In this way, we will eventually be able to understand the interactions and interdependencies among the different morphological systems that cooperate in particular functions, for example, seed-eating during which the jaw apparatus opens the seed shells and crushes the seeds, the lingual apparatus manipulates and holds the seeds, and the laryngeal and lingual apparatus cooperate in swallowing the food. The eventual functional analysis of an integrated system, such as the feeding apparatus, could provide a model that would represent a major step towards a better understanding of "coevolutionary" mechanisms affecting the various parts of an organism as an integrated entity.

ACKNOWLEDGMENTS

A special debt is owed to Dr. Walter J. Bock for introducing me to the joys and sorrows of functional-anatomical studies of muscle-bone systems, for crucial comments and suggestions, for well-taken criticisms, and linguistic improvements of various drafts of the manuscript, and for his tolerant attitude towards helvetic thoroughness. The late Professores emeriti Dr. Hans Höfliger and Dr. Eugen Seiferle, Department of Anatomy, School of Veterinary Medicine, University of Zürich, deserve credit for having taught me the art of descriptive anatomy. Drs. Mercedes S. Foster, Carl Gans, Wolfgang F. Gutmann, Kathleen K. Smith, Richard L. Zusi, and Gart A. Zweers courageously and selflessly volunteered to read the manuscript and provided valuable and gratefully acknowledged criticisms. The illustrations were prepared with dedication and perseverance by Mr. Taro Suzuki (Figs. 1, 2, 15–30, 31B–33, 35A, 36A, 37, 39A, 41A, 42, 44, 45), Ms. Yvette Ledoux (Figs. 3–14, 34, 35B, 36B, 38, 39B, 40, 41B, 43, 46–52; as well as the labeling of these and T.S.'s figures and their assembly into plates), and Ms. Michelle J. Cave (Figs. 31A, 53–64; as well as the labeling of these figures and their assembly into plates and various corrections of the other figures). The final copy of the manuscript and innumerable drafts were typed by Ms. Lynnda Halbrook with expertise and unending patience. The photostatic reductions of the ink illustrations were produced by the Printing Office of Louisiana State University. The photographs of the half-tone illustrations were taken by Ron Bouchard. Catherine L. Cummins helped at various stages of the proof reading chore.

Anatomical specimens for this study were made available from the collections of the Zoological Museum at the University of Zürich (Switzerland) through Prof. Dr. Vinzenz Ziswiler, of the American Museum of Natural History in New York through Dr. Wesley E. Lanyon, and of the Museum of Natural Science at Louisiana State University in Baton Rouge through Mr. John J. Morony. The present study was initiated when I held a post-doctoral fellowship from the Swiss National Foundation for the Promotion of Scientific Research at Columbia University in New York City. Funds for the final preparation of the illustrations were provided by grants from the Frank M. Chapman Memorial Fund and from the American Philosophical Society and by the Department of Zoology and Physiology of Louisiana State University. Part of the study was done when I held a Summer Faculty Research Grant from the Louisiana State University Council on Research.

SUMMARIES

ENGLISH: SUMMARY

The present study of the lingual apparatus of the African Grey Parrot, *Psittacus erithacus,* consists of two interdependent parts: a descriptive part, which significantly amends and completes previous anatomical descriptions of the psittacine lingual apparatus, and a functional-analytical part, in which a theoretical mechanical model is formulated to explain the functions and interactions of the different parts of the apparatus.

The anatomical description covers only mechanically functioning elements (i.e., skeletal elements, ligaments and fasciae, muscles, synovial bursae, epithelial structures, cavernous vascular tissue, salivary glands) and excludes nerves and blood vessels. Special attention was given to the range of variation exhibited by the various structures. To facilitate future use of the anatomical nomenclature, a Latin name was suggested for each newly described structure. Whenever possible, already established names were adopted for the described structures, and synonymies were provided.

The hyoid skeleton consists of the unpaired basihyale (fused to the unpaired urohyale) and of the paired epibranchialia, ceratobranchialia, and paraglossalia. Cartilaginous processes are located at the caudal ends of the urohyale and epibranchialia. A cartilaginous bridge unites the anterior ends of the paraglossalia. Most hyoid joints, viz., the Art. paraglosso-basihyalis, Art. cerato-basihyalis, and Art. epi-ceratobranchialis, are true diarthroses. The shape of the articular facets and the position of the articular ligaments determine to a large degree the direction and extent of the movements at particular joints. Attachment sites of muscles and ligaments of the skeletal elements are often, but not always, marked by ridges, rugosities or depressions on the skeletal surface.

The mandible and the laryngeal and tracheal skeletons are described only as far as they are relevant for the description of the lingual apparatus.

The ligaments can be classified into three categories. The six intraosseal ligaments connect two different areas of the same skeletal element and have various functions, such as connecting different tissues or serving as attachment sites for muscles. The four articular ligaments connect two articulating skeletal elements and limit the possible movements at the particular articulation. Linkage ligaments are represented only by the Lig. nodulo-ceratobranchiale, which contains a sesamoid bone (Nodulus) at the point where it crosses the ventral surface of the urohyale. This ligament spans two articulations, viz., the paired Artt. cerato-basihyales, thus coupling the movements of the ceratobranchialia and basihyale. In addition, six fasciae serve to envelop or connect different structures. The Fascia vaginalis hyoidei forms a sleeve-like sheath in which the caudal part of the hyoid horn (the epibranchiale and part of the ceratobranchiale) glides back and forth during retraction and protraction of the lingual apparatus.

The lingual musculature consists of six extrinsic and seven intrinsic pairs of muscles which are described in a standardized way to facilitate comparison. Each description is supplemented by tables listing the synonymies of homologous muscles described for other birds and comparing previously published functional interpretations of each psittacine lingual muscle with my own analysis. The lingual muscles of parrots differ in many ways from those of other birds; some of their most distinctive characteristics are summarized. The M. serpihyoideus and M. stylohyoideus merge towards their origins on the mandible. The M. branchiomandibularis is strongly developed and consists of three parts. The M. genioglossus is prominent and runs more or less vertically between the paraglossalia and the mandibular symphysis. The M. tracheohyoideus originates from the sternum and inserts on the basihyale and ceratobranchiale. This muscle is clearly distinct from the M. tracheolateralis, and a special effort was made to homologize these two muscles with previously described avian muscles. The configuration of the M. ceratohyoideus does not differ much from that found in other birds. The M. ceratoglossus is extremely complex with up to five different parts, each with a somewhat different function. The large M. hypoglossus obliquus consists of two different parts and contributes to the formation of the tendinous envelope surrounding the lingual salivary gland (Glandula lingualis). The M. supraglossus is present only in parrots and is derived from the M. hypoglossus obliquus. It is an extraordinarily complex muscle with two muscular and five, roughly distinguishable, tendinous parts. The M. hypoglossus anterior is strongly developed and forms a sling passing around the apical end of the paraglossale to attach to the ventro-apical tip of the basihyale. The M. hypoglossus transversus has not been described previously. It was found in only one specimen and is presumably derived from the M. hypoglossus anterior. The M. mesoglossus occurs only in parrots. It does not attach to bony skeletal elements, but to the aponeurosis of the M. hypoglossus anterior, to the lingual nail, and to the connective tissue supporting the central area of the dorsal lingual epithelium. It is derived from the M. hypoglossus anterior.

Synovial bursae are described for the first time for an avian lingual apparatus. They are found at pressure points between skeletal ridges and overcrossing tendons. The Bursa synovialis urohyalis lies between the urohyale and the Nodulus. The paired Bursae synoviales paraglossales lie between the apical border of the Synchondrosis paraglossalis and the tendon of the M. hypoglossus anterior.

The epithelium covering the lingual surface is differentiated into various specialized structures depending on its location on the tongue. One of the most striking structures is the lingual nail, a heavily keratinized cuticle on the lateroventral surface of the lingual tip, which interacts with the underlying vascular cavernous tissue.

Surface structures of the tongue, most of which have already been described by Homberger (1980a), were correlated with the underlying tissues and other internal structures. One of the most complex surface structures is found on the dorsal surface of the lingual tip which is differentiated into a central area and the flanking V-shaped mound. The shape of these surface structures can be modified by the underlying muscles and cavernous vascular tissue.

Corpora cavernosa (cavernous vascular tissue) underlie the lateral parts and the extreme end of the lingual tip. They act as "active hydraulic structures" and interact with the M. supraglossus, M. hypoglossus anterior, and M. mesoglossus in broadening, flattening, or narrowing the lingual tip.

Three salivary glands, the Gl. lingualis, Gl. sublingualis, and Gl. mandibularis, are present in the psittacine lingual apparatus. As an exception among parrots, a Gl. praeglottalis is not present in *Psittacus*. The glandular bodies can function as hydrostatic skeletal elements or as cushions between two contracting muscles when external forces act upon them.

A method was formulated for the functional analysis of complex systems such as the lingual apparatus. It is based on a synthesis of anatomical, physiological, and mechanical data. This approach leads to the construction of a theoretical mechanical model with which the accuracy of the anatomical description can be evaluated and which explains the functioning of the lingual apparatus as a result of the interactions of the constituent structural elements. The model itself can be tested by testing experimentally certain predictions made by the model. With this approach to functional analysis, the processes of hypothesis formulation and hypothesis testing are clearly separated into two steps.

Movements of the hyoid relative to the mandible are brought about by coordinated contractions of the extrinsic lingual muscles and the intrinsic M. ceratohyoideus. These movements are linked with an up-down rotation of the basihyale.

Movements of the epibranchiale relative to the ceratobranchiale are controlled by the M. branchiomandibularis during protraction of the hyoid, but are passive during retraction.

Movements of the paraglossale relative to the basihyale are based on coordinated contractions of the M. ceratoglossus, M. hypoglossus obliquus and M. supraglossus. Movements of the paraglossale have to be synchronized with those of the hyoid apparatus relative to the mandible.

The shape and relief of the dorsal epithelial surface of the lingual tip can be modified through coordinated contractions of the M. supraglossus, M. hypoglossus anterior, M. hypoglossus transversus, and M. mesoglossus, in cooperation with the Corpora cavernosa. These surface modifications must also be synchronized with movements of the paraglossale relative to the basihyale and with those of the hyoid relative to the mandible.

This study of the lingual apparatus of *Psittacus erithacus* was intended to serve several functions: as a reference for future comparative anatomical studies on the avian lingual apparatus, as a basis for evolutionary and adaptive studies on the psittacine and avian feeding apparatus, and as an example of a theoretical functional analysis of a complex apparatus in which a mechanical model that can be tested by using experimental methods is constructed.

German: Zusammenfassung

Die vorliegende Untersuchung des Zungenapparates des Graupapageis, *Psittacus erithacus,* setzt sich aus zwei Teilen zusammen, nämlich aus einem beschreibenden Teil, der frühere anatomische Beschreibungen des psittacinen Zungenapparates berichtigt und bedeutend ergänzt, und einem funktionell-analytischen Teil, in welchem ein theoretisches, mechanisches Modell entwickelt wird, das die Funktionen und Interaktionen der verschiedenen Bauteile des Zungenapparates erklärt.

Der beschreibend-anatomische Teil behandelt nur die mechanisch wirkenden Bauteile, wie knöcherne und knorpelige Skeletteile, Bänder, Faszien, Muskeln, Schleimbeutel, epitheliale Strukturen, Schwellgewebe und Speicheldrüsen. Besondere Aufmerksamkeit wurde der strukturellen Variabilität der einzelnen Bauteile geschenkt. Um die künftige Anwendung der hier geprägten anatomischen Nomenklatur zu erleichtern, wurde für jede neu beschriebene Struktur ein lateinischer Name vorgeschlagen. Soweit wie möglich wurden bereits bestehende anatomische Namen für die beschriebenen Strukturen übernommen und deren Synonyme angegeben.

Das Zungenbein besteht aus einem unpaaren Basihyale, das mit dem ebenfalls unpaaren Urohyale verwachsen ist, und aus den paarigen Epibranchialia, Ceratobranchialia und Paraglossalia. Knorpelige Fortsätze befinden sich an den kaudalen Enden des Urohyale und der Epibranchialia. Eine knorpelige Brücke (Synchondrosis paraglossalis) verbindet die apikalen Enden der Paraglossalia. Die Gelenke des Zungenbeins (Paraglossobasihyalgelenk, Ceratobasihyalgelenk und Epiceratobranchialgelenk) sind echte Gelenke (Diarthrosen). Das Oberflächenrelief der Gelenkflächen bestimmt, zusammen mit den Bändern, die Richtung und das Ausmass der möglichen Bewegungen der Gelenke. Die Ansatzflächen für die Muskeln und Bänder an den knöchernen Skeletteilen sind oft durch Kämme, Rauhheiten oder Vertiefungen auf der Knochenoberfläche gekennzeichnet.

Der Unterkiefer und das Kehlkopf- und Trachealskelett sind nur soweit beschrieben wie sie für die Beschreibung des Zungenapparates von Bedeutung sind.

Die Bänder des Zungenbeinapparates können in drei Kategorien eingestuft werden. Die insgesamt sechs intraossealen Bänder verbinden jeweils zwei verschiedene Regionen desselben Skelettelementes und dienen der Verbindung verschiedener Gewebe oder als Ansatzstellen für Muskeln. Die insgesamt vier Gelenkbänder verbinden jeweils zwei verschiedene Skelettelemente, die gelenkig miteinander verbunden sind, und schränken die möglichen Bewegungen der Gelenke ein. Kupplungsbänder sind nur durch das Ligamentum nodulo-ceratobranchiale vertreten, das einen Sesamknochen, den Nodulus, an der Stelle, wo es die Ventralseite des Urohyale überquert, enthält. Dieses Band überspannt beide Ceratobasihyalgelenke und koppelt auf diese Art die Bewegungen der Ceratobranchialia mit denjenigen des Basihyale. Ausser den typischen Bändern sind noch sechs verschiedene Faszien vorhanden, die verschiedenen Aufgaben dienen, z.B. dem Umhüllen von Strukturen oder der Verbindung verschiedener Strukturen. Die Fascia vaginalis hyoidei bildet eine bindegewebige Scheide, in der die Zungenbeinhörner, die aus den Ceratobranchialia und Epibranchialia bestehen, während des Vor- und Rückziehens des Zungenbeinapparates vor- und rückwärts gleiten.

Die Zungenmuskulatur besteht aus sechs extrinsischen und sieben intrinsischen Muskelpaaren. Deren Beschreibung ist in einer standartisierten Weise in mit Untertiteln versehene Abschnitte unterteilt, um den Vergleich zwischen den einzelnen Muskeln zu erleichtern. Jede Muskelbeschreibung ist von einer Synonymieliste und von einer vergleichenden Aufstellung der funktionellen Interpretationen früherer Autoren begleitet. Die Zungenmuskeln des Graupapageis unterscheiden sich in mannigfaltiger Weise von denen anderer Vögel. Die hervorstechendsten Merkmale der psittacinen Zungenmuskeln sind im folgenden zusammengefasst aufgeführt. Der M. serpihyoideus entspringt dem kaudo-ventralen Ende des Unterkieferastes und setzt am apikalen Ende des Ceratobranchiale, am Lig. nodulo-ceratobranchiale und entlang einer sagittalen Raphe an. Der M. serpihyoideus und der M. stylohyoideus verschmelzen in der Nähe ihres gemeinsamen Ursprungs. Der M. branchiomandibularis ist stark entwickelt und besteht aus drei Teilen. Er entspringt dem Vorderende des Unterkieferastes und setzt am Epibranchiale und kaudalen Ende des Ceratobranchiale an. Der kräftige M. genioglossus verbindet die Paraglossalia mit der Unterkiefersymphyse. Der M. tracheohyoideus entspringt dem Brustbein und setzt am Basihyale und Ceratobranchiale an. Er ist deutlich vom M. tracheolateralis zu unterscheiden; eine korrekte Homologisierung der beiden Muskeln mit den in der Literatur beschriebenen Zungen- und Trachealmuskeln verschiedener Vögel ist vorgeschlagen worden. Der M. ceratohyoideus unterscheidet sich kaum von demjenigen anderer Vögel. Er entspringt dem Ceratobranchiale und setzt am Nodulus an. Der M. ceratoglossus ist äusserst komplex und besteht aus bis zu fünf verschiedenen Teilen mit je verschiedener Funktion. Er setzt sich sehnig am Paraglossale an und entspringt dem Lig. nodulo-ceratobranchiale und verschiedenen Stellen am Vorderende des Ceratobranchiale. Der M. hypoglossus obliquus besteht aus zwei Teilen und trägt zur Bildung der sehnigen Hülle der paarigen Zungendrüse (Glandula lingualis) bei. Er entspringt dem Ceratobranchiale und Basihyale und setzt am Paraglossale an. Der M. supraglossus ist nur bei Papageien vorhanden und ist vom M. hypoglossus obliquus abgeleitet. Er ist ein ausserordentlich komplexer Muskel mit zwei paarigen Muskelportionen und einem sehnigen Teil, in dem fünf verschiedene Sehnengruppen unterschieden werden können. Der Muskelteil entspringt dem Ceratobranchiale und Basihyale, und die sehnigen Abschnitte setzen am Paraglossale, an der epithelialen Zungenoberfläche und an der Aponeurosis des M. hypoglossus anterior an. Der kräftig ausgebildete M. hypoglossus anterior entspringt dem Paraglossale und bildet eine Schlinge, die um die Vorderkante der Synchondrosis paraglossalis zieht, um an der ventro-apikalen Spitze des Basihyale anzusetzen. Der M. hypoglossus transversus, der die beiden Paraglossalhälften verbindet, ist hier zum ersten Mal beschrieben. Er wurde nur in einem von sieben Exemplaren gefunden und ist wahrscheinlich vom M. hypoglossus anterior abgeleitet. Der M. mesoglossus kommt nur bei Papageien vor. Er setzt nicht an knöchernen oder knorpeligen Skeletteilen an, sondern entspringt der Aponeurosis des M. hypoglossus anterior und dem Zungennagel (siehe unten) und setzt am Bindegewebe, das die Schleimhautoberfläche der Zungenspitze unterlagert, an. Er leitet sich vom M. hypoglossus anterior ab.

Schleimbeutel (Bursae synoviales) sind hier zum ersten Mal in einer Vogelzunge beschrieben. Sie befinden sich an Druckstellen zwischen Skelettkanten und darüberziehenden Sehnen. Die unpaare Bursa synovialis urohyalis liegt zwischen dem Urohyale und Nodulus, und die paarige Bursa synovialis paraglossalis liegt zwischen der Vorderkante der Synchondrosis paraglossalis und der Sehne des M. hypoglossus anterior.

Das Zungenepithel hat sich regional in verschiedene Spezialstrukturen weiterdifferenziert. Die auffälligste dieser Strukturen ist der Zungennagel, der als stark verhorntes Häutchen die Unter- und Seitenflächen der Zungenspitze hufeisenartig umfasst und der mit dem unterlagernden Schwellgewebe zusammenwirkt (siehe unten).

Die Oberflächenstrukturierung der Zunge, die zum grossen Teil bereits von Homberger (1980a) beschrieben worden war, ist mit der inneren Struktur der Zunge in Beziehung gebracht worden. Eine der kompliziertesten Oberflächenstrukturierungen befindet sich auf der Dorsalseite der Zungenspitze.

Diese hat sich in eine von einem lippenartigen, V-förmigen Wulst eingerahmte, zentrale Fläche differenziert, deren Form und Relief von den unterlagernden Muskeln und Schwellgeweben verändert werden kann.

Schwellgewebe (Corpora cavernosa) sind am Vorderende der Zungenspitze und seitlich zwischen den Paraglossalia und dem Zungennagel gelegen. Sie funktionieren als "aktive hydraulische Strukturen" und wirken mit dem M. supraglossus, M. hypoglossus anterior und M. mesoglossus zusammen, wenn die Zungenspitze verbreitert, verschmälert oder verflacht wird.

Drei verschiedene Speicheldrüsen, nämlich die paarige Zungendrüse (Glandula lingualis), die Unterzungendrüse (Glandula sublingualis) und die Unterkieferdrüse (Glandula mandibularis), sind im Zungenapparat des Graupapageis enthalten. Als Ausnahme unter den Papageien besitzt der Graupapagei keine unpaare Zungendrüse (Glandula praeglottalis). Die Speicheldrüsen sind passive hydrostatische Strukturen. Wenn äussere Kräfte auf sie einwirken, können sie als hydrostatische Skelettelemente (z.B. paarige Zungendrüse und Unterzungendrüse) oder als Polster zwischen zwei sich kreuzende kontrahierende Muskeln (z.B. Unterkieferdrüse und Unterzungendrüse) wirken.

Für die funktionelle Analyse des Zungenapparates ist eine Methode entwickelt worden, die auf der Synthese von anatomischen, physiologischen und biomechanischen Daten beruht. Diese Methode führt zur Konstruktion eines theoretischen mechanischen Modells, mit dem die Richtigkeit und Vollständigkeit der anatomischen Bestandesaufnahme überprüft werden kann und welches das Funktionieren des Zungenapparates als ein Zusammenwirken seiner Bestandteile erklärt. Das Modell selbst kann geprüft werden, indem gewisse Voraussagen des Modells experimentell geprüft werden. Mit dieser Methode wird der logische Ablauf der Hypothesenformulation und Hypothesenprüfung in zwei klar getrennte Schritte aufgeteilt.

Die Bewegungen des Zungenbeins gegenüber dem Unterkiefer unterliegen der Kontrolle der extrinsischen Zungenmuskeln und des intrinsischen M. ceratohyoideus. Diese Bewegungen sind mit einer Vertikalrotation des Basihyale gekoppelt.

Die Epibranchialbewegungen gegenüber dem Ceratobranchiale werden während der Protraktion des Hyoids vom M. branchiomandibularis kontrolliert und verlaufen während der Retraktion des Hyoids passiv.

Die Paraglossalbewegungen gegenüber dem Basihyale unterliegen der Kontrolle des M. ceratoglossus, M. hypoglossus obliquus und M. supraglossus. Die Paraglossalbewegungen müssen mit den Hyoidbewegungen synchronisiert werden.

Die Form der Zungenspitze und das Relief der Zungenspitzenoberfläche können durch koordinierte Kontraktionen des M. supraglossus, M. hypoglossus anterior, M. hypoglossus transversus und M. mesoglossus, in Zusammenarbeit mit dem Schwellgewebe, modifiziert werden. Die Form- und Reliefveränderungen der Zungenspitze müssen ebenfalls mit den Paraglossal- und Hyoidbewegungen synchronisiert werden.

Die vorliegende Studie des Graupapageis erfüllt mehrere Zwecke. Sie kann als Referenz für zukünftige, vergleichend-anatomische Untersuchungen der Vogelzunge und als Basis für Studien zur Evolution und adaptiven Radiation des Nahrungsaufnahmeapparates der Papageien und anderer Vögel dienen. Sie kann auch als Beispiel einer theoretischen funktionellen Analyse eines komplexen biomechanischen Apparates dienen, mit der ein experimentell prüfbares mechanisches Modell konstruiert werden kann, welches das Funktionieren des Apparates erklärt.

LITERATURE CITED

ALEXANDER, R. MCN. 1968. Animal mechanics. Univ. of Washington Press, Seattle.

ANTONY, M. 1920. Ueber die Speicheldrüsen der Vögel. Zool. Jahrb. Anat. 41:547–660.

BARGMANN, W. 1962. Histologie und mikroskopische Anatomie des Menschen, 4th ed. Thieme Verlag, Stuttgart.

BARNIKOL, A. 1951. Ueber einige Gesetzmässigkeiten im Aufbau der Trigeminusäste bei Vögeln. Anat. Anz. 98:217–223.

BASMAJIAN, J. V. 1978. Muscles alive. Their functions revealed by electromyography. Williams and Wilkins, Baltimore.

BASMAJIAN, J. V. 1980. Electromyography—dynamic gross anatomy: A review. Amer. J. Anat. 159: 245–260.

BAUMEL, J. J. 1975. Aves nervous system. Pp. 2019–2062, *In* R. Getty (ed.), Sisson and Grossman's The Anatomy of the Domestic Animals, Vol. 2, 5th ed. Saunders, Philadelphia.

BAUMEL, J. J. 1979. Osteologia. Pp. 53–121, *In* J. J. Baumel, A. S. King, A. M. Lucas, J. E. Breazile, and H. E. Evans (eds.), Nomina Anatomica Avium. Academic Press, London.

BAUMEL, J. J., A. S. KING, A. M. LUCAS, J. E. BREAZILE, AND H. E. EVANS (EDS.). 1979. Nomina anatomica avium. Academic Press, New York.

BEAMENT, J. W. L. 1960. Physical models in biology. Pp. 83–101, *In* J. W. L. Beament (ed.), Models and Analogues in Biology. Symp. Soc. Exp. Biol. No. 14. Cambridge Univ. Press, Cambridge.

BHATTACHARYYA, B. N. 1980. The morphology of the jaw and tongue musculature of the common pigeon, *Columba livia,* in relation to its feeding habit. Proc. Zool. Soc., Calcutta 31:95–127 (1978).

BLOOM, W., AND D. W. FAWCETT. 1975. A textbook of histology. Saunders, Philadelphia.

BOCK, W. J. 1961. Salivary glands in the Gray Jays (*Perisoreus*). Auk 78:355–365.

BOCK, W. J. 1972. Morphology of the tongue apparatus of *Ciridops anna* (Drepanididae). Ibis 114: 61–78.

BOCK, W. J. 1974. The avian skeletomuscular system. Pp. 119–257, *In* D. S. Farner and J. R. King (eds.), Avian Biology, Vol. 4. Academic Press, New York.

BOCK, W. J. 1978. Tongue morphology and affinities of the Hawaiian Honeycreeper *Melamprosops phaeosoma.* Ibis 120:467–479.

BOCK, W. J., AND R. S. HIKIDA. 1969. Turgidity and function of the hatching muscle. Amer. Midl. Natur. 81:99–106.

BOCK, W. J., AND H. MORIOKA. 1971. Morphology and evolution of the ectethmoidmandibular articulation in the Meliphagidae (Aves). J. Morph. 135:13–50.

BOCK, W. J., AND J. J. MORONY. 1978a. Relationships of the passerine finches (Passeriformes: Passeridae). Bonn. Zool. Beitr. 29:122–147.

BOCK, W. J., AND J. J. MORONY. 1978b. The preglossale of *Passer* (Aves: Passeriformes)—A skeletal neomorph. J. Morph. 155:99–110.

BOCK, W. J., AND C. R. SHEAR. 1972. A staining method for gross dissection of vertebrate muscles. Anat. Anz. 130:222–227.

BOCK, W. J., AND G. VON WAHLERT. 1965. Adaptation and the form-function complex. Evolution 19:269–299.

BOCK, W. J., AND H. WINKLER. 1978. Mechanical analysis of the external forces on climbing mammals. Zoomorphologie 91:49–61.

BOCK, W. J., R. P. BALDA, AND S. B. VANDER WALL. 1973. Morphology of the sublingual pouch and tongue musculature in Clark's Nutcracker. Auk 90:491–519.

BODINE, S. C., R. R. ROY, D. A. MEADOWS, R. F. ZERNICKE, R. D. SACKS, M. FOURNIER, AND V. R. EDGERTON. 1982. Architectural, histochemical, and contractile characteristics of a unique biarticular muscle: the cat semitendinosus. J. Neurophysiol. 48:192–201.

BONIK, K., M. GRASSHOFF, W. F. GUTMANN, AND W. MAIER. 1978. Hydraulik als Grundlage der Morphologie aller tierischen Lebewesen. Natur u. Museum 108:162–174.

BOULLION, T. L., AND D. G. HOMBERGER. 1982. The lingual apparatus of *Gallus gallus.* Am. Zool. 22:943.

BRACKENBURY, J. 1980. Control of sound production in the syrinx of the fowl *Gallus gallus.* J. Exp. Biol. 85:239–251.

BREDER, C. M. 1947. An analysis of the geometry of symmetry with especial reference to the squamation of fishes. Bull. Am. Mus. Nat. Hist. 88:327–412.

BROWN, R. H. J. 1960. Mechanical models in zoology. Pp. 69–82, *In* J. W. L. Beament (ed.), Models and Analogues in Biology. Symp. Soc. Exp. Biol. No. 14. Cambridge Univ. Press, Cambridge.

BRUSH, A. H. 1980. Chemical heterogeneity in keratin proteins of avian epidermal structures: possible relations to structure and function. Pp. 87–109, *In* R. I. C. Spearman and P. A. Riley (eds.), The Skin of Vertebrates. Academic Press, London.

BÜHLER, P. 1970. Schädelmorphologie und Kiefermechanik der Caprimulgidae (Aves). Z. Morph. Tiere 66:337–399.

BÜHLER, P. 1980. Zur Methodik funktions-morphologischer Untersuchungen. Pp. 185–189, *In* R. Nöhring (ed.), Acta XVII Congr. Internat. Ornithol. Deutsche Ornithologen-Gesellschaft, Berlin.

BÜHLER, P. 1981. Functional anatomy of the avian jaw apparatus. Pp. 439–468, *In* A. S. King and J. McLelland (eds.), Form and Function of Birds, Vol. II. Academic Press, London.

BURTON, P. J. K. 1974a. Feeding and feeding apparatus in waders. Brit. Mus. (Nat. Hist.), London.

BURTON, P. J. K. 1974b. Anatomy of head and neck in the Huia (*Heteralocha acutirostris*) with comparative notes on other Callaeidae. Bull. Brit. Mus. (Nat. Hist.) Zool. 27:3–48.

BURTON, P. J. K. 1974c. Jaw and tongue features in Psittaciformes and other orders with special reference to the anatomy of the Tooth-billed Pigeon (*Didunculus strigirostris*). J. Zool. Lond. 174:255–276.

CARLSON, F. D., AND D. R. WILKIE. 1974. Muscle physiology. Prentice-Hall, Englewood Cliffs, New Jersey.

CARLSÖÖ, S., AND S. MOLBECH. 1966. The functions of certain two-joint muscles in a closed muscular chain. Acta Morphol. Neerl. Scand. 6:368–377.

CHAINE, J. 1904. Remarques sur la musculature de la langue des oiseaux. C. R. Soc. Biol. Paris:991–992.

CHAINE, J. 1905. La langue des oiseaux. Etude de myologie comparative. Bull. Sci. Fr. Belg. 39:487–504.

CHAMBERLAIN, F. W. 1943. Atlas of avian anatomy. Michigan State Coll. Agr. Exp. Station, East Lansing.

CHAPMAN, G. 1975. Versatility of hydraulic systems. J. Exp. Zool. 194:249–270.

CHOLODKOWSKY, N. 1892. Zur Kenntnis der Speicheldrüsen der Vögel. Zool. Anz. 15:250–254.

CLARK, R. B. 1964. Dynamics in metazoan evolution. The origin of the coelom and segments. Clarendon Press, Oxford.

CRACRAFT, J. 1971. The functional morphology of the hind limb of the domestic pigeon, *Columba livia*. Bull. Am. Mus. Nat. Hist. 144:171–268.

CROWNINSHIELD, R. D., AND R. D. BRAND. 1981. The prediction of forces in joint structures: distribution of intersegmented resultants. Exercise Sport Sciences Reviews 9:159–181.

DEMPSTER, W. T. 1961. Free-body diagrams as an approach to the mechanics of human posture and motion. Pp. 81–135, *In* F. G. Evans (ed.), Biomechanical Studies of the Musculoskeletal System. Thomas, Springfield, Illinois.

DENKER, A. 1907. Das Gehörorgan und die Sprechwerkzeuge der Papageien. Bergmann, Wiesbaden.

DUBALE, M. S., AND U. M. RAWAL. 1965. A morphological study of the cranial muscles associated with the feeding habit of *Psittacula krameri* (Scopoli). Pavo 3:1–13.

DU BRUL, E. L. 1976. Biomechanics of speech sounds. Pp. 631–642, *In* S. R. Harnad, H. D. Steklis, and J. Lancaster (eds.), Origins and Evolution of Language and Speech. Ann. N.Y. Acad. Sci. Vol. 200. N.Y. Acad. Sci., New York.

DULLEMEIJER, P. 1974. Concepts and approaches in animal morphology. Van Gorcum, Assen.

DULLEMEIJER, P. 1980. Functional morphology and evolutionary biology. Acta Biotheor. 29:151–250.

DUVERNOY, G. L. 1835. Mémoire sur quelques particularités des organes de la déglutition de la classe des oiseaux et des reptiles. Mém. Soc. Nat. Strasbourg 2:1–24.

DYM, M. 1983. The male reproductive system. Pp. 1000–1053, *In* L. Weiss (ed.), Histology. Elsevier Biomedical, New York, Amsterdam, Oxford.

EDGEWORTH, F. H. 1935. The cranial muscles of vertebrates. Cambridge Univ. Press, Cambridge.

ENGELS, W. L. 1938. Tongue musculature of passerine birds. Auk 55:642–650.

EVANS, H. E. 1969. Anatomy of the Budgerigar. Pp. 45–112, *In* M. L. Petrak (ed.), Diseases of Cage and Aviary Birds. Lea and Febiger, Philadelphia.

FAHRENHOLZ, C. 1937. Drüsen der Mundhöhle. Pp. 115–210, *In* L. Bolk, E. Göppert, E. Kallius, and W. Lubosch (eds.), Handbuch der Vergleichenden Anatomie der Wirbeltiere, Vol. 3. Urban und Schwarzenberg, Berlin, Wien. Reprinted 1967: Asher, Amsterdam.

FÄHRMANN, W. 1978. Zur Histochemie und Ultrastruktur der mittleren Zungendrüsen des Haussperlings (*Passer domesticus* L.). Zool. Jahrb., Abt. Anat. Ontog. Tiere 99:468–482.

FEDER, F. H. 1969. Beitrag zur makroskopischen und mikroskopischen Anatomie des Verdauungsapparates beim Wellensittich. Anat. Anz. 125:232–255.

FISHER, H. I. 1956. The landing forces of pigeons. Auk 73:87–105.

FISHER, H. I., AND D. C. GOODMAN. 1955. The myology of the Whooping Crane, *Grus americana*. Ill. Biol. Monogr. 24:1–127.

FITZGERALD, T. C. 1969. The *Coturnix* quail. Anatomy and histology. Iowa State Univ. Press, Ames.

FOELIX, R. F. 1970. Vergleichend-morphologische Untersuchungen an den Speicheldrüsen körnerfressender Singvögel. Zool. Jahrb. Anat. 87:523–587.

FORSHAW, J. M. 1978. Parrots of the world, 2nd ed. Landsdowne, Melbourne.

FRASER, R. D. B., AND T. P. MACRAE. 1980. Molecular structure and mechanical properties of keratin. Pp. 211–246, *In* The Mechanical Properties of Biological Materials. Symp. Soc. Exp. Biol. No. 34. Cambridge Univ. Press, Cambridge.

FUJIOKA, T. 1963. Comparative and topographical anatomy of the fowl. IV. On the origins and insertions of muscles of the head and neck in the fowl. Part I. Muscles of the head. Jap. J. Vet. Sci. 25:219–226.

FUNG, Y. C. 1981. Biomechanics. Mechanical properties of living tissues. Springer-Verlag, New York, Heidelberg, Berlin.

FÜRBRINGER, M. 1888. Untersuchungen zur Morphologie und Systematik der Vögel, zugleich ein Beitrag zur Anatomie der Stütz- und Bewegungsorgane. T. J. van Holkema, Amsterdam.

FÜRBRINGER, M. 1922. Das Zungenbein der Wirbeltiere insbesondere der Reptilien und Vögel. Pp. 79–109, *In* Abhandl. Heidelberg. Akad. Wissensch., B, No. 11.

GADOW, H. 1891. Vögel. Pp. 294–325, *In* Bronn's Klassen und Ordnungen des Thier-Reichs, Vol. 6. C. F. Winter, Leipzig.

GANS, C. 1967. The chamaeleon. Natural History 76(4):52–59.

GANS, C. 1974. Biomechanics. An approach to vertebrate biology. Lippincott, Philadelphia.

GANS, C. 1982. Fiber architecture and muscle function. Exercise Sport Sciences Reviews 10:160–207.

GANS, C., AND W. J. BOCK. 1965. The functional significance of muscle architecture—a theoretical analysis. Ergeb. Anat. Entwicklungsgesch. 38:115–142.

GANS, C., AND G. C. GORNIAK. 1982a. How does the toad flip its tongue? Test of two hypotheses. Science 216:1335–1337.

GANS, C., AND G. C. GORNIAK. 1982b. Functional morphology of lingual protrusion in marine toads (*Bufo marinus*). Am. J. Anat. 163:195–222.

GAUNT, A. S., AND S. L. L. GAUNT. 1977. Mechanics of the syrinx in *Gallus gallus.* II. Electromyographic studies of ad libitum vocalizations. J. Morph. 152:1–20.

GEORGE, J. C., AND A. J. BERGER. 1966. Avian myology. Academic Press, New York.

GIEBEL, C. 1858. Die Zunge der Vögel und ihr Gerüst. Zeitschr. f. Naturwiss. 11:19–51.

GIEBEL, C. 1862. Zur Anatomie der Papageien, nach Chr. Nitzsch's Untersuchungen mitgetheilt. Zeitschr. Ges. Naturw. 19:133–152.

GOODMAN, D. C., AND H. I. FISHER. 1962. Functional anatomy of the feeding apparatus in waterfowl. Southern Illinois Univ. Press, Carbondale.

GRESCHIK, E. 1913. Histologische Untersuchungen der Unterkieferdrüse (Glandula mandibularis) der Vögel. Aquila 20:331–371.

GRESCHIK, E. 1921. Die Zunge von *Plegadis falcinellus* L. Aquila 28:90–96.

GRESCHIK, E. 1928. Morphologische Studien an der Zunge einiger Tagraubvögel aus den Gattungen *Accipiter, Buteo* und *Circus.* Ann. Musei Nationalis Hungarici 25:39–59.

GRIFFITHS, T. A. 1978. Muscular and vascular adaptation for nectar-feeding in the glossophagine bats *Monophyllus* and *Glossophaga.* J. Mammal. 59:414–418.

GROEBBELS, F. 1932. Der Vogel, Vol. 1. Bornträger, Berlin.

GÜNTERT, M. 1981. Morphologische Untersuchungen zur adaptiven Radiation des Verdauungstraktes bei Papageien (Psittaci). Zool. Jahrb. Anat. 106:471–526.

GÜNTERT, M., AND V. ZISWILER. 1972. Konvergenzen in der Struktur von Zunge und Verdauungstrakt nektarfressender Papageien. Rev. Suisse Zool. 79:1017–1026.

GUTMANN, W. F. 1972. Die Hydroskelett-Theorie. Aufsätze u. Reden Senckenberg. Naturf. Ges. No. 21. Waldemar Kramer, Frankfurt am Main.

GUTMANN, W. F. 1975. Das Schuppenhemd der niederen Wirbeltiere und seine mechanische Bedeutung. Natur u. Museum 105:169–185.

GUTMANN, W. F. 1981. Die praktische Bedeutung von Theorien in der Wissenschaft. Pp. 27–37 *In* D. Starck, K. Fiedler, P. Harth, and J. Richter (eds.), Biologie. Verlag Chemie, Weinheim, Deerfield Beach, Basel.

HARGRAVE, L. L. 1970. Mexican macaws. Comparative osteology and survey of remains from the Southwest. Anthropological Papers Univ. Arizona No. 20, 67 pp.

HEIDRICH, K. 1908. Die Mund-Schlundkopfhöhle der Vögel und ihre Drüsen. Morphol. Jahrb. 37: 10–69.

HERRING, S. W., A. F. GRIMM, AND B. R. GRIMM. 1979. Functional heterogeneity in a multipinnate muscle. Amer. J. Anat. 154:563–576.

HODGES, R. D. 1974. The histology of the fowl. Academic Press, New York.

HOLYOAK, D. T. 1973. Comments on taxonomy and relationships in the parrot subfamilies Nestorinae, Loriinae and Platycercinae. Emu 73:157–176.

HOMBERGER, D. G. 1978. Functional morphology of the parrot tongue (*Psittacus erithacus*). Am. Zool. 18:623.

HOMBERGER, D. G. 1979a. Review of "Parrots of the world," 2nd ed., by J. M. Forshaw. Auk 96: 829–830.

HOMBERGER, D. G. 1979b. Functional morphology of the larynx in the parrot *Psittacus erithacus*. Am. Zool. 19:988.

HOMBERGER, D. G. 1980a. Funktionell-morphologische Untersuchungen zur Radiation der Ernährungs- und Trinkmethoden der Papageien (Psittaci). Bonn. Zool. Monographien No. 13, 192 pp.

HOMBERGER, D. G. 1980b. Functional morphology and evolution of the feeding apparatus in parrots, with special reference to the Pesquet's Parrot, *Psittrichas fulgidus* (Lesson). Pp. 471–485, *In* R. F. Pasquier (ed.), Conservation of New World Parrots. International Council for Bird Preservation Technical Paper No. 1. Smithsonian Institution Press, Washington, D.C.

HOMBERGER, D. G. 1980c. Review of "Nomina anatomica avium" by J. J. Baumel, A. S. King, A. M. Lucas, J. E. Breazile, and H. E. Evans (eds.). Auk 97:912–914.

HOMBERGER, D. G. 1981. Morphological foundations of the bill honing behavior in parrots (Psittaci). Am. Zool. 21:1039.

HOMBERGER, D. G. 1982. Hydraulic structures in the avian lingual apparatus. Am. Zool. 22:943.

HOMBERGER, D. G. 1983. Nonadaptive evolution of avian drinking methods. Am. Zool. 23:894.

HOMBERGER, D. G. 1985. Parrot. Pp. 437–439, *In* B. Campbell and E. Lack (eds.), A Dictionary of Birds. British Ornithologists' Union, London.

HOMBERGER, D. G., AND A. H. BRUSH. 1986. Functional-morphological and biochemical correlations in the keratinized structures of the African Grey Parrot (*Psittacus erithacus* Linné). Zoomorphology: in press.

HOMBERGER, D. G., AND V. ZISWILER. 1972. Funktionell-morphologische Untersuchungen am Schnabel von Papageien. Rev. Suisse Zool. 79:1038–1048.

JANSEN, D. W., AND R. C. FOEHRING. 1983. The mechanism of venom secretion from Duvernoy's gland of the snake *Thamnophis sirtalis*. J. Morph. 175:271–277.

KALLIUS, E. 1906. Beiträge zur Entwickelung der Zunge. II. Teil. Vögel. 3. *Melopsittacus undulatus*. Anat. Hefte 31:603–651.

KASAI, N. 1957. A comparative anatomical study on the tongue muscles of birds. Acta Med. Fukuoka 27:1403–1431.

KASTELIC, J., AND E. BAER. 1980. Deformation in tendon collagen. Pp. 397–435, *In* The Mechanical Properties of Biological Materials. Symp. Soc. Exp. Biol. No. 34. Cambridge Univ. Press, Cambridge.

KATZ, J. L. 1980. The structure and biomechanics of bone. Pp. 137–168, *In* The Mechanical Properties of Biological Materials. Symp. Soc. Exp. Biol. No. 34. Cambridge Univ. Press, Cambridge.

KIER, W. M., AND K. K. SMITH. 1983. The biomechanics of movement in muscular-hydrostats. Am. Zool. 24:904.

KILHAM, L. 1979. Chestnut-colored woodpeckers feeding as a pair on ants. Wilson Bull. 91:149–150.

KING, A. S. 1975. Aves respiratory system. Pp. 1883–1918, *In* R. Getty (ed.), Sisson and Grossman's The Anatomy of the Domestic Animals, Vol. 2. Saunders, Philadelphia.

KING, A. S. 1979. Systema respiratorium. Pp. 227–265, *In* J. J. Baumel, A. S. King, A. M. Lucas, J. E. Breazile, and H. E. Evans (eds.), Nomina Anatomica Avium. Academic Press, London.

KRULIS, V. 1978. Struktur und Verteilung von Tastrezeptoren im Schnabel-Zungenbereich von Singvögeln, im besonderen der Fringillidae. Rev. Suisse Zool. 85:385–447.

KUTORGA, S. 1832. De organis vocis et loquelae Psittaci erithaci. Inaugural Dissertation. J. C. Schünmann, Dorpati Livonorum.

LEIBER, A. 1907. Vergleichende Anatomie der Spechtzunge. Zoologica, Stuttgart 20:1–79.

LOCKNER, F. R., AND O. M. YOUNGREN. 1976. Functional syringeal anatomy of the mallard. I. In situ electromyograms during ESB elicited calls. Auk 93:324–342.

LOMBARD, R. E., AND D. B. WAKE. 1976. Tongue evolution in the lungless salamanders, family Plethodontidae. I. Introduction, theory and a general model of dynamics. J. Morph. 148:265–286.

LOMBARD, R. E., AND D. B. WAKE. 1977. Tongue evolution in the lungless salamanders, family Plethodontidae. II. Function and evolutionary diversity. J. Morph. 153:39–79.

LOWERY, G. H., AND D. A. TALLMAN. 1976. A new genus and species of nine-primaried oscine of uncertain affinities from Peru. Auk 93:415–428.

LUBOSCH, W. 1933. Untersuchungen über die Visceralmusculatur der Sauropsiden. Der Untersuchungen über die Kaumuskulatur der Wirbeltiere 3. Teil. Morphol. Jahrb. 72:584–666.

LUCAS, A. M., AND P. R. STETTENHEIM. 1972. Avian anatomy—integument, Vol. 2. U.S. Dept. of Agric., Washington, D.C.

MCCONNELL, J., G. S. BENSON, AND W. A. SCHMIDT. 1982. The vasculature of the human penis: a reexamination of the morphological basis for the polster theory of erection. Anat. Rec. 203: 475–484.

MCLELLAND, J. 1965. The anatomy of the rings and muscles of the trachea of *Gallus domesticus.* J. Anat. 99:651–656.

MCLELLAND, J. 1968. The hyoid muscles of *Gallus gallus.* Acta Anat. 69:81–86.

MCLELLAND, J. 1975. Aves digestive system. Pp. 1857–1882, *In* R. Getty (ed.), Sisson and Grossman's The Anatomy of the Domestic Animals, Vol. 2. Saunders, Philadelphia.

MCLELLAND, J. 1979a. Systema digestiorum. Pp. 267–295, *In* J. J. Baumel, A. S. King, A. M. Lucas, J. E. Breazile, and H. E. Evans (eds.), Nomina Anatomica Avium. Academic Press, London.

MCLELLAND, J. 1979b. Digestive system. Pp. 69–181, *In* A. S. King and J. McLelland (eds.), Form and Function in Birds, Vol. I. Academic Press, London.

MEDWAY, L. 1962. The relation between the reproductive cycle, moult and changes in the sublingual salivary glands of the Swiftlet *Collocalia maxima* Hume. Proc. Zool. Soc. Lond. 138:305–315.

MERCER, E. H. 1981. The foundations of biological theory. Wiley, New York.

MIVART, ST. J. 1895. On the hyoid bone of certain parrots. Proc. Zool. Soc. Lond.:162–174.

MIVART, ST. J. 1896. On the hyoid bones of *Nestor meridionalis* and *Nanodes discolor.* Proc. Zool. Soc. Lond.:236–240.

MOLBECH, S. 1966. On the paradoxical effect of some two-joint muscles. Acta Morphol. Neerl. Scand. 6:171–178.

MOLLER, W. 1930. Ueber die Schnabel- und Zungenmechanik blütenbesuchender Vögel. I. Ein Beitrag zur Biologie des Blumenvogels. Biol. Generalis 6:651–726.

MOLLER, W. 1931. Ueber die Schnabel- und Zungenmechanik blütenbesuchender Vögel. II. Ein Beitrag zur Biologie des Blumenvogels. Biol. Generalis 7:99–154.

MUDGE, G. P. 1903. On the morphology of the tongue of parrots with a classification of the order, based upon the structure of the tongue. Trans. Zool. Soc. Lond. 16:211–278.

NACHTIGALL, W. 1971. Biotechnik. Quelle und Meyer, Heidelberg.

NAVAS, P., C. BUENO, J. PINERO, J. AIJON, AND J. L. LOPEZ-CAMPOS. 1980a. Histochemical analysis on salivary glands of *Columba columba*—anterior mandibular gland. Morf. Normal Patol., Sec. A. Histol. 4:525–531.

NAVAS, P., J. HIDALGO, C. BUENO, AND J. L. LOPEZ-CAMPOS. 1980b. Histochemical analysis of the lingual glands in *Columba columba.* Arch. Biol. 90:437–444.

NICKEL, R., A. SCHUMMER, E. SEIFERLE, W. G. SILLER, AND P. A. L. WRIGHT. 1977. Anatomy of the domestic birds. Paul Parey, Berlin, and Springer-Verlag, New York.

NOTTEBOHM, F. 1976. Phonation in the Orange-winged Amazon Parrot, *Amazona amazonica.* J. Comp. Physiol. 108A:157–170.

PORTMANN, A. 1950. Le tube digestif. Pp. 270–284, *In* P.-P. Grassé (ed.), Traité de Zoologie, Vol. 15: Oiseaux. Masson, Paris.

PREUSS, F., D. DONAT, AND G. LUCKHAUS. 1969. Funktionelle Studie über die Zunge der Hausvögel. Berl. Münch. Tierärztl. Wochenschr. 82:45–48.

RAWAL, U. M. 1970. A comparative account of the lingual myology of some birds. Proc. Indian Acad. Sci. 71:36–47.

RICHARDS, L. P., AND W. J. BOCK. 1973. Functional anatomy and adaptive evolution of the feeding apparatus of the Hawaiian Honeycreeper genus *Loxops* (Drepanididae). Ornithol. Monogr. No. 15, 173 pp.

ROSENBERG, H. I. 1967. Histology, histochemistry, and emptying mechanism of the venom glands of some elapid snakes. J. Morph. 123:133–150.

RUSSELL, A. P. 1975. A contribution to the functional analysis of the foot of the Tokay, *Gekko gecko* (Reptilia: Gekkonidae). J. Zool. Lond. 176:437–476.

RUSSELL, A. P. 1981a. Arteries of the antebrachium and manus of the Tokay (*Gekko gecko*) (Reptilia, Gekkonidae). Can. J. Zool. 59:573–582.

RUSSELL, A. P. 1981b. Descriptive and functional anatomy of the digital vascular system of the Tokay, *Gekko gecko*. J. Morph. 169:293–323.

SCHARNKE, H. 1931. Beitrag zur Morphologie und Entwicklungsgeschichte der Zunge der Trochilidae, Meliphagidae und Picidae. J. Ornithol. 79:425–491.

SHUFELDT, R. W. 1886. Osteology of *Conurus carolinensis*. J. Anat. Physiol. 20:407–425.

SHUFELDT, R. W. 1890. The myology of the Raven (*Corvus corax sinuatus*). Macmillan, New York.

SMITH, G. A. 1975. Systematics of parrots. Ibis 117:18–67.

SMITH, K. K. 1983. The morpholpogy and function of the tongue in lizards. Am. Zool. 24:1028.

STADTMÜLLER, F. 1938. Zunge, Mundhöhlenboden. Pp. 971–976, *In* L. Bolk, E. Göppert, E. Kallius, and W. Lubosch (eds.), Handbuch der Vergleichenden Anatomie der Wirbeltiere, Vol. 5. Urban und Schwarzenberg, Berlin. Reprinted 1967: Asher, Amsterdam.

STARCK, D. 1959. Neuere Ergebnisse der vergleichenden Anatomie und ihre Bedeutung für die Taxonomie erläutert an der Trigeminus-Muskulatur der Vögel. J. Ornithol. 100:47–59.

STEINBACHER, G. 1935. Zur Anatomie von *Micropsitta*. Ornith. Monatsber. 43:139–144.

STEINBACHER, G. 1951. Die Zungenborsten der Loris. Zool. Anz. 146:57–65.

STRAUS, W. L. 1946. The concept of nerve-muscle specificity. Biol. Rev. 21:75–91.

SUZUKI, M., AND S. NOMURA. 1975. Electromyographic studies on the deglutition movements in the fowl. Jap. J. Vet. Sci. 37:289–293.

THUET, M. J. 1838. Disquisitiones anatomicae psittacorum. Dissertatio Universitatis Turicensis. Orelli et Füssli, Turicum (Zürich).

TUCKER, V. A. 1975. The energetic cost of moving about. Am. Scientist 63:413–419.

VANDEN BERGE, J. C. 1975. Aves myology. Pp. 1802–1848, *In* R. Getty (ed.), Sisson and Grossman's The Anatomy of the Domestic Animals, Vol. II. Saunders, Philadelphia.

VANDEN BERGE, J. C. 1979. Myologia. Pp. 175–219, *In* J. J. Baumel, A. S. King, A. M. Lucas, J. E. Breazile, and H. E. Evans (eds.), Nomina Anatomica Avium. Academic Press, London.

VANDEN BERGE, J. C. 1982. Notes on the myology of the pelvic limb in Kiwi (*Apteryx*) and in other birds. Auk 99:309–315.

VINCENT, J. F. V. 1982. Structural biomaterials. Halstead Press, New York.

WAINWRIGHT, S. A., W. D. BIGGS, J. D. CURREY, AND J. M. GOSLINE. 1976. Mechanical design in organisms. Halstead Press, New York.

WAINWRIGHT, S. A., F. VOSBURG, AND J. HEBRANK. 1980. Shark skin and locomotion. Science 207: 1004–1005.

WEISS, L. 1983. Histology. Elsevier Biomedical, New York, Amsterdam, Oxford.

WEYMOUTH, R. D., R. C. LASIEWSKI, AND A. J. BERGER. 1964. The tongue apparatus in hummingbirds. Acta Anat. 58:252–270.

WINKLER, H., AND W. J. BOCK. 1976. Analyse der Kräfteverhältnisse bei Klettervögeln. J. Ornithol. 117:398–418.

WINOKUR, R. M. 1977. The integumentary tentacles of the snake *Erpeton tentaculatum*: Structure, function, evolution. Herpetologica 33:247–253.

WINOKUR, R. M. 1981. Erectile tissue and smooth muscle in the snouts of turtles. Am. Zool. 21: 959.

WINOKUR, R. M. 1982. Erectile tissue and smooth muscle in snouts of *Carettochelys insculpta*, Trionychids and other Chelonia. Zoomorphology 101:83–93.

YOUNG, J. A., AND E. W. VAN LENNEP. 1978. The morphology of salivary glands. Academic Press, London, New York.

YOUNG, J. Z. 1963. The life of mammals. Oxford University Press, Oxford.

ZISWILER, V. 1979. Zungenfunktionen und Zungenversteifung bei granivoren Singvögeln. Rev. Suisse Zool. 86:823–831.

ZISWILER, V. 1980. Use of adaptational analysis in evolutionary and phylogenetic study. Pp. 209–213, *In* R. Nöhring (ed.), Acta XVII Congr. Internat. Ornithol. Deutsche Ornithologen-Gesellschaft, Berlin.

ZISWILER, V., AND D. S. FARNER. 1972. Digestion and digestive system. Pp. 343–430, *In* D. S. Farner and J. R. King (eds.), Avian Biology, Vol. II. Academic Press, London, New York.

ZUSI, R. L., AND R. W. STORER. 1969. Osteology and myology of the head and neck of the Pied-billed grebe (*Podilymbus*). Misc. Publ. Mus. Zool. Univ. Mich. No. 139, 49 pp.

ZWEERS, G. A. 1974. Structure, movement, and myography of the feeding apparatus of the mallard (*Anas platyrhynchos* L.). A study in functional anatomy. Netherl. J. Zool. 24:323–467.

ZWEERS, G. A. 1979. Explanation of structure through optimization and systemization. Netherl. J. Zool. 29:418–440.

ZWEERS, G. A. 1980. Experimental functional analysis and formulation of causal models. Pp. 195–201, *In* R. Nöhring (ed.), Acta XVII Congr. Internat. Ornithol. Deutsche Ornithologen-Gesellschaft, Berlin.

ZWEERS, G. A. 1982. The feeding system of the pigeon (*Columba livia* L.). Adv. Anat. Embryol. Cell Biol. Vol. 73, 108 pp. Springer-Verlag, Berlin, Heidelberg, New York.

ZWEERS, G. A., A. F. C. GERRITSEN, AND P. J. VAN KRAANENBURG-VOOGD. 1977. Mechanics of feeding of the mallard (*Anas platyrhynchos* L., Aves, Anseriformes). *In* M. K. Hecht and F. S. Szalay (eds.), Contributions to Vertebrate Evolution, Vol. 3. Karger, Basel.

ZWEERS, G. A., H. C. VAN PELT, AND A. BECKERS. 1981. Morphology and mechanics of the larynx of the pigeon (*Columba livia* L.): A drillchuck system (Aves). Zoomorphology 99:37–69.

APPENDIX I

GLOSSARY OF ABBREVIATIONS

Ap	aponeurosis
A. CP	area of attachment for the cartilaginous process of the epibranchiale (Area cartilaginis epibranchialis)
A. CUH	Area of attachment for the cartilaginous process of the urohyale (Area cartilaginis urohyalis)
A. lig	Area ligamentosa (basihyalis or paraglossalis), attachment site of articular ligaments
ant.	anterior
Art. cb	Articulatio cerato-basihyalis
Art. ec	Articulatio epi-ceratobranchialis
Art. pb	Articulatio paraglosso-basihyalis
A. SPG	area of attachment for the Synchondrosis paraglossalis
A. synd. pg	area of attachment for the syndesmosis between the paraglossalia
ATW	Anterior lingual wing, Ala linguae anterior ("apikaler Abschnitt des Zungenflügels" of Homberger 1980a)
BH	basihyale
BL	blood lacunae, cavernous veins
BTpg	Bursa synovialis paraglossalis; usually shown opened
BTuh	Bursa synovialis urohyalis; usually only its roof on the urohyale shown
C	cartilage
C. bh	Corpus basihyale, body of the basihyale
Cart	Capsula articularis, articular capsule
CB	ceratobranchiale
Cca; C. c. ap.	Corpus cavernosum apicale
Ccl; C. c. lat.	Corpus cavernosum laterale
Conn. T	connective tissue
CP	Cartilago epibranchialis, cartilaginous process of epibranchiale
Cr. bm	Crista branchiomandibularis (ceratobranchialis)
Cr. d. bh	Crista dorsalis basihyalis
Cr. d-l	Crista dorso-lateralis (paraglossalis)
Cr. d-m	Crista dorso-medialis (ceratobranchialis)
Cr. l. bh	Crista lateralis basihyalis
Cr. l. cb	Crista lateralis ceratobranchialis
Cr. l. pg	Crista lateralis paraglossalis
Cr. m	Crista medialis (paraglossalis)
Cr. v. bh	Crista ventralis basihyalis
Cr. v. cb	Crista ventralis ceratobranchialis
CUH	Cartilago urohyalis, cartilaginous process of urohyale
CWBTpg	wall of Bursa synovialis paraglossalis, cut open
CW Env. Gl	wall of tendinous envelope of Glandula lingualis, cut open

APPENDIX I

CONTINUED

dors.	dorsalis, respectively dorsale
EB	epibranchiale
Env. Gl Mho	tendinous envelope of Glandula lingualis from M. hypoglossus obliquus
Env. Gl PG	tendinous envelope of Glandula lingualis from M. hypoglossus obliquus, supported by the paraglossale
EP	epithelium (Tunica mucosa, mucosa)
ep	cut edge of epithelium (only in Fig. 20)
Fac. d	Facies dorsalis (basihyalis), dorsal surface of broadened anterior end of basihyale
Fac. med	Facies medialis (ceratobranchialis), medial surface of ceratobranchiale
F. art. cb lat	Facies articularis cerato-basihyalis, pars lateralis; lateral part of articular facet of Art. cerato-basihyalis
F. art. cb med	Facies articularis cerato-basihyalis, pars medialis; medial part of articular facet of Art. cerato-basihyalis
F. art. ec	Facies articularis epi-ceratobranchialis, articular facet of Art. epi-ceratobranchialis
F. art. pb	Facies articularis paraglosso-basihyalis, articular facet of Art. paraglosso-basihyalis
F. bh	Fossa basihyalis
F. d. cb	Fossa dorsalis ceratobranchialis
For. ip	Foramen interparaglossale
Fpgd	Fascia paraglossalis dorsalis
G	glottis ("Larynxspalte" of Homberger 1980a)
Gl	Glandula lingualis, lingual salivary gland ("paarige Zungendrüse" of Homberger 1980a)
Gl.	Glandula
Gsl	Glandula sublingualis, sublingual salivary gland ("Unterzungendrüse" of Homberger 1980a)
Ins.	insertion
Lacb	Ligamentum articulare cerato-basihyale
Lapbv	Ligamentum articulare paraglosso-basihyale ventrale
Lapbd	Ligamentum articulare paraglosso-basihyale dorsale
lat.	lateralis, respectively laterale
Lcrb	Ligamentum crico-basihyale
LF	laryngeal flap, Papilla laryngeales ("Larynxpapillenreihe" of Homberger 1980a)
Licb	Ligamentum interarticulare cerato-basihyale
Lig.	Ligamentum, ligament
Ligpa	Ligamentum interparaglossale apicale
Lipgc	Ligamentum interparaglossale caudale
Liph	Ligamentum interparahyale
LM	laryngeal mound, Mons laryngealis ("Larynxplatte" of Homberger 1980a)
Lnc	Ligamentum nodulo-ceratobranchiale
Lpga	Ligamentum paraglossale apicale
M.	Musculus, muscle
Ma. d	Margo dorsalis (paraglossalis), dorsal edge of paraglossale
Ma. d. F. art. ec	Margo dorsalis faciei articularis epi-ceratobranchialis, dorsal rim of articular facet of Art. epi-ceratobranchialis
Ma. d. F. art. pb	Margo dorsalis faciei articularis paraglosso-basihyalis, dorsal rim of articular facet of Art. paraglosso-basihyalis
Ma. d-m	Margo dorso-medialis (epibranchialis), dorso-medial edge of epibranchiale
Ma. l	Margo lateralis (basihyalis), lateral edge of body of basihyale
Mamev	M. adductor mandibulae externus ventralis
Ma. v. F. art. cb	Margo ventralis faciei articularis cerato-basihyalis, ventral rim of articular facet of Art. cerato-basihyalis

APPENDIX I

CONTINUED

Ma. v. F. art. pb	Margo ventralis faciei articularis paraglosso-basihyalis, ventral rim of articular facet of Art. paraglosso-basihyalis
Ma. v-l	Margo ventro-lateralis (epibranchialis), ventro-lateral edge of epibranchiale
Mbma; M. bm. ant.	M. branchiomandibularis anterior
Mbmal; M. bm. ant. lat.	M. branchiomandibularis anterior lateralis
Mbmam; M. bm. ant. med.	M. branchiomandibularis anterior medialis
Mbmp; M. bm. post.	M. branchiomandibularis posterior
Mcg; M. cg.	M. ceratoglossus
Mcgl; M. cg. lat.	M. ceratoglossus lateralis
Mcgm; M. cg. med.	M. ceratoglossus medialis
Mcgma; M. cg. med. acc.	M. ceratoglossus medialis accessorius
Mcgmd; M. cg. med. dors.	M. ceratoglossus medialis dorsalis
Mcgmv; M. cg. med. ventr.	M. ceratoglossus medialis ventralis
Mcgp; M. cg. prof.	M. ceratoglossus profundus
Mcgs; M. cg. supf.	M. ceratoglossus superficialis
Mcgsl; M. cg. supf. lig.	M. ceratoglossus superficialis ligamenti
Mcgsn; M. cg. supf. nod.	M. ceratoglossus superficialis noduli
Mcgsp; M. cg. supf. lig. prof.	M. ceratoglossus superficialis ligamenti profundus
Mcgss; M. cg. supf. lig. supf.	M. ceratoglossus superficialis ligamenti superficialis
Mch; M. ch.	M. ceratohyoideus
Mchp; M. ch. prof.	M. ceratohyoideus profundus
Mchs; M. ch. supf.	M. ceratohyoideus superficialis
Mcrdi	M. cricohyoideus dorsalis intermedialis
Mcrdp	M. cricohyoideus dorsalis profundus
Mcrds	M. cricohyoideus dorsalis superficialis
Mcrv	M. cricohyoideus ventralis
MD	mandible
Mdm	M. depressor mandibulae
med.	medialis, respectively mediale
Mem	M. ethmomandibularis
Mgg; M. gg.	M. genioglossus
Mhga; M. hg. ant.	M. hypoglossus anterior
Mhgt; M. hg. trans.	M. hypoglossus transversus
Mho; M. ho.	M. hypoglossus obliquus
Mhol; M. ho. lat.	M. hypoglossus obliquus lateralis
Mhom; M. ho. med.	M. hypoglossus obliquus medialis
Mmg; M. mg.	M. mesoglossus
Mmga; M. mg. ant.	M. mesoglossus, pars anterior
Mmgp; M. mg. post.	M. mesoglossus, pars posterior
Mmha; M. mh. ant.	M. mylohyoideus anterior
Mmhp; M. mh. post.	M. mylohyoideus posterior
Mpvl	M. pterygoideus ventralis lateralis
MS	meniscus or articular pad of articular cavity
Msg; M. sg.	M. supraglossus
Msgb; M. sg. bas.	M. supraglossus basihyalis
Msgc; M. sg. cer.	M. supraglossus ceratobranchialis
Msh; M. sh.	M. serpihyoideus
Mst; M. st.	M. stylohyoideus
Mth; M. th.	M. tracheohyoideus
Mthb; M. th. bas.	M. tracheohyoideus basihyalis
Mthc; M. th. cer.	M. tracheohyoideus ceratobranchialis
Mthp; M. th. para.	M. tracheohyoideus parahyalis
Mtl; M. tl.	M. tracheolateralis
N	nerve
NL	lingual nail, Cuticula linguae ("Nagel" of Homberger 1980a)
ND	Nodulus, sesamoid bone of Lig. nodulo-ceratobranchiale
OE	oesophagus
Or.	origin

APPENDIX I

CONTINUED

P. ant. pg	Processus anterior paraglossalis
P. centr. bh	Processus centralis basihyalis
P. centr. pg	Pars centralis paraglossalis
P. post. pg	Processus posterior paraglossalis
Pch	perichondrium
PG	paraglossale
PL	palatal lobe ("Gaumenlappen" of Homberger 1980a)
post.	posterior
Proc.	Processus
PTW	posterior lingual wing, Ala linguae posterior ("caudaler Abschnitt des Zungenflügels" of Homberger 1980a)
RNL	root of nail, Radix cuticulae ("Nagelwurzel" of Homberger 1980a)
s.	*sive,* or
SPG	Synchondrosis paraglossalis, cartilaginous bridge between the anterior processes of paraglossalia
Sr	Sulcus radialis ("Radiärrinne" of Homberger 1980a)
S. v-m	Sulcus ventro-medialis (ceratobranchialis)
Synd. pg	Syndesmosis paraglossalis
TA	angle between the lingual wings, angulus alarum linguae ("Zungenflügelwinkel" of Homberger 1980a)
TMcg	tendon of M. ceratoglossus
TMhga	tendon and tendinous fibers of M. hypoglossus anterior
TMsg	tendinous fibers of M. supraglossus
TMsg A–TMsg E	part A–part E of the tendinous fibers of M. supraglossus
TR	trachea
TT	central area of the tongue's tip, Area centralis apicis linguae ("Zungenspitze" of Homberger 1980a)
Tub. cb	Tuberositas ceratobranchialis
Tub. pg	Tuberositas paraglossalis
UH	urohyale
ventr.	ventralis, respectively ventrale
vFpgd	ventral extension of the Fascia paraglossalis dorsalis, attaching on ventral side of paraglossale
VSM	V-shaped mound on dorsal side of lingual tip ("lippenartige Wülste" of Homberger 1980a)

APPENDIX II

SYNONYMIES

A: SELECTED SYNONYMIES OF THE ELEMENTS OF THE AVIAN HYOID SKELETON

Present study: *Psittacus erithacus*	Paraglossale	Basihyale	Parahyale	Urohyale	Nodulus (Os sesamoideum urohyale)	Ceratobranchiale	Epibranchiale	Cartilaginous process of epibranchiale (Cartilago epibranchialis)
						hyoid horn		
Baumel (1979): avian	Os entoglossum	Os basibranchiale rostrale	Processus parahyalis	Os basibranchiale caudale	Os sesamoideum	Os ceratobranchiale	Os epibranchiale	—
						Cornu branchiale		
Zweers (1982): *Columba livia*	paraglossal	basihyal	—	urohyal	—	ceratobranchial	epibranchial	pharyngobranchial
Denker (1907): *Amazona amazonica*	Paraglossale	Copula, s. Entoglossum	Fortsätze des Entoglossum	Zungenbeinstiel, s. Stiel, s. Kiel	—	Ceratobranchiale, Keratobranchiale	ceratohyale, s. Keratohyale	—
						Zungenbeinhorn		
Dubale and Rawal (1965); Rawal (1970): *Psittacula krameri*	entoglossum	basihyal	parahyal process	urohyal	—	hypobranchial	ceratobranchial	epibranchial
Duvernoy (1835): psittacine	os lingual	hyoïde	—	queue de l'hyoïde	—	première pièce	pièce intermédiaire	deuxième pièce
						corne hyoïde		
Evans (1969): *Melopsittacus undulatus*	entoglossal	basihyal	parahyal	urohyal	sesamoid plate	ceratobranchial	epibranchial	—
						hyoid cornu		
Fürbringer (1922): psittacine	Paraglossum, Paraglossale	Copula I (corpus + processus lingualis)	Processus parahyalis	Copula II	—	proximales Glied, s. Hypobranchiale	distales Glied, s. Keratobranchiale	Knorpelige Endepiphyse
						cornu branchiale I		
Gadow (1891, p. 239, Pl. 30): psittacine	Os entoglossum, s. Kern	Basihyal, s. Copula, s. unpaares Mittelstück	—	Stiel, s. 2. Copula, s. Os urohyale, s. Zungenstiel	—	1. Glied	2. Glied	—
						Zungenbeinhorn		

A: CONTINUED

Giebel (1858, p. 42): psittacine	Kern	Körper	Vorderecken	Stiel	—	1. Glied Zungenhorn	2. Glied	—
Hargrave (1970): *Ara*	—	basihyal	—	basibranchial	—	—	—	—
Kallius (1906): *Melopsittacus undulatus*	Paraglossalia	Copula (entoglossum = tip of Copula)	—	Stiel	supcopularer Knorpel, s. Cartilago subcopularis	proximaler Abschnitt Zungenbeinhorn	distaler Abschnitt	—
Kutorga (1832): *Psittacus erithacus*	Os linguale	Os hyoideum	Processus ossis hyoidei anticus lateralis	Processus styloideus ossis hyoidei	—	Pars cornuum anterior ossea	Pars cornuum posterior ossea	Pars cornuum fibroso-cartilaginea
Mivart (1895, 1896): *Psittacus erithacus*, etc.	entoglossum	basihyal	parahyal process	urohyal	—	hypobranchial	ceratobranchial	—
Mudge (1903): psittacine	entoglossum	basihyal	parahyal process	urohyal	nodule	hypobranchial	ceratobranchial	—
Nitzsch (in Giebel 1862): psittacine	Zungenkern, s. Zungenbeinkernstück	Zungenbeinkörper	Flügeldorn s. seitliche Flügelspitze	Zungenstiel	—	1. Glied Zungenbeinhorn	2. Glied	—
Shufeldt (1886): *Conurus carolinensis*	ceratohyal	first basibranchial	osseus outgrowth	second basibranchial	—	ceratobranchial	epibranchial	thyrohyal element
Steinbacher (1951): *Lorius* sp.	Paraglossalis	Entoglossum	Parahyalia	—	—	Keratobranchiale	Epibranchiale	—
Thuet (1838): *Psittacus erithacus*	Os hyoides accessorium	Corpus ossis hyoidis	—	Cornu medium	—	—		—
						Cornu magnum		

B: Synonymies of the Avian M. serpihyoideus

Name	Author
M. serpihyoideus	Greschik (1921, 1928), Scharnke (1931), Engels (1938), Goodman and Fisher (1962), Weymouth et al. (1964), George and Berger (1966), Zusi and Storer (1969), Bock and Morioka (1971), Bock (1972, 1978), Bock et al. (1973), Richards and Bock (1973), Burton (1974a, b, c), Zweers (1974, 1982), Lowery and Tallman (1976), Zweers et al. (1977), Bock and Morony (1978a, b), Vanden Berge (1979), Bhattacharyya (1980)
M. serpi-hyoidien, portion postérieure	Duvernoy (1835)
M. serpihyoideus (posterior part of the M. mylohyoideus posterior)	Gadow (1891), Mudge (1903), Leiber (1907)
M. mylohyoideus	Kasai (1957)
M. mylohyoideus obliquus	Kutorga (1832)
M. mylohyoideus obliquus posterior	Nitzsch (in Giebel 1862)
M. hyomandibularis medialis	Kallius (1906), Denker (1907), Moller (1930, 1931), Fujioka (1963)
M. basibranchialis mandibularis	Vanden Berge (1975)
M. articulohyoideus	Fitzgerald (1969)
M. articulohyoideus ventralis	Suzuki and Nomura (1975)
M. articulohyoideus caudalis	Chamberlain (1943)
M. gularis posterior	Edgeworth (1935), Dubale and Rawal (1965), Rawal (1970)
Caudal part of the oral portion of the Constrictor secundus	Lubosch (1933)
M. stylohyoideus	Nickel et al. (1977)

C: Synonymies of the Avian M. stylohyoideus

Name	Author
M. stylohyoideus	Shufeldt (1890), Greschik (1921, 1928), Engels (1938), Fisher and Goodman (1955), Kasai (1957), Goodman and Fisher (1962), Weymouth et al. (1964), George and Berger (1966), Zusi and Storer (1969), Bock and Morioka (1971), Bock (1972, 1978), Bock et al. (1973), Richards and Bock (1973), Burton (1974a, b, c), Zweers (1974, 1982), Lowery and Tallman (1976), Zweers et al. (1977), Nickel et al. (1977), Bock and Morony (1978a, b), Vanden Berge (1979), Bhattacharyya (1980)
Anterior part of M. mylohyoideus posterior (M. stylohyoideus)	Gadow (1891), Mudge (1903), Leiber (1907)
M. mylohyoideus obliquus anterior	Nitzsch (in Giebel 1862)
M. hyomandibularis lateralis	Kallius (1906), Denker (1907), Moller (1930, 1931), Scharnke (1931), Fujioka (1963)
M. serpihyoidien, portion antérieure	Duvernoy (1835)
M. gularis anterior	Edgeworth (1935), Dubale and Rawal (1965), Rawal (1970)
M. basibranchialis mandibularis	McLelland (1968)
M. basibranchialis mandibularis pars lateralis	Vanden Berge (1975)
M. articulohyoideus dorsalis	Suzuki and Nomura (1975)
M. articulohyoideus caudalis	Chamberlain (1943)
Rostral part of the oral portion of the Constrictor secundus	Lubosch (1933)

D: Synonymies of the Avian M. branchiomandibularis

Name	Author
M. branchiomandibularis	Edgeworth (1935), Engels (1938), Dubale and Rawal (1965), Zusi and Storer (1969), Rawal (1970), Bock and Morioka (1971), Bock (1972, 1978), Bock et al. (1973), Richards and Bock (1973), Burton (1974b, c), Lowery and Tallman (1976), Bock and Morony (1978a, b), Vanden Berge (1979), Bhattacharyya (1980)
M. branchiomandibularis visceralis	Lubosch (1933)
M. mandibularis epibranchialis	McLelland (1968), Vanden Berge (1975)
M. ceratomandibularis	Kallius (1906), Denker (1907), Greschik (1921, 1928), Fujioka (1963)
M. keratomandibularis	Moller (1930, 1931), Scharnke (1931)
M. hyomandibularis	Chamberlain (1943), Fitzgerald (1969)
M. stylomandibularis	Suzuki and Nomura (1975)
M. mylo-cératoidien	Duvernoy (1835)
M. geniohyoideus (M. mylo-ceratoideus anterior and posterior)	Nitzsch (in Giebel 1862)
M. geniohyoideus	Shufeldt (1890), Gadow (1891), Mudge (1903), Leiber (1907), Fisher and Goodman (1955), Kasai (1957), Goodman and Fisher (1962), Weymouth et al. (1964), George and Berger (1966), Burton (1974a), Zweers (1974, 1982), Zweers et al. (1977), Nickel et al. (1977)
M. conici ossis hyoidei	Kutorga (1832)

E: Synonymies of the Avian M. mylohyoideus

Name	Author
M. mylohyoideus (undivided)	Shufeldt (1890), Greschik (1921, 1928), Moller (1930, 1931), Engels (1938), Chamberlain (1943), Kasai (1957), Weymouth et al. (1964), George and Berger (1966), Zusi and Storer (1969), Fitzgerald (1969), Bock and Morioka (1971), Bock (1972, 1978), Richards and Bock (1973), Burton (1974a, b), Lowery and Tallman (1976), Bock and Morony (1978a, b), Bhattacharyya (1980)
M. mylohyoideus anterior	Leiber (1907), Nickel et al. (1977)
M. mylohyoideus anterior et posterior	Duvernoy (1835), Bock et al. (1973)
M. mylohyoideus anterior, anterior and posterior portion	Gadow (1891), Mudge (1903)
M. mylohyoideus transversus, anterior and posterior portion	Nitzsch (in Giebel 1862)
M. mylohyoideus and M. hyomandibularis transversus	Kallius (1906), Denker (1907)
M. intermandibularis (undivided)	Fujioka (1963), Suzuki and Nomura (1975), Vanden Berge (1975)
M. intermandibularis anterior et posterior	Dubale and Rawal (1965)
M. intermandibularis ventralis et dorsalis	Goodman and Fisher (1962), Rawal (1970), Vanden Berge (1979)
M. intermandibularis, M. intermandibularis dorsalis	Edgeworth (1935)
M. intermandibularis ventralis rostralis	Zweers (1974, 1982), Zweers et al. (1977)

E: Continued

Name	Author
M. intermandibularis, s. Constrictor secundus mandibularis secundus, s. M. hyomandibularis transversus, s. M. mylohyoideus anterior	Lubosch (1933)
M. genioglossus (probably misidentified M. mylohyoideus)	Fisher and Goodman (1955)

F: Synonymies of the Avian M. genioglossus

Name	Author
M. genioglossus	Kutorga (1832), Nitzsch (in Giebel 1862), Gadow (1891), Mudge (1903), Chaine (1905), Moller (1930, 1931), Lubosch (1933), Edgeworth (1935), Steinbacher (1951), George and Berger (1966), Bock and Morioka (1971), Bock (1972, 1978), Bock et al. (1973), Richards and Bock (1973), Burton (1974a, b, c), Bock and Morony (1978a, b), Vanden Berge (1979), Bhattacharyya (1980)
M. geniohyoideus	Denker (1907)
M. geniohyoideus/M. genioglossus	Kallius (1906), Greschik (1921, 1928), Engels (1938), Dubale and Rawal (1965), Rawal (1970)
M. geniopharyngealis	Zweers (1982)
M. myloglossus	Duvernoy (1835)

G: Synonymies of the Avian M. tracheohyoideus and M. tracheolateralis

Author	M. tracheohyoideus	M. tracheolateralis (see Table 1)	Comments
Bock and Morioka (1971), Bock (1972, 1978), Bock et al. (1973), Richards and Bock (1973), Burton (1974b), Lowery and Tallman (1976), Gaunt and Gaunt (1977), Bock and Morony (1978a, b), Brackenbury (1980), Bhattacharyya (1980)	M. tracheohyoideus	M. tracheolateralis	Descriptions of passerine birds, *Gallus,* and *Columba.*
Moller (1931)	M. tracheohyoideus	M. tracheohyoideus	The "M. tracheohyoideus" in *Zosterops* is described as consisting of two parts. (These two parts correspond to the M. th. and M. tracheolateralis in the present nomenclature.) The "M. tracheohyoideus" is undivided in *Dacnis.*

G: Continued

Author	M. tracheohyoideus	M. tracheolateralis (see Table 1)	Comments
Weymouth et al. (1964)	M. tracheohyoideus, lateral part	M. tracheohyoideus, ventral part	Description is vague
Duvernoy (1835), Moller (1930)	M. tracheohyoideus	not mentioned	
George and Berger (1966)	M. tracheohyoideus, M. sternohyoideus, M. ypsilotrachealis	M. tracheolateralis	Various non-homologous muscles of different authors are assembled under the name of "M. tracheohyoideus" and "M. ypsilotrachealis." "M. tracheolateralis" is included with various syringeal muscles, the homologies of which are unclear.
Vanden Berge (1979)	M. tracheohyoideus, M. sternohyoideus, M. cleidohyoideus	M. trachealis lateralis, M. sternotrachealis, M. sternotracheolaryngealis	Taken from various authors. The homologies are often only partially correct.
Fisher and Goodman (1955)	M. tracheohyoideus	M. tracheolaryngeus superior et inferior	
Kallius (1906)	M. tracheolaryngohyoideus (M. sternohyoideus)	not mentioned	
Denker (1907)	M. tracheo-laryngohyoideus and M. keratohyoideus II	not mentioned	The "M. keratohyoideus II" consists actually of the parahyal and ceratobranchial heads of the M. tracheohyoideus in the present nomenclature
Dubale and Rawal (1965)	M. sternohyoideus	M. tracheocricoideus	
Nitzsch (in Giebel 1862)	M. sternohyoideus	M. sternotrachealis	
Greschik (1921, 1928)	not mentioned	M. sternotrachealis	
Mudge (1903), Kasai (1957), Lubosch (1933)	M. sternohyoideus	not mentioned	
Burton (1974c)	M. sterynohyoideus (sic)	not mentioned	
Gadow (1891)	M. sternohyoideus, M. claviculohyoideus, M. cleidohyoideus	M. sterno-thyreoideus, M. coraco-thyreoideus, M. tracheolaryngeus	
Chamberlain (1943)	M. sternothyrohyoideus	M. ypsilotrachealis	
Fitzgerald (1969)	M. sternothyrohyoideus	M. sternotrachealis	
Nickel et al. (1977)	M. sternothyrohyoideus	M. sternotrachealis	
McLelland (1965), King (1975)	M. sternotracheolaryngeus lateralis	M. sternotracheolaryngeus medialis	
Scharnke (1931)	M. sternotrachealis	not mentioned	The "M. tracheohyoideus" of Scharnke is the M. cricohyoideus in the present nomenclature (Table 1)

G: CONTINUED

Author	M. tracheohyoideus	M. tracheolateralis (see Table 1)	Comments
Fujioka (1963)	M. sternolaryngeus	M. trachealis lateralis	
Burton (1974a)	M. cleidohyoideus	M. sternotrachealis	The "M. sternotrachealis" consists of both the M. sternotrachealis and M. tracheolateralis in the present nomenclature
Leiber (1907)	M. cleidothyreoideus	M. trachealis	The "M. tracheohyoideus" of Leiber is the M. cricohyoideus in the present nomenclature (Table 1)
Shufeldt (1890)	M. cleidotrachealis	M. tracheolateralis	
Zweers (1982)	M. claviculohyoideus, M. claviculoglandularis	M. trachealis lateralis	
Lockner and Youngren (1976)	M. ypsilotrachealis	M. tracheolateralis	
Edgeworth (1935)	M. rectus cervicis, posterior part, external fasciculus; M. sternohyoideus	M. rectus cervicis, posterior part, internal fasciculus; M. cleidohyoideus 1) M. sternotrachealis 2) M. tracheohyoideus	
Engels (1938)	M. rectus cervicis, external fasciculus	not mentioned	The "M. tracheohyoideus" of Engels is the M. cricohyoideus in the present nomenclature (Table 1)
Rawal (1970)	not mentioned	M. tracheocricoideus	
Kutorga (1832)	not mentioned	M. sternotrachealis	
Zusi and Storer (1969)	absent	not mentioned	Fig. 7-B, p. 21, shows two unidentified muscles in the position of the M. tracheohyoideus and M. tracheolateralis

H: Synonymies of the Avian M. ceratohyoideus

Name	Author
M. ceratohyoideus	Kutorga (1832), Duvernoy (1835), Nitzsch (in Giebel 1862), Gadow (1891), Mudge (1903), Leiber (1907), Greschik (1921, 1928), Engels (1938), Kasai (1957), Weymouth et al. (1964), George and Berger (1966), Zusi and Storer (1969), Bock and Morioka (1971), Bock (1972, 1978), Bock et al. (1973), Richards and Bock (1973), Burton (1974a, b, c), Zweers et al. (1977), Bock and Morony (1978a, b), Bhattacharyya (1980)
M. keratohyoideus	Nickel et al. (1977)
M. keratohyoideus III	Denker (1907)
M. interceratoideus	Kallius (1906), Fujioka (1963)
M. interkeratoideus	Edgeworth (1935), Dubale and Rawal (1965), Rawal (1970)
M. interceratobranchialis	McLelland (1968), Vanden Berge (1975, 1979)
M. interbranchialis ventralis	Lubosch (1933)
M. transversus hyoideus	Chamberlain (1943), Fitzgerald (1969), Suzuki and Nomura (1975)
M. ceratoglossus	Goodman and Fisher (1962)

I: Synonymies of the Avian M. ceratoglossus

Name	Author
M. ceratoglossus	Kutorga (1832), Duvernoy (1835), Gadow (1891), Mudge (1903), Chaine (1904), Greschik (1921, 1928), Kasai (1957), Fujioka (1963), Weymouth et al. (1964), George and Berger (1966), Bock and Morioka (1971), Bock (1972, 1978), Bock et al. (1973), Richards and Bock (1973), Burton (1974a, b, c), Zweers (1974, 1982), Lowery and Tallman (1976), Zweers et al. (1977), Bock and Morony (1978a, b), Vanden Berge (1979), Bhattacharyya (1980)
M. keratoglossus	Nickel et al. (1977)
M. ceratoglossus posterior	Engels (1938), Zusi and Storer (1969)
M. ceratoglossus inferior	Leiber (1907)
M. ceratoglossus inferior, pars lateralis	Fisher and Goodman (1955)
M. ceratoglossus inferior et lateralis	Nitzsch (in Giebel 1862), Steinbacher (1951)
M. ceratoglossus ventralis inferior, s. superficialis	Chaine (1905)
M. keratoglossus II and M. keratohyoideus IV	Denker (1907)
M. ceratohyoideus	Shufeldt (1890)
M. keratohyoideus	Moller (1930, 1931)
M. hyoglossus	Edgeworth (1935), Rawal (1970)
M. hyoglossus posterior, portio d	Kallius (1906)
M. hyoglossus lateralis	Dubale and Rawal (1965)
M. styloentoglossus	Chamberlain (1943), Fitzgerald (1969), Suzuki and Nomura (1975)
M. paraglossoceratobranchialis	McLelland (1968), Vanden Berge (1975)
M. branchiocopuloglossus, M. branchioparaglossalis	Lubosch (1933)

J: Synonymies of the Avian M. hypoglossus obliquus

Name	Author
M. hypoglossus obliquus	Nitzsch (in Giebel 1862), Gadow (1891), Mudge (1903), Greschik (1921, 1928), Steinbacher (1951), Fisher and Goodman (1955), Kasai (1957), Bock and Morioka (1971), Bock (1972, 1978), Bock et al. (1973), Richards and Bock (1973), Burton (1974b, c), Lowery and Tallman (1976), Nickel et al. (1977), Homberger (1978), Bock and Morony (1978a, b), Vanden Berge (1979), Bhattacharyya (1980)
M. hyoglossus obliquus	Weymouth et al. (1964), George and Berger (1966), Burton (1974a), Zweers (1974, 1982), Zweers et al. (1977)
M. hypoglossus obliquus and M. ceratoglossus ventralis profundus	Chaine (1905)
M. hypoglossus obliquus and M. hypoglossus posterior	Scharnke (1931)
M. hypoglossus posterior	Edgeworth (1935), Engels (1938), Dubale and Rawal (1965), Zusi and Storer (1969), Rawal (1970)
M. hypoglossus posterior, portio c	Kallius (1906)
M. hyoglossus posterior and M. keratoglossus I	Denker (1907)
M. hyoglossus lateralis	Fujioka (1963)
M. hyoglossus transversus et rectus	Duvernoy (1835)
M. hyoglossus	Suzuki and Nomura (1975)
M. basioglossus inferior	Kutorga (1832)
M. ceratoglossus superior	Leiber (1907)
M. copuloentoglossum	Chamberlain (1943)
M. paraglossobasibranchialis lateralis	McLelland (1968), Vanden Berge (1975)
M. branchioparaglossalis dorsalis	Lubosch (1933)

K: Synonymies on the Psittacine M. supraglossus

Name	Author
M. supraglossus	Homberger (present study)
M. hyoglossus superior	Dubale and Rawal (1965), Rawal (1970)
M. hyoglossus posterior, portio b	Kallius (1906)
M. ceratoglossus superior	Nitzsch (in Giebel 1862), Gadow (1891), Mudge (1903), Steinbacher (1951), George and Berger (1966), Burton (1974c)
M. ceratoglossus dorsalis	Chaine (1905)
M. basioglossus superior	Kutorga (1832)
M. keratohyoideus I	Denker (1907)

L: SYNONYMIES OF THE AVIAN M. HYPOGLOSSUS ANTERIOR

Name	Author
M. hypoglossus anterior	Scharnke (1931), Edgeworth (1935), Dubale and Rawal (1965), Rawal (1970), Bock and Morioka (1971), Bock et al. (1973), Richards and Bock (1973), Burton (1974b, c), Lowery and Tallman (1976), Homberger (1978), Bock and Morony (1978a, b)
M. hyoglossus anterior	Denker (1907), Weymouth et al. (1964), George and Berger (1966), Zweers (1974, 1982), Zweers et al. (1977)
M. hyoglossus anterior, portio b and c	Kallius (1906)
M. hypoglossus rostralis	Vanden Berge (1979)
M. hypoglossus rectus	Nitzsch (in Giebel 1862), Gadow (1891), Mudge (1903), Greschik (1921, 1928), Steinbacher (1951), Fisher and Goodman (1955), Nickel et al. (1977), Bhattacharyya (1980)
M. hyoglossus rectus	Chaine (1904, 1905)
M. hyoglossus medialis	Kasai (1957)
M. hyoglossus anterior and medialis, and M. ceratoglossus anterior	Burton (1974a)
M. ceratoglossus anterior	Engels (1938), Zusi and Storer (1969)
M. paraglossobasibranchialis medialis	McLelland (1968), Vanden Berge (1975)
M. paraglosso-glossus	Lubosch (1933)
M. depressor glossus	Shufeldt (1890)
M. lingualis	Duvernoy (1835)
M. lingualis inferior	Kutorga (1832)

M: SYNONYMIES OF THE PSITTACINE M. MESOGLOSSUS

Name	Author
M. mesoglossus	Mudge (1903), Steinbacher (1951), George and Berger (1966), Burton (1974c), Vanden Berge (1975, 1979)
M. hyoglossus anterior, portio a	Kallius (1906)
M. lingualis	Chaine (1905)
M. lingualis impar	Kutorga (1832)

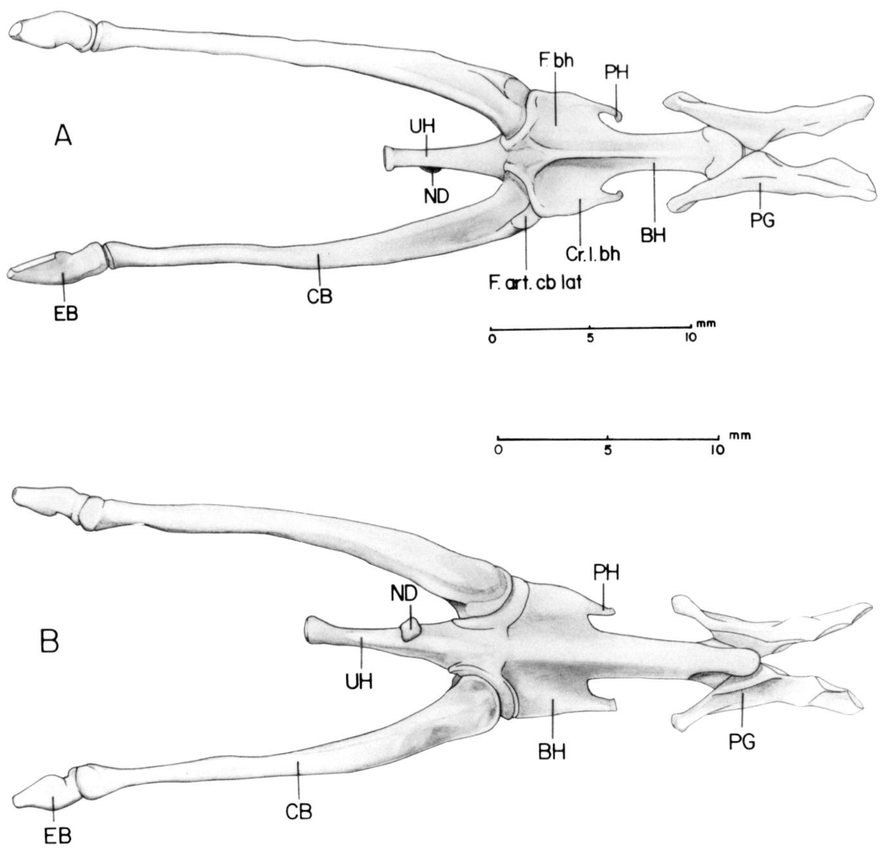

FIG. 1. The bony elements of the articulated hyoid skeleton of *Psittacus erithacus* in protracted position, AMNH 3448. A: Dorsal view, with horizontal ceratobranchialia. B: Ventral view, with horizontal basihyale. Abbreviations for all labels are found in Appendix I.

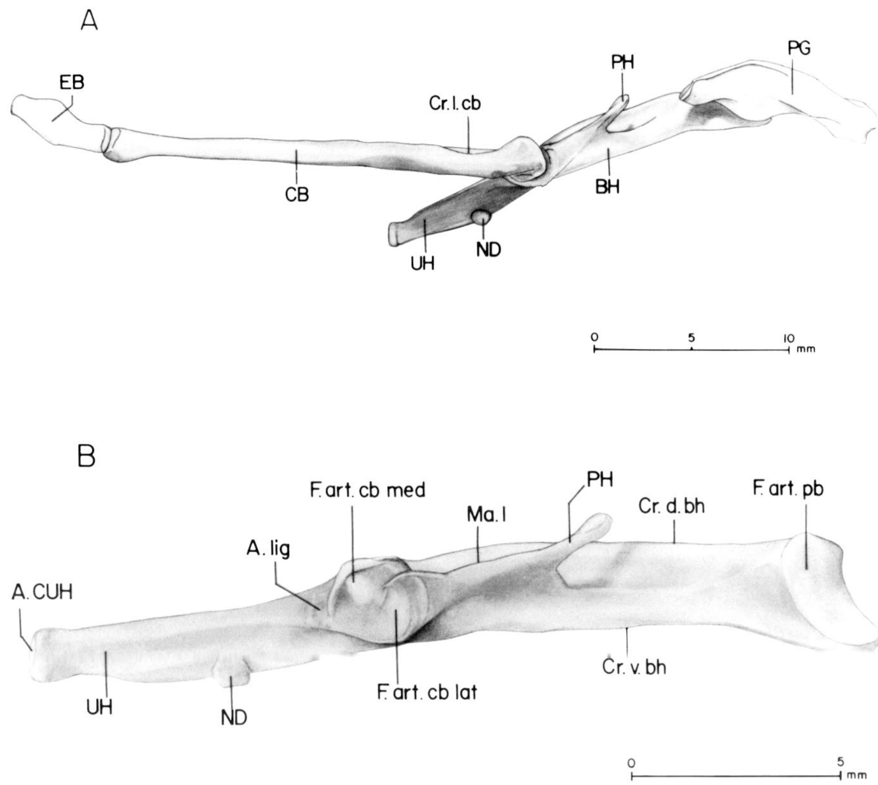

FIG. 2. A: Lateral view of the bony elements of the articulated hyoid skeleton of *Psittacus erithacus* in protracted position, AMNH 3448. B: Lateral view of the basihyale and Nodulus of *Psittacus erithacus,* AMNH 3448.

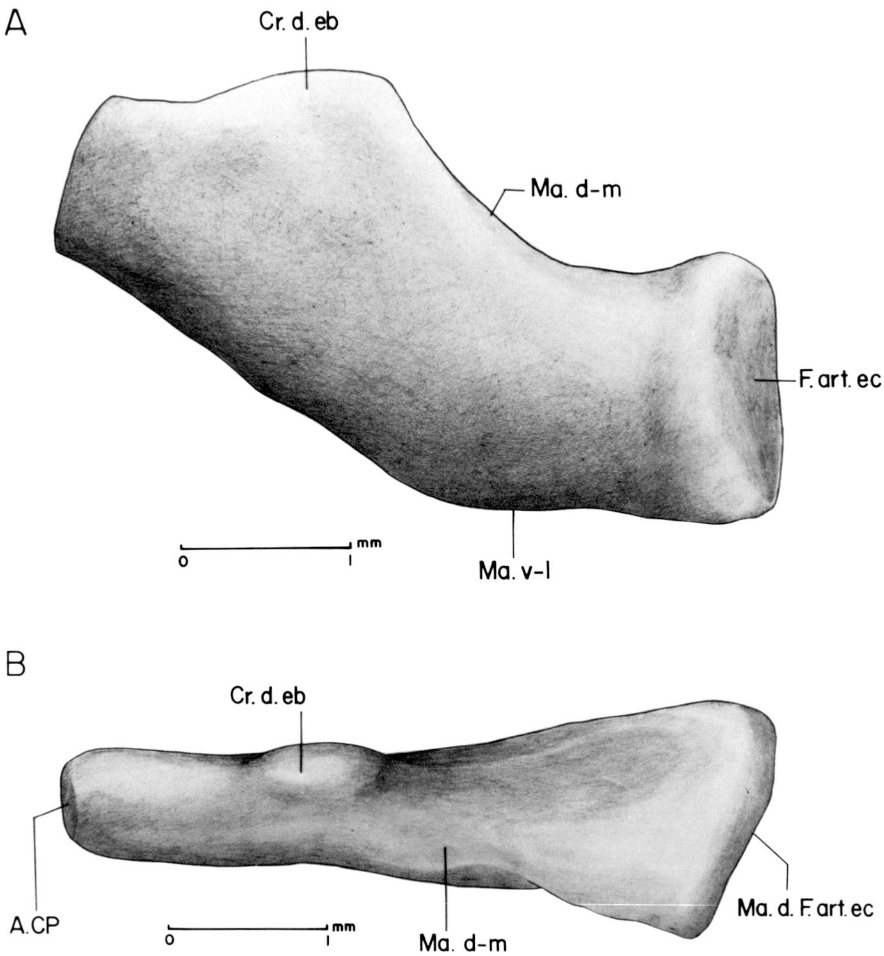

FIG. 3. Left epibranchiale of *Psittacus erithacus*. A: Dorso-medial view, AMNH 3448. B: Medial view, AMNH 2354.

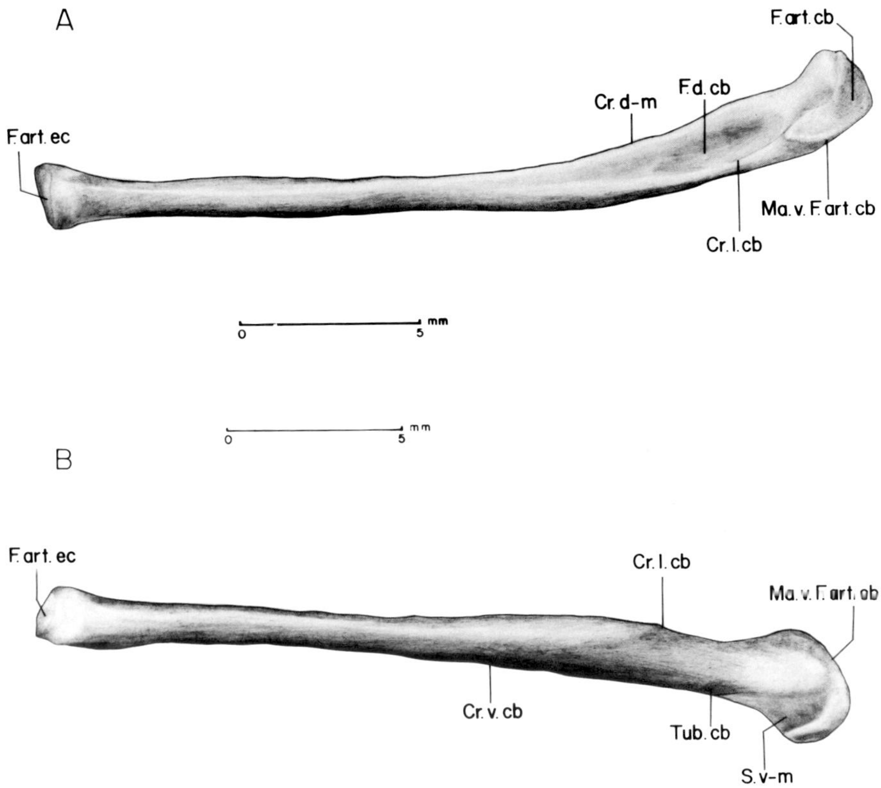

FIG. 4. Right ceratobranchiale of *Psittacus erithacus,* AMNH 2354. A: Dorsal view. B: Ventro-lateral view.

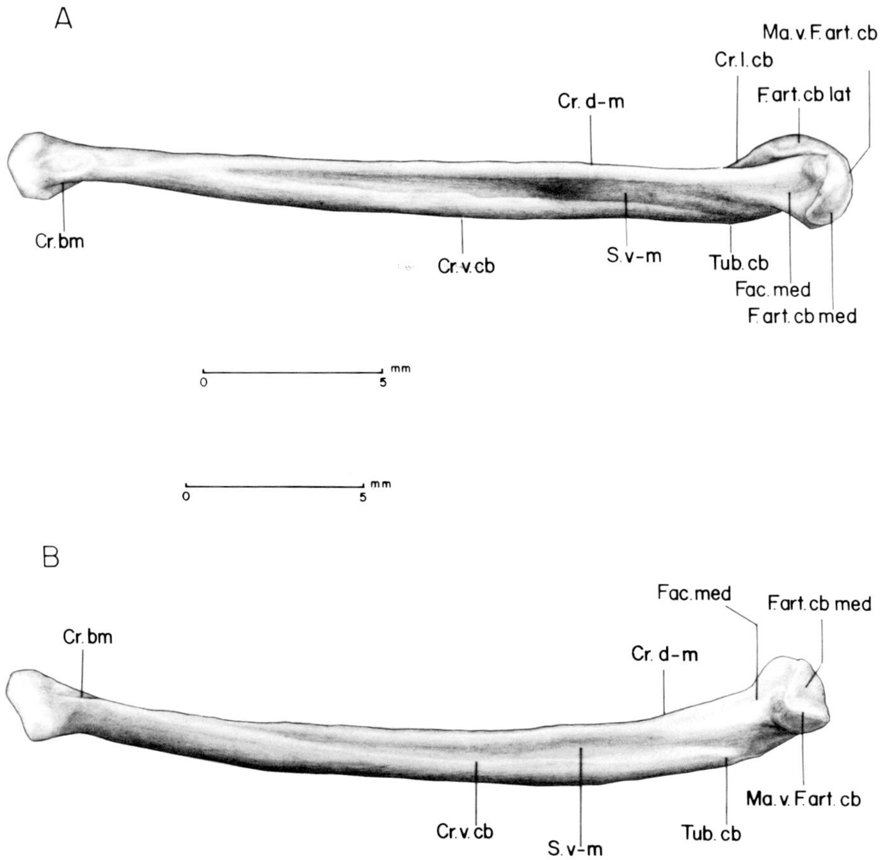

FIG. 5. Left ceratobranchiale of *Psittacus erithacus,* AMNH 2354. A: Medial view. B: Ventromedial view.

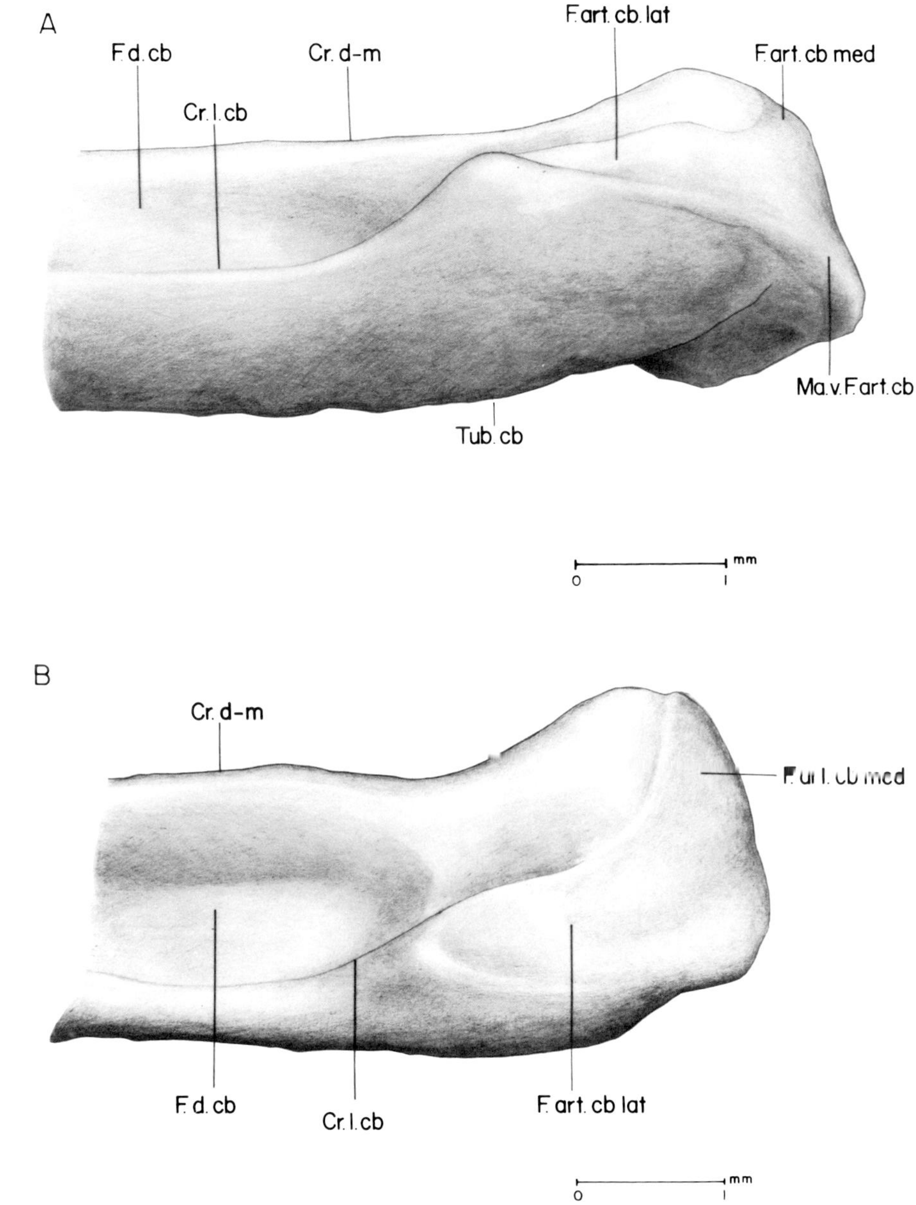

FIG. 6. Apical end of the right ceratobranchiale of *Psittacus erithacus,* AMNH 2354. A: Lateral view. B: Dorsal view.

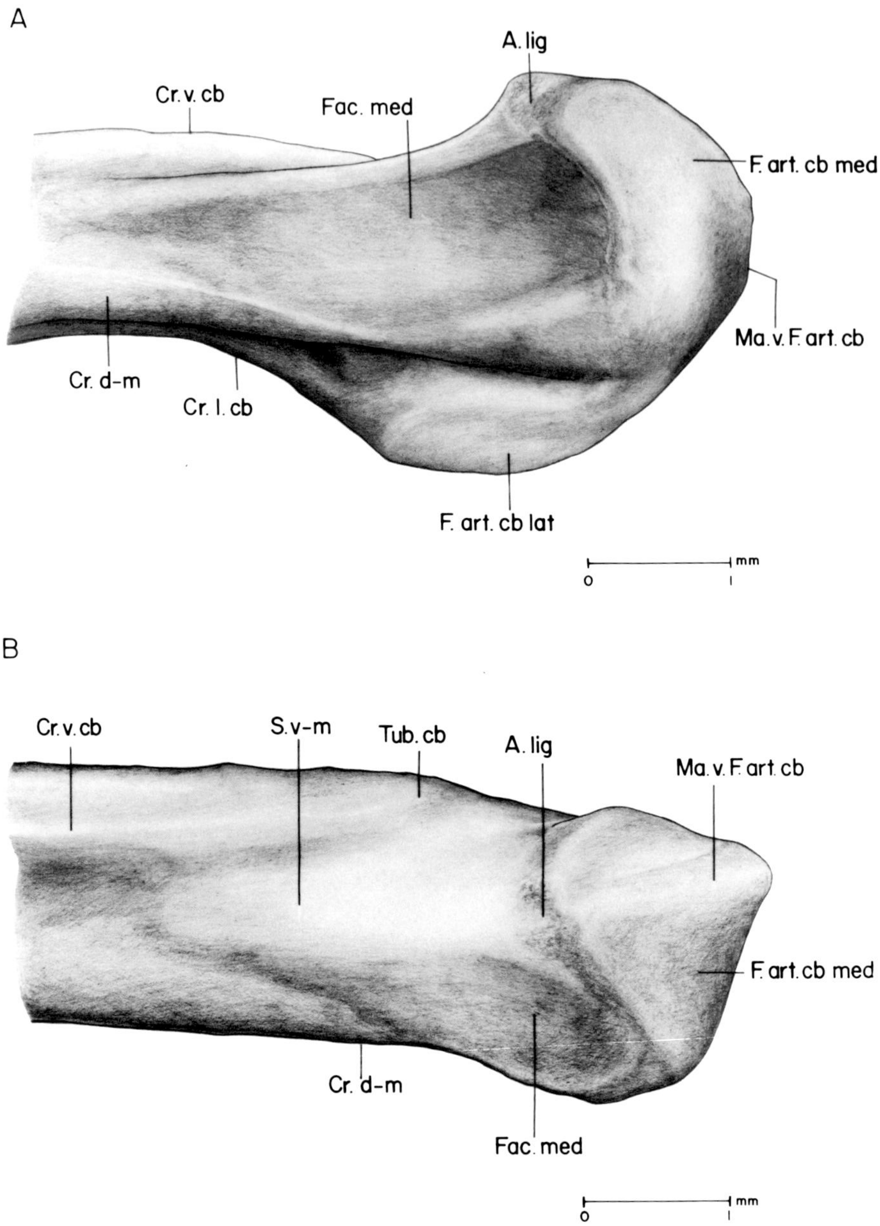

FIG. 7. Apical end of the right ceratobranchiale of *Psittacus erithacus,* AMNH 2354. A: Dorsomedial view. B: Medial view.

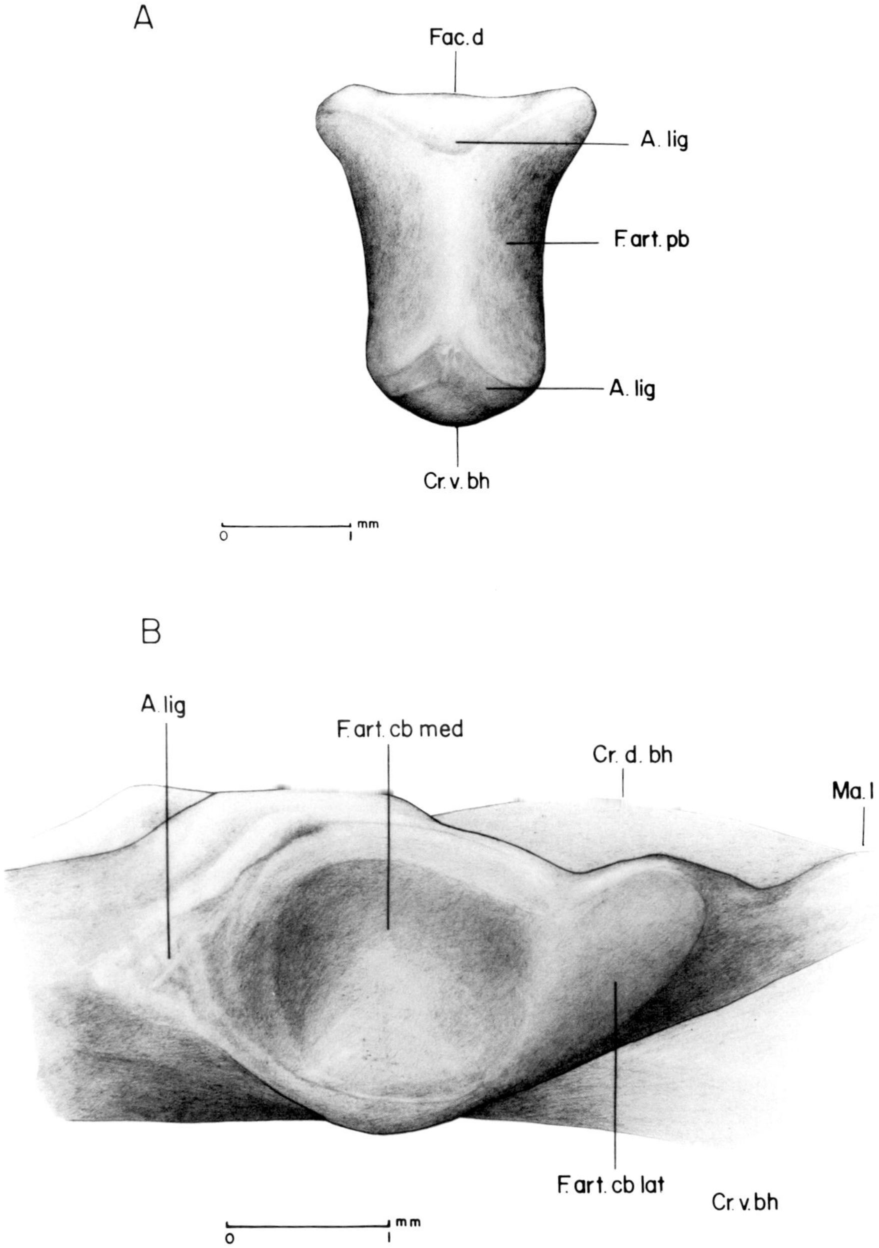

FIG. 8. A: Frontal view of the apical end of the basihyale of *Psittacus erithacus,* AMNH 3448. B: Caudo-lateral view of the right articular facet of the Articulatio cerato-basihyalis on the basihyale of *Psittacus erithacus,* AMNH 3448.

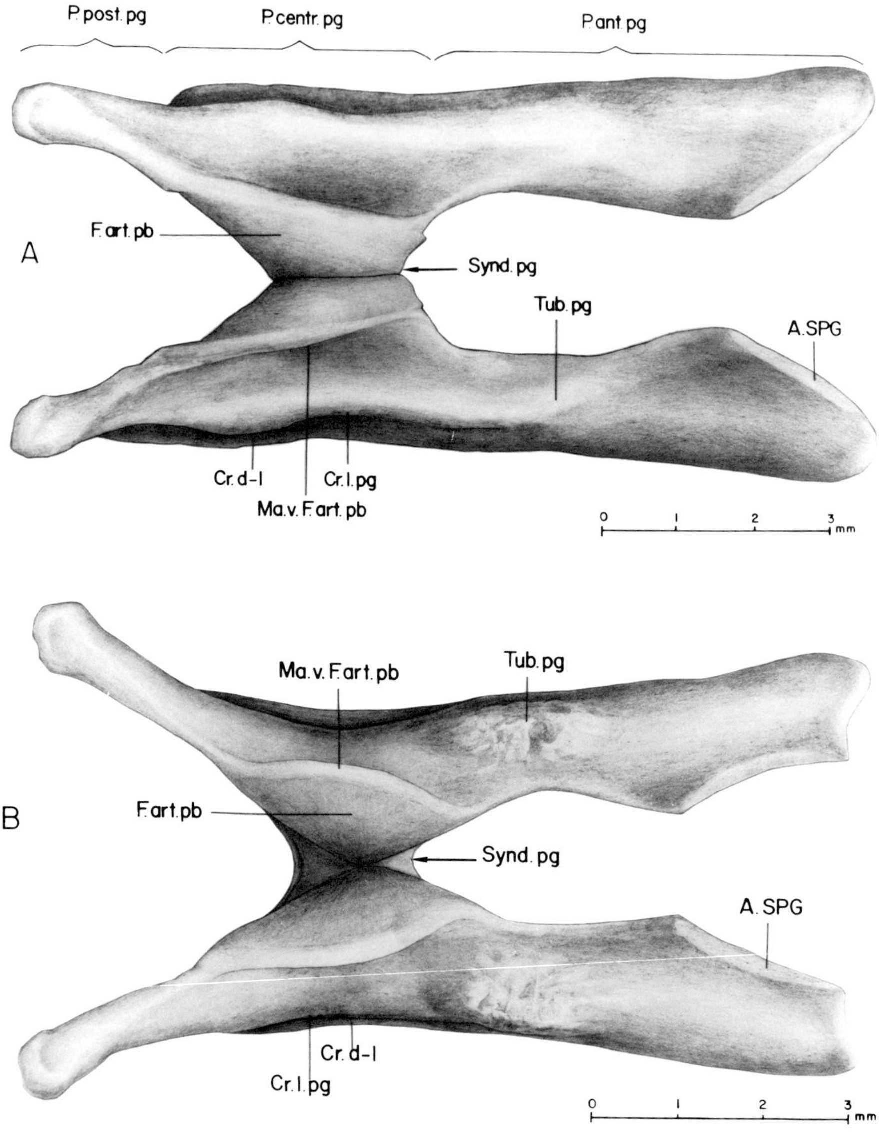

FIG. 9. Ventral views of the bony portion of the articulated paraglossalia of *Psittacus erithacus*. A: AMNH 2354. B: AMNH 3448.

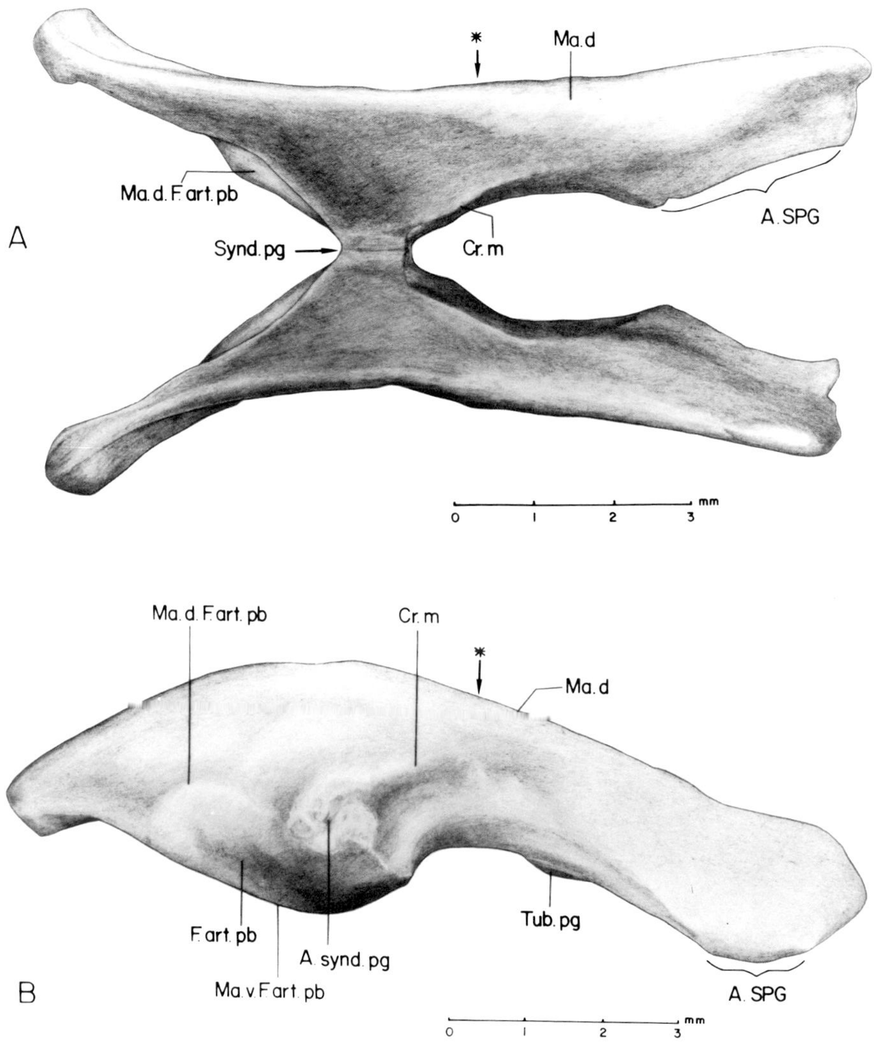

FIG. 10. A: Dorsal view of the bony portion of the articulated paraglossalia of *Psittacus erithacus*, AMNH 3448. B: Medial view of the left paraglossale of *Psittacus erithacus*, AMNH 2354.

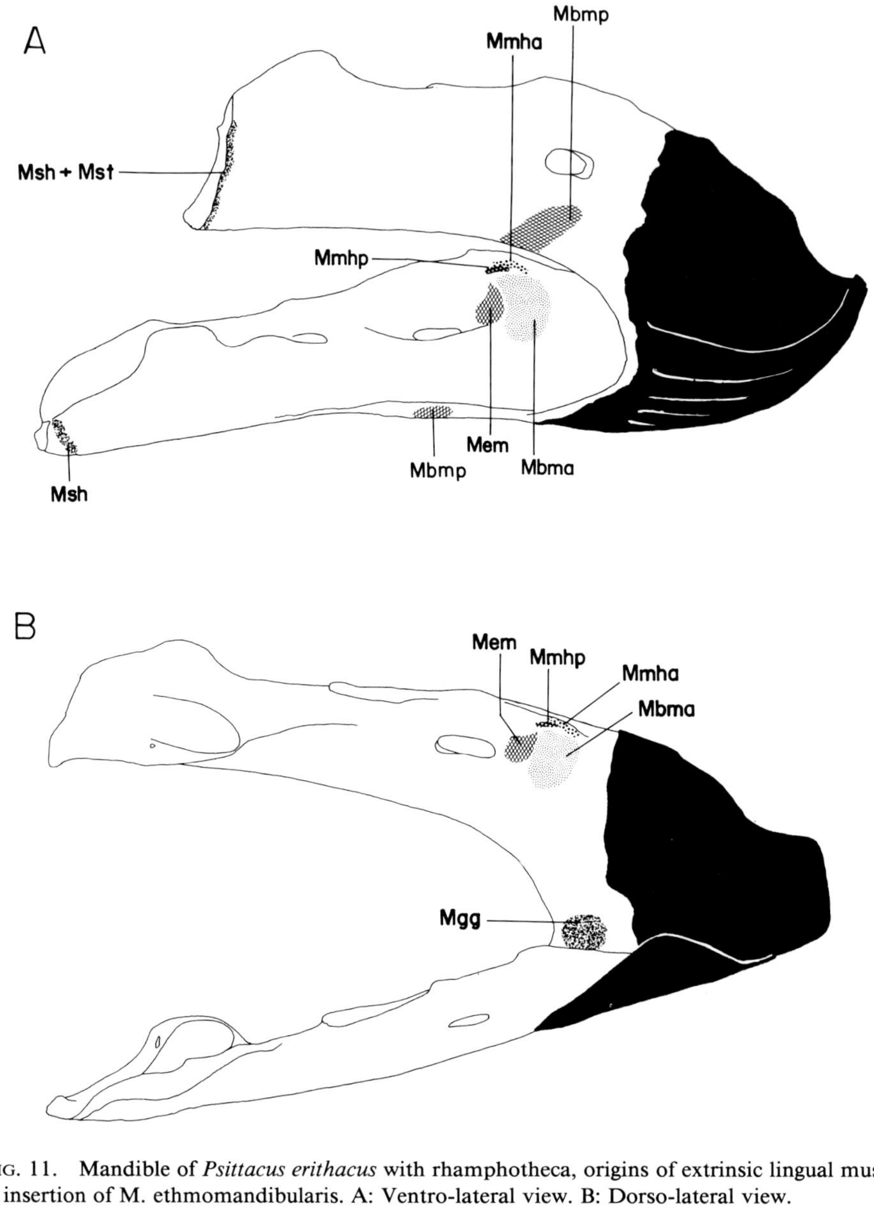

FIG. 11. Mandible of *Psittacus erithacus* with rhamphotheca, origins of extrinsic lingual muscles, and insertion of M. ethmomandibularis. A: Ventro-lateral view. B: Dorso-lateral view.

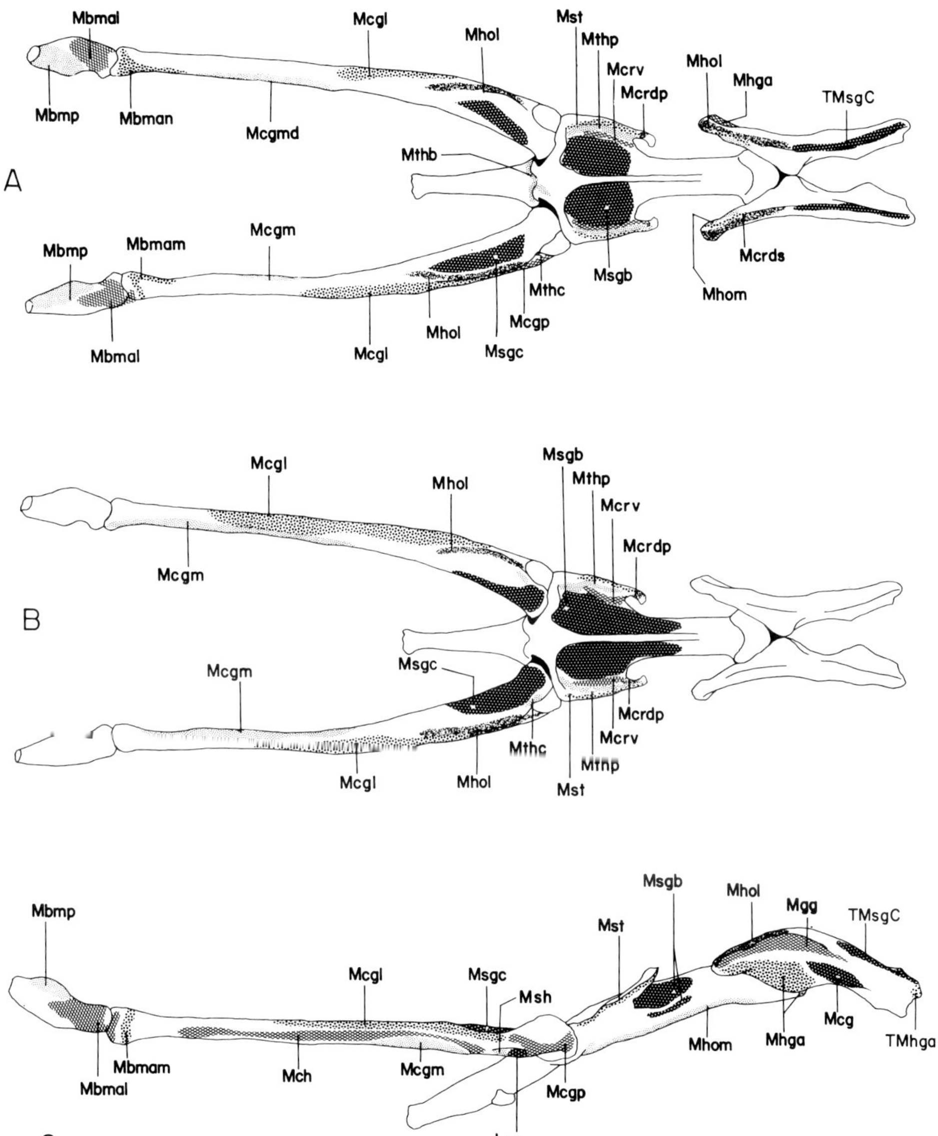

FIG. 12. Hyoid skeleton and attachment sites of lingual muscles of *Psittacus erithacus*. A–B: Dorsal views (A: Complete muscle attachments compiled from specimens Z3 and AMNH 7203; B: Selected variations of muscle attachments compiled from specimens Z2 and Z4, shown only for the cerato-branchialia and basihyale). C: Lateral view.

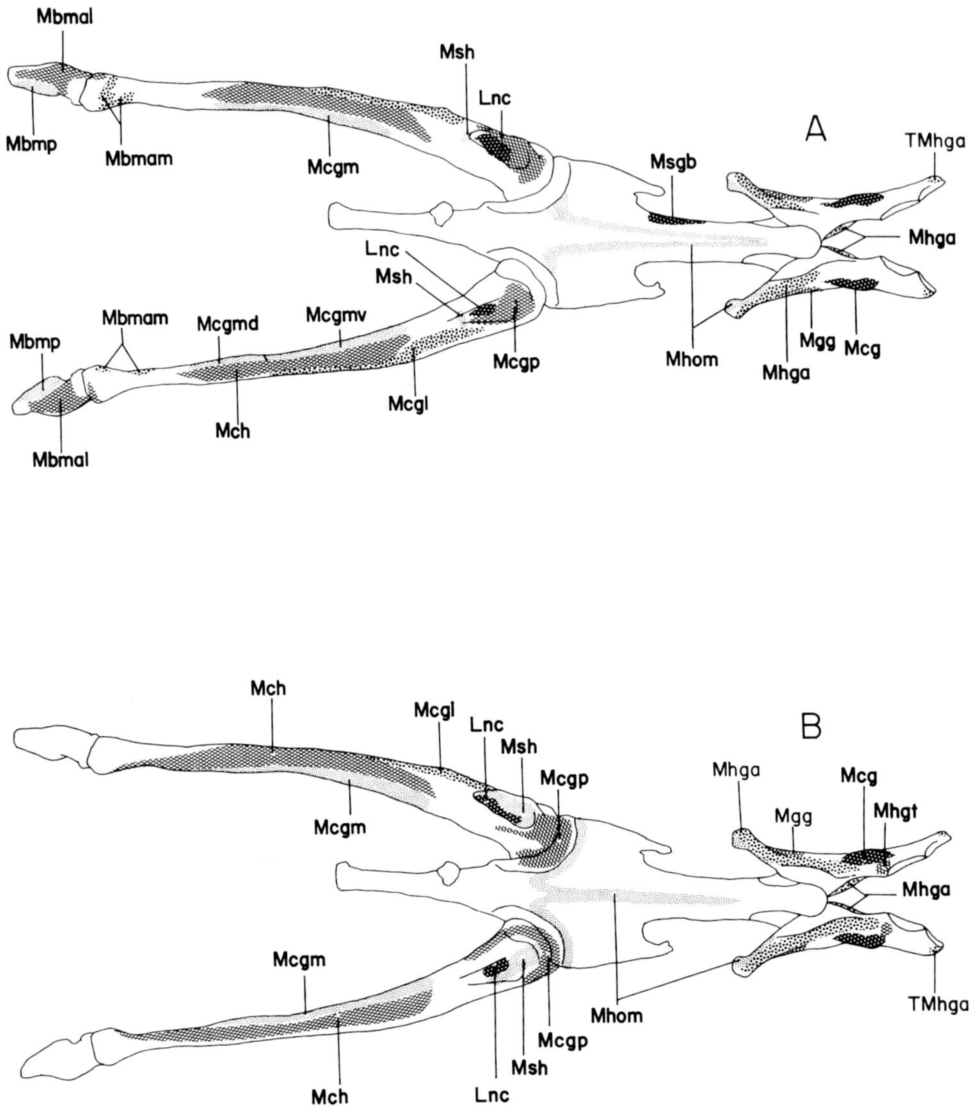

FIG. 13. Ventral view of the hyoid skeleton with attachment sites of lingual muscles of *Psittacus erithacus*. A: Complete muscle attachments compiled from specimens Z3 and AMNH 7203. B: Selected variations of muscle attachments compiled from specimens Z2, Z4, and Z6, shown only for the paraglossalia, basihyale, and ceratobranchialia.

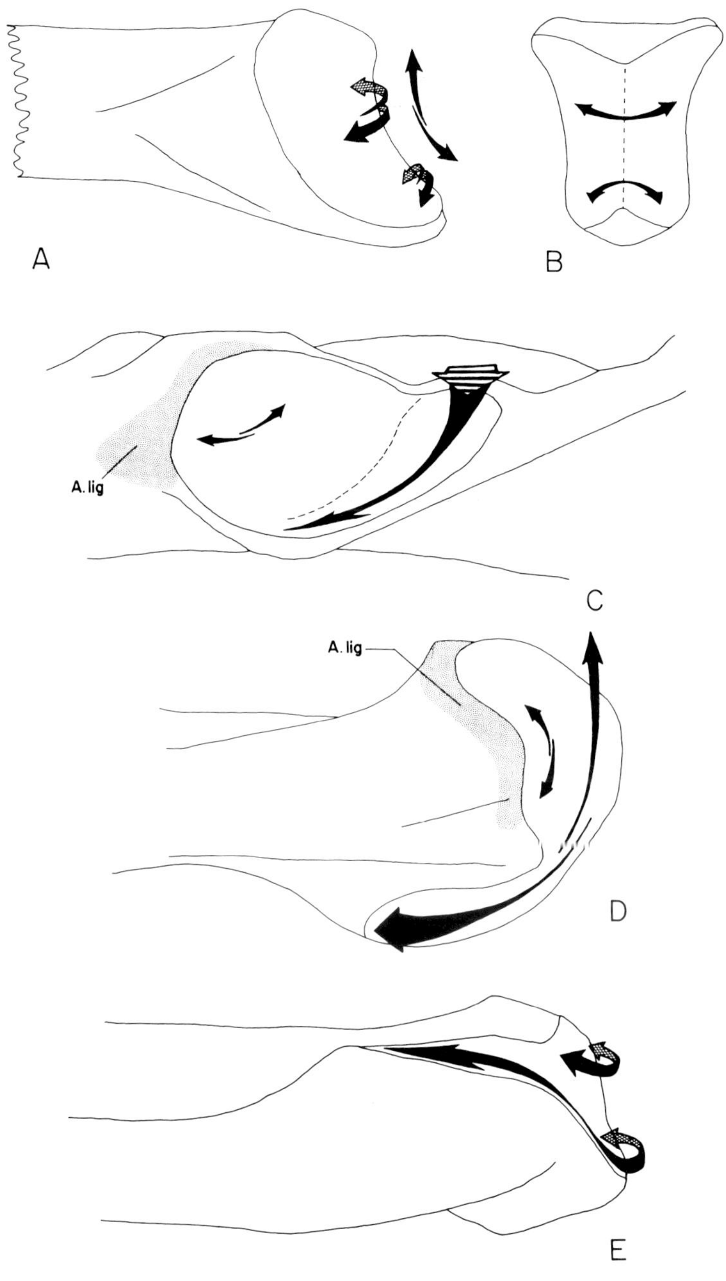

FIG. 14. Articular facets of the hyoid skeleton of *Psittacus erithacus* showing trajectories of the movements of the complementary articular facets. A–B: Articular facet of the apical tip of the basihyale with trajectories of the movements of the articular facet of the paraglossale (A: Lateral view; B: Apical view). C: Caudo-lateral view of the articular facet of the basihyale with trajectories of the movements of the right ceratobranchiale. D–E: Articular facet of the apical end of the right ceratobranchiale with trajectories of the movements of the articular facet of the basihyale (D: Dorso-medial view; E: Lateral view).

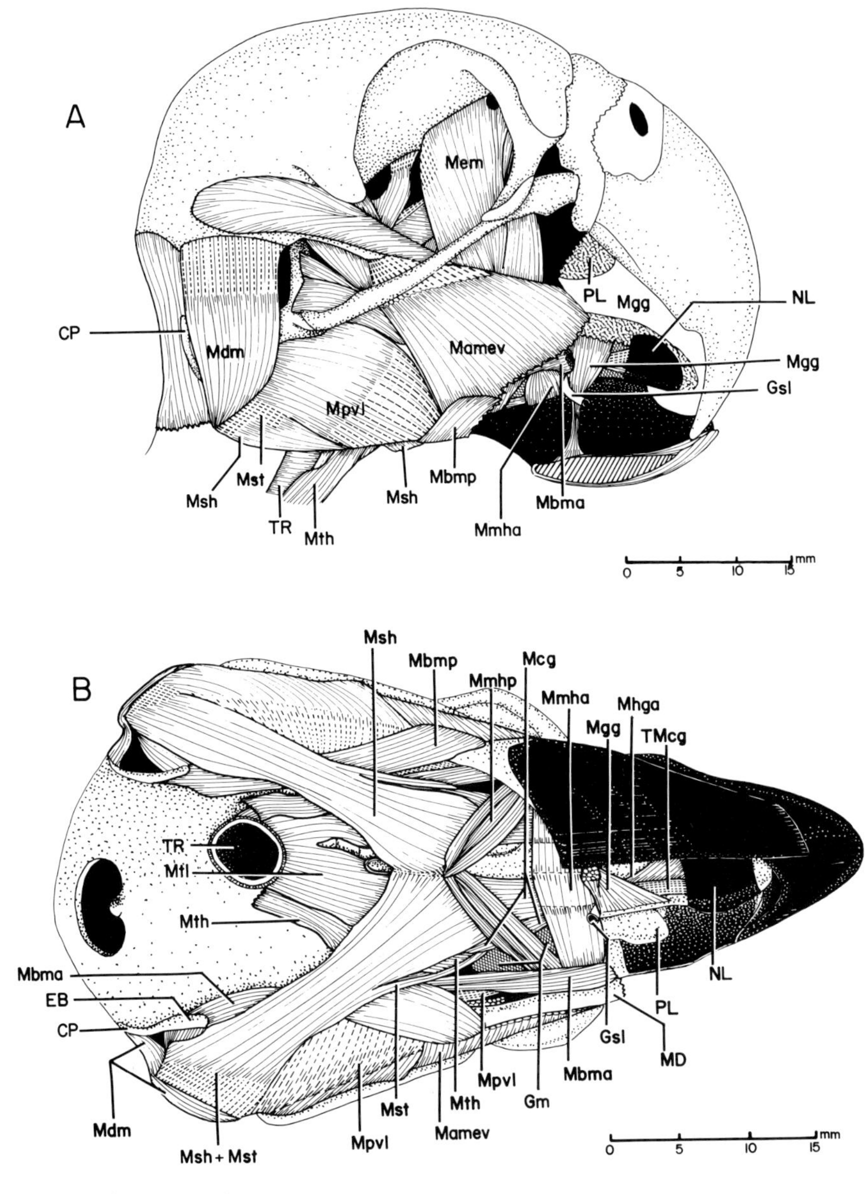

FIG. 15. In situ views of the superficial jaw and lingual musculature of *Psittacus erithacus,* Z3. Left half of the lower rhamphotheca and apical part of the left mandibular ramus removed; lingual epithelium removed on the ventral side of the tongue up to the lingual nail. A: Lateral view (left–right inverted). B: Ventral view.

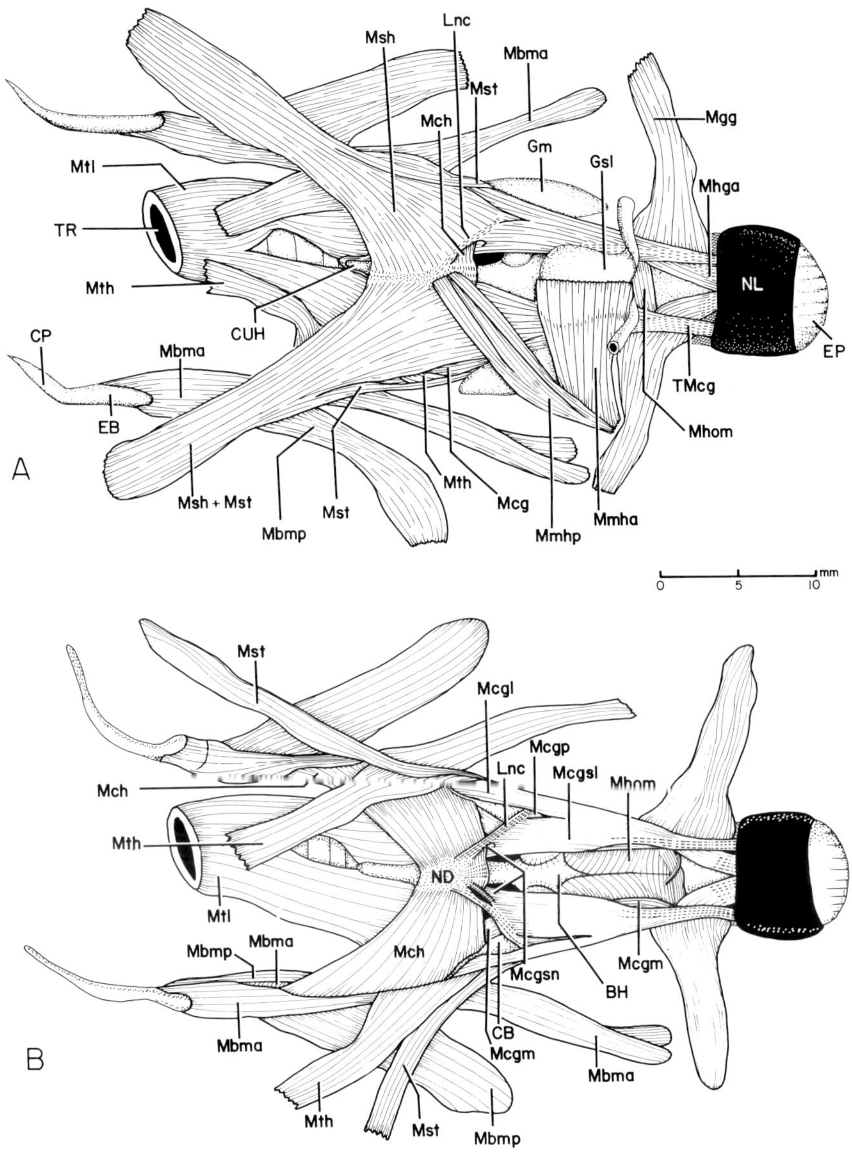

FIG. 16. Ventral views of the lingual apparatus of *Psittacus erithacus,* Z3, after separation from the mandible. A: Modified from Fig. 15B (M. genioglossus laterally reflected; right M. mylohyoideus removed). B: Modified from Fig. 16A (M. serpihyoideus, Gl. mandibularis, and Gl. sublingualis removed; M. stylohyoideus and M. tracheohyoideus laterally reflected).

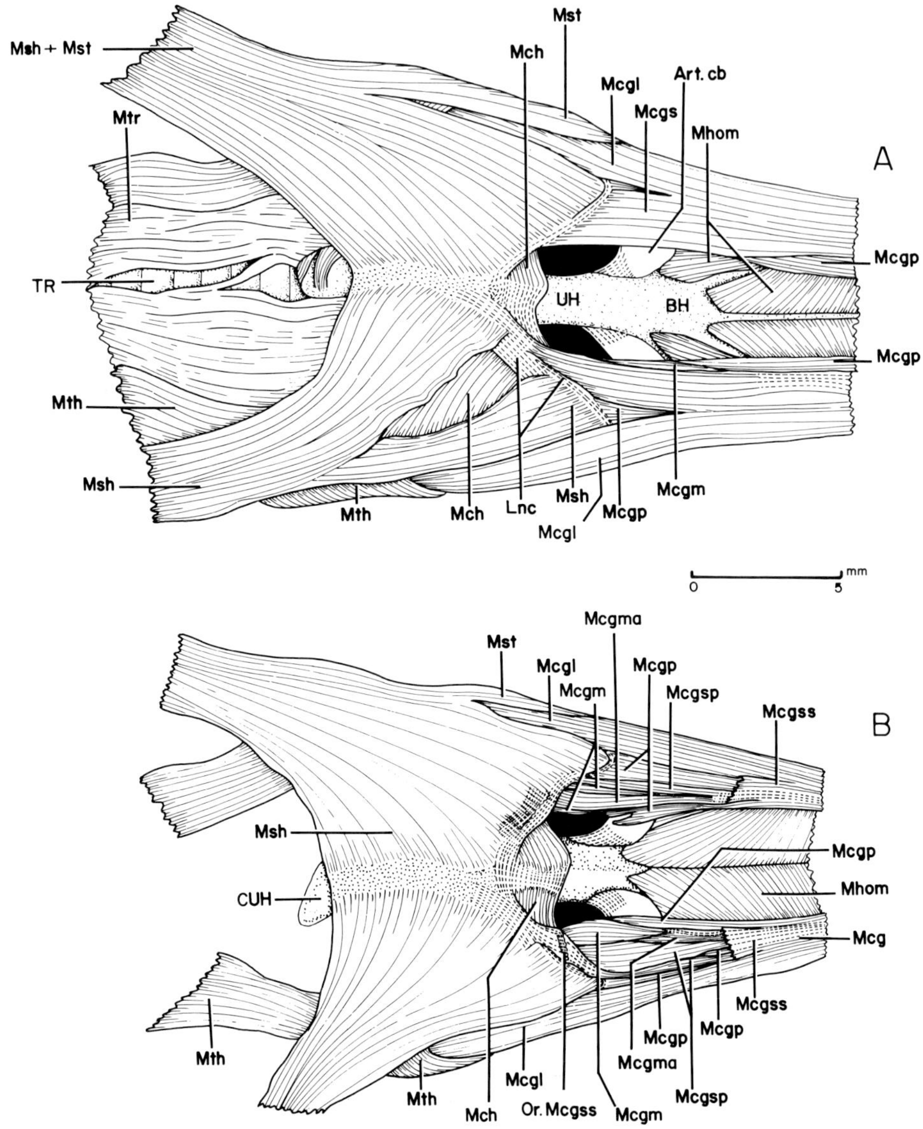

FIG. 17. Enlarged ventral views of the central portion of the lingual apparatus of *Psittacus erithacus*. A: Z4 (right side: superficial view; left side: M. stylohyoideus removed). B: Z2 (M. ceratoglossus superficialis ligamenti superficialis, i.e., Mcgss, and left M. stylohyoideus removed).

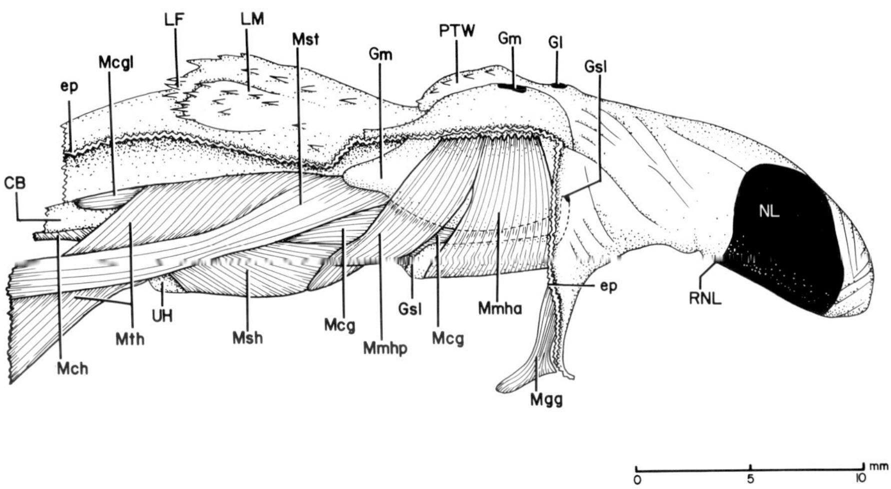

FIG. 20. Lateral view of the antero-central portion of the lingual apparatus of *Psittacus erithacus*, Z4. Connections of the lingual epithelium to the rhamphotheca of the mandible and to the epithelium of the mouth cavity cut; M. mylohyoideus incomplete near its origin.

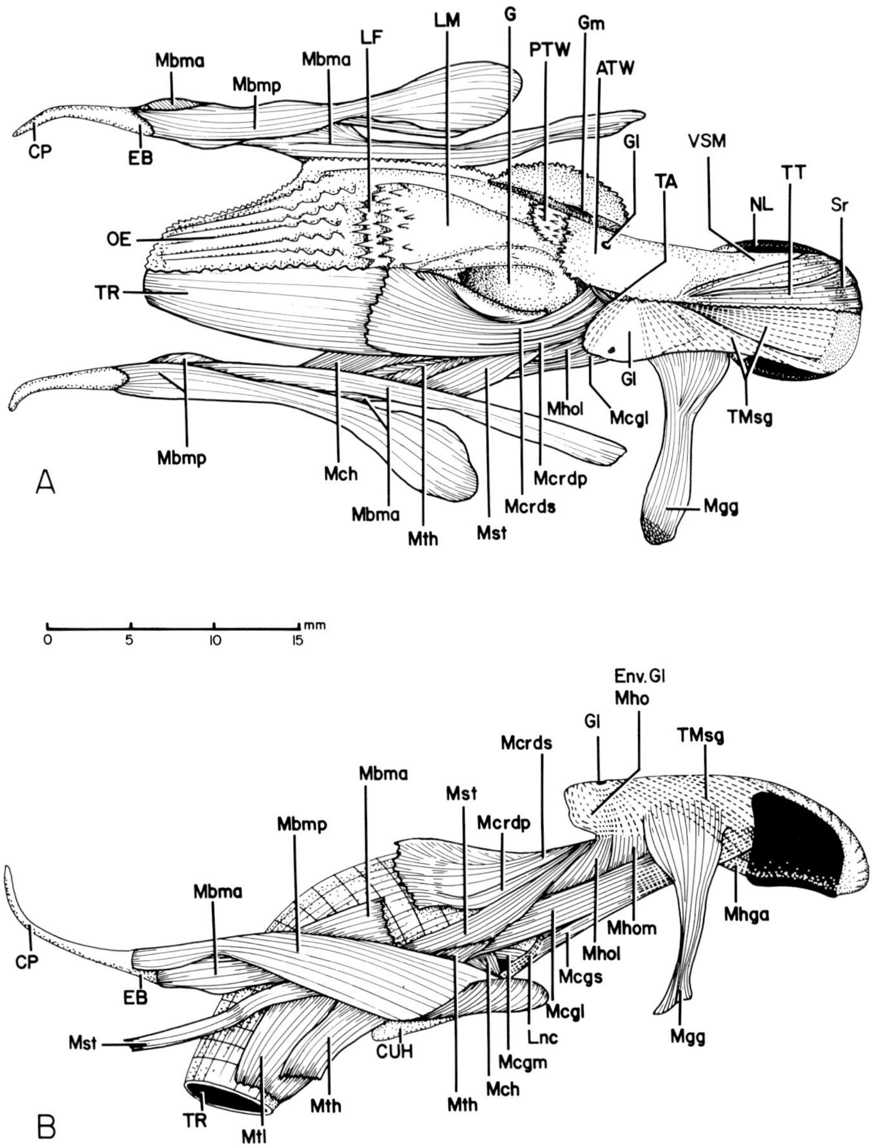

FIG. 21. Superficial lingual musculature and epithelial structures of the lingual apparatus of *Psittacus erithacus,* Z3; epithelium removed on the right and ventral sides of the lingual apparatus. A: Dorsal view, modified from Fig. 16A (M. serpihyoideus and right Gl. mandibularis removed). B: Lateral view (apical half of M. branchiomandibularis anterior removed and insertion of M. tracheolateralis not shown).

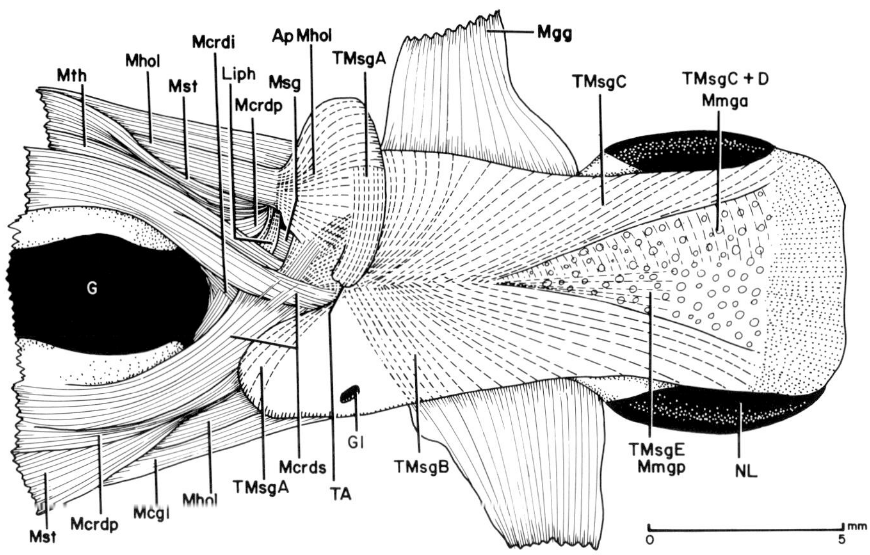

FIG. 22. Enlarged dorsal view of anterior half of the tongue of *Psittacus erithacus,* Z3. Modified from Fig. 21A (lingual epithelium, except lingual nail, and left Gl. lingualis removed; anterior lingual wing apically reflected).

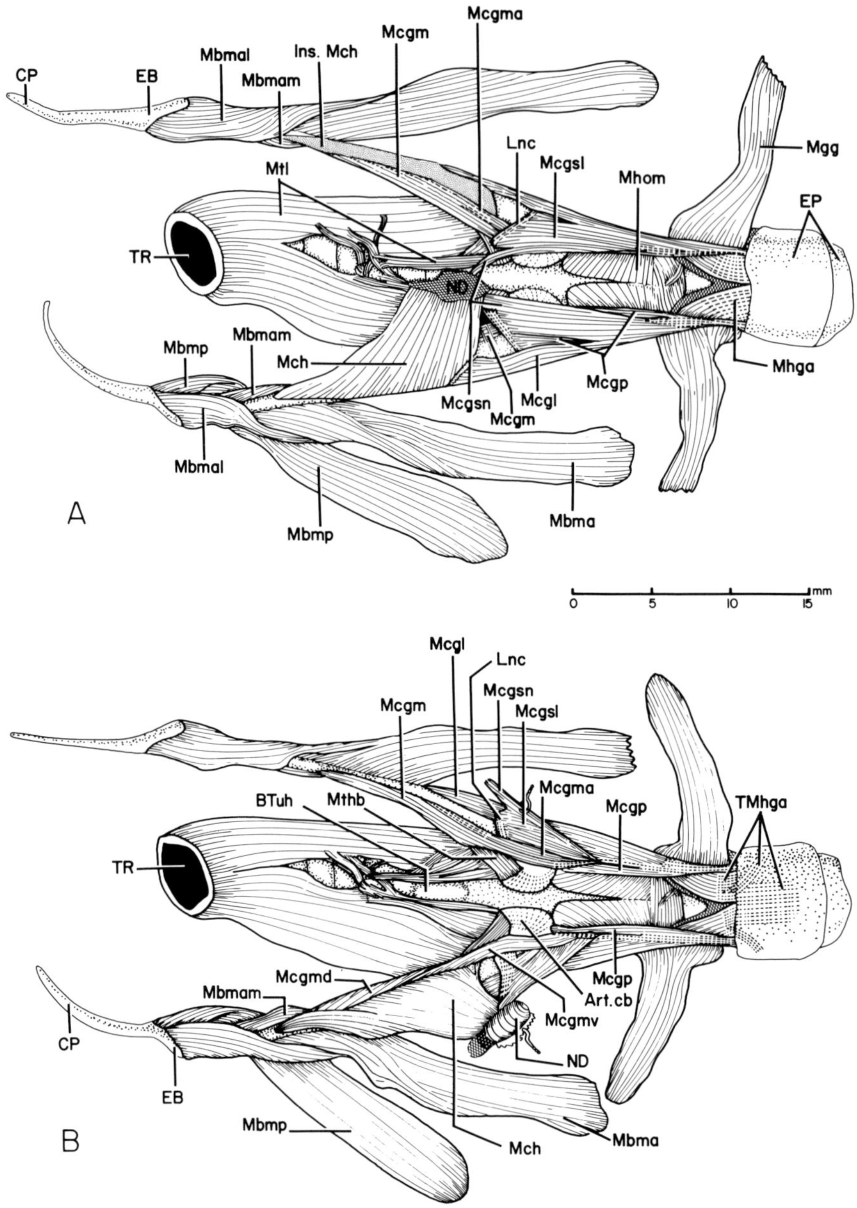

FIG. 23. Ventral views of the lingual apparatus of *Psittacus erithacus*, Z3. A: Modified from Fig. 16B (lingual nail, M. stylohyoideus, M. tracheohyoideus, left M. ceratohyoideus, and left M. branchiomandibularis posterior removed; half of the right segment of the Lig. nodulo-ceratobranchiale attaching to the Nodulus removed; M. branchiomandibularis anterior unfolded). B: Modified from Fig. 23A (right M. ceratoglossus superficialis reflected to one side, and left M. ceratoglossus superficialis, M. ceratohyoideus, and Nodulus reflected together to the other side).

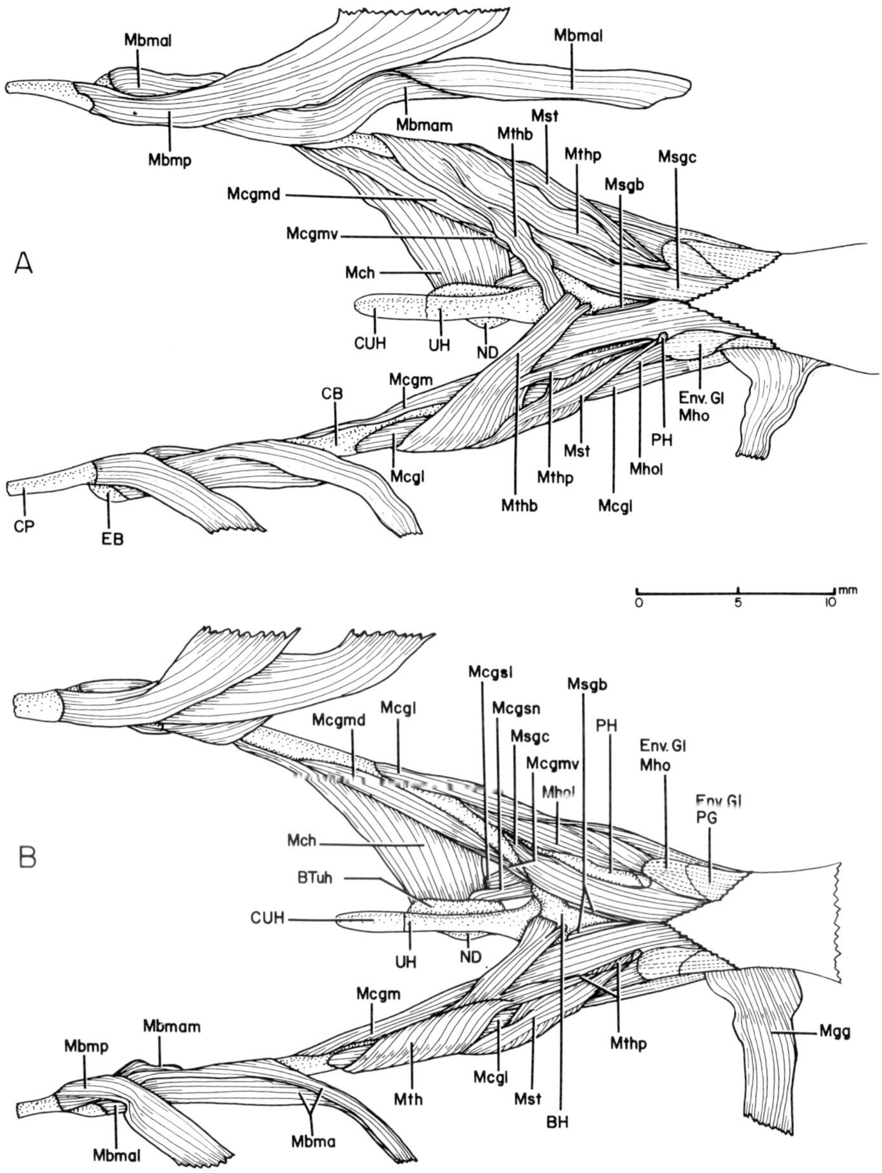

FIG. 24. Dorsal views of the lingual apparatus of *Psittacus erithacus,* Z3, excluding the tongue's tip. A: Modified from Figs. 21A and 22 (laryngeal apparatus with the extrinsic laryngeal musculature removed; paired Gll. linguales and the superficial parts of their tendinous envelopes removed, leaving only the floor of these envelopes; tendinous part of the M. supraglossus not shown; right M. ceratohyoideus and M. ceratoglossus superficialis removed). B: Modified from Fig. 24A (left M. stylohyoideus and M. tracheohyoideus removed).

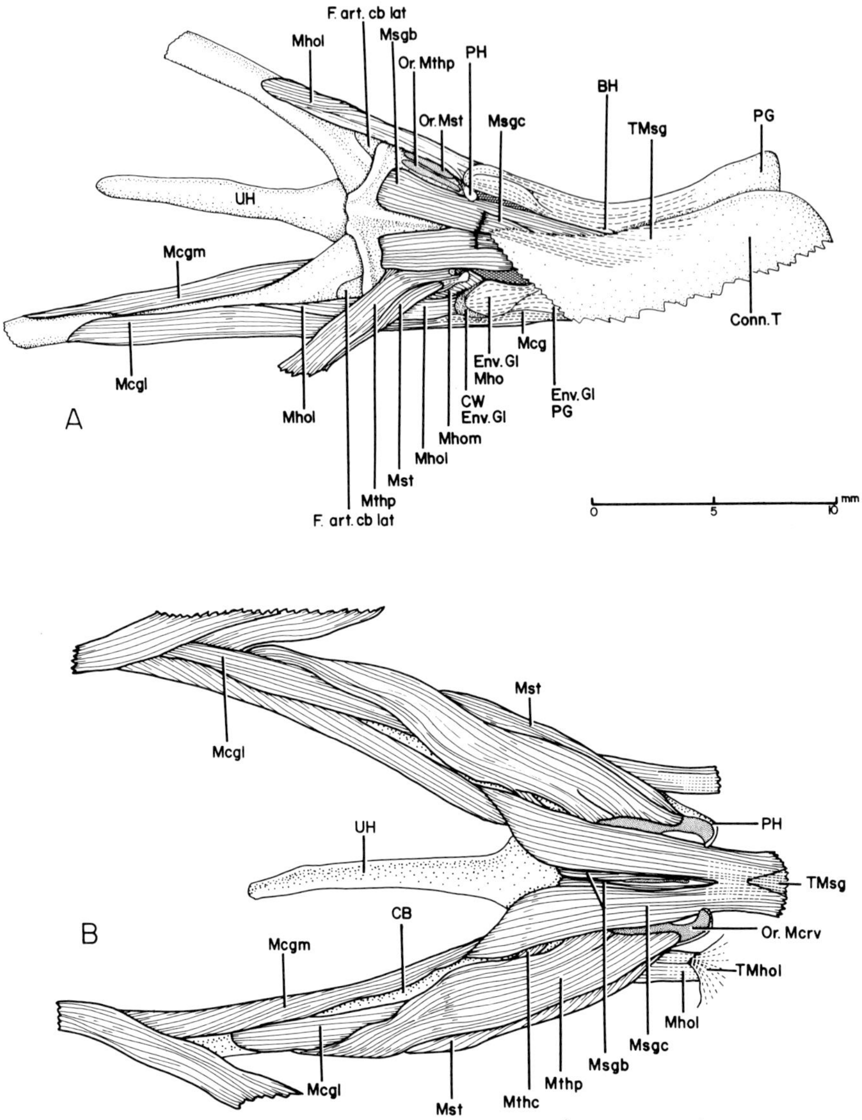

FIG. 25. A: Dorsal view of the apical three-quarters of the lingual apparatus of *Psittacus erithacus,* Z3; modified from Fig. 24B (M. supraglossus ceratobranchialis, right M. tracheohyoideus basihyalis, left M. ceratoglossus, and M. ceratohyoideus removed; left side of tendinous part of M. supraglossus reflected medially; right side of tendinous part of M. supraglossus not shown; right M. tracheohyoideus paraglossalis and M. stylohyoideus spread laterally). B: Dorsal view of the central part of the lingual apparatus of *Psittacus erithacus,* Z4, excluding the paraglossale and lingual tip (laryngeal apparatus and extrinsic laryngeal muscles, M. ceratohyoideus, M. hypoglossus obliquus medialis, and apical ends of M. ceratoglossus, M. hypoglossus obliquus lateralis and M. supraglossus removed; M. tracheohyoideus basihyalis absent).

A

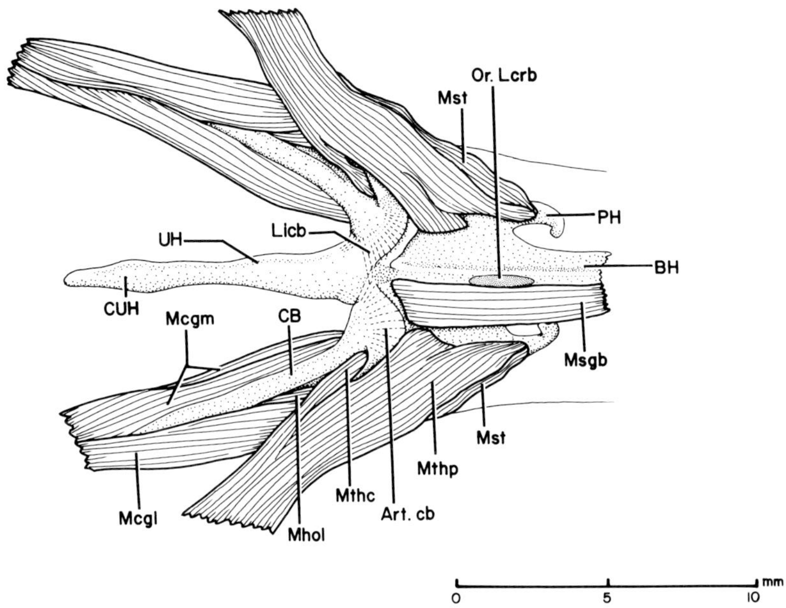

B

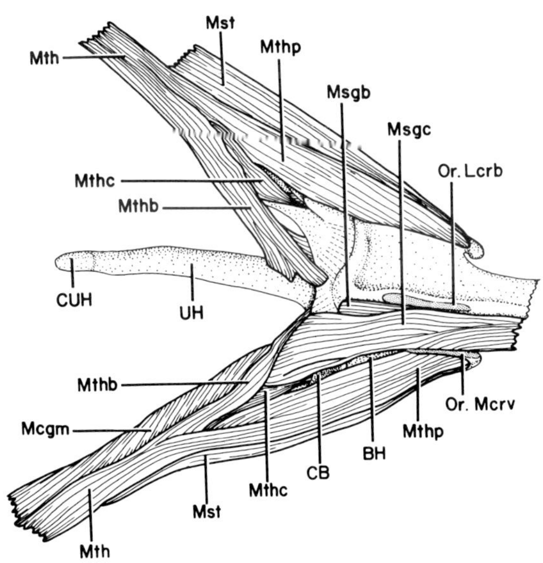

FIG. 26. Dorsal views of central portions of the lingual apparatus of *Psittacus erithacus*. A: Z4; modified from Fig. 25B (M. supraglossus ceratobranchialis, M. hypoglossus obliquus lateralis, and left M. supraglossus basihyalis removed; M. tracheohyoideus spread laterally). B: Z1 (muscles removed as in Fig. 26A, except right M. supraglossus ceratobranchialis).

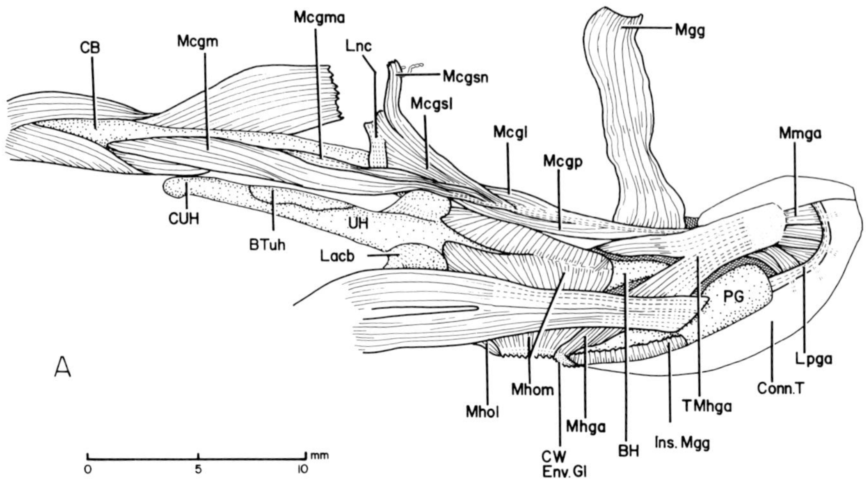

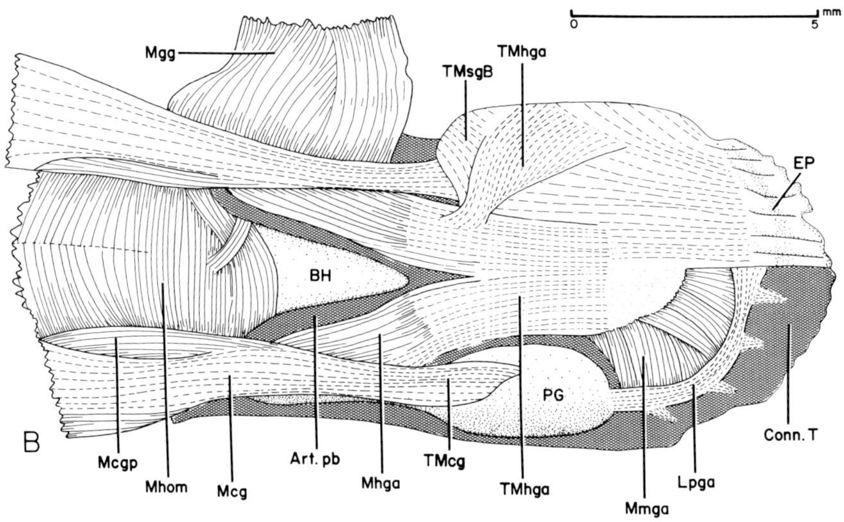

FIG. 27. A: Ventro-lateral view of the apical three-quarters of the lingual apparatus of *Psittacus erithacus,* Z3; modified from Fig. 23B (laryngeal apparatus, M. genioglossus, and connective tissue covering the left half of the ventral side of the lingual tip removed). B: Enlarged ventral view of the tongue's tip of *Psittacus erithacus,* Z3, unmodified from Fig. 27A.

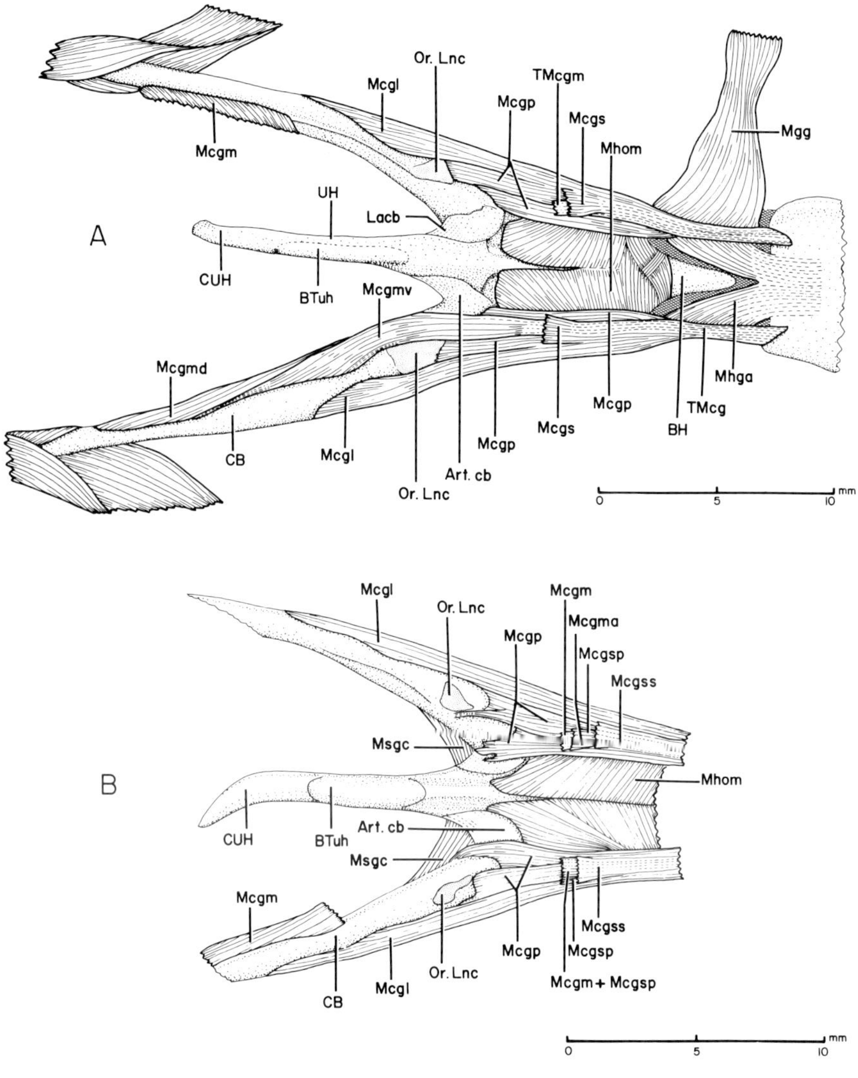

FIG. 28. Ventral views of central portions of the lingual apparatus of *Psittacus erithacus*. A: Z3; modified from Figs. 23B and 27A (M. ceratoglossus superficialis, right M. ceratoglossus medialis, and left M. ceratohyoideus, Lig. nodulo-ceratobranchiale, and Nodulus removed). B: Z2; modified from Fig. 17B (M. serpihyoideus, M. ceratoglossus superficialis et medialis, M. tracheohyoideus, and right M. stylohyoideus removed).

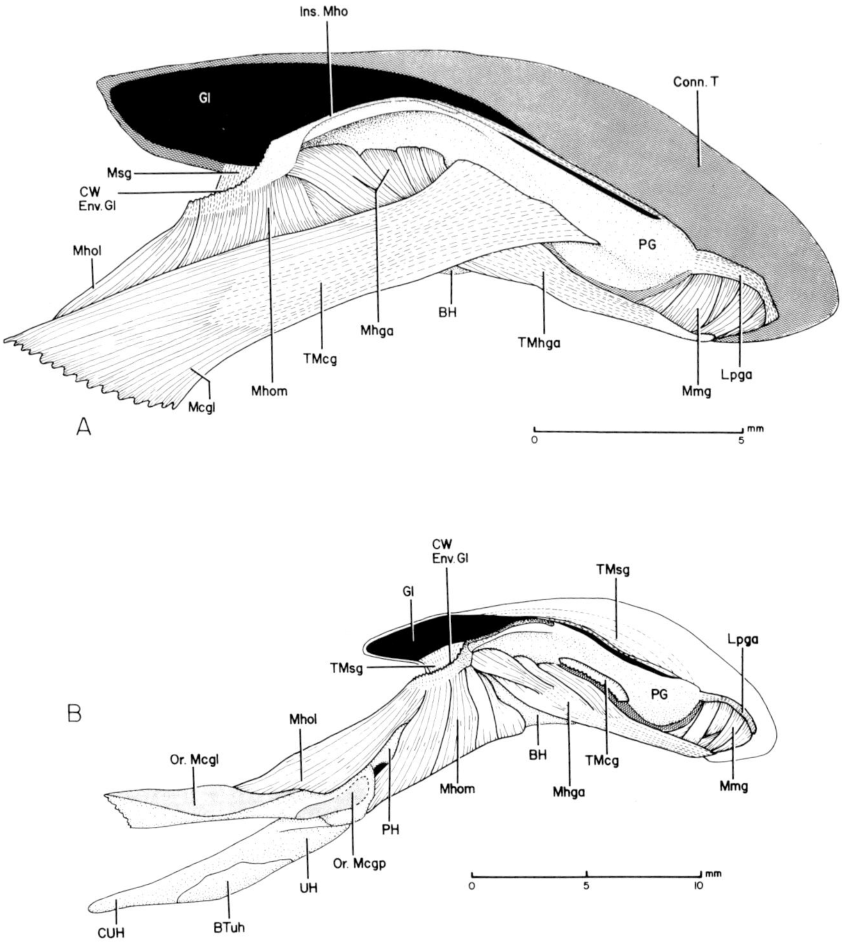

FIG. 29. Lateral views of the apical portion of the lingual apparatus of *Psittacus erithacus,* Z3. A: Modified from Fig. 21B (right M. genioglossus, right Gl. lingualis, lateral wall of tendinous envelope of right Gl. lingualis, and lingual nail and connective tissue covering the ventro-lateral surface of the lingual tip removed. The internal surface of the medial wall of the right glandular envelope is shown in black. The part of the right M. supraglossus shown will continue along the external surface of the medial wall of the right glandular envelope). B: Modified from Fig. 29A (M. ceratoglossus removed).

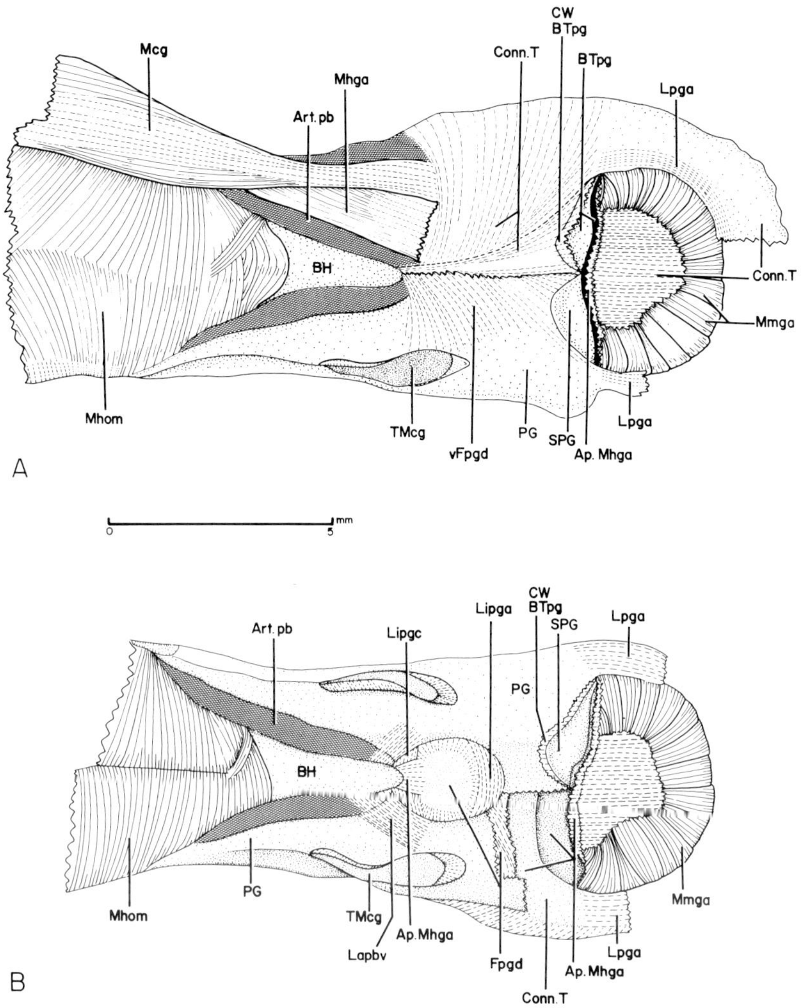

FIG. 30. Ventral views of the tongue of *Psittacus erithacus,* Z3. A: Modified from Fig. 27B (left side: M. ceratoglossus, M. hypoglossus anterior, capsule of Bursa synovialis paraglossalis, Lig. paraglossale apicale, and connective tissue covering the ventral side of the lingual tip removed; right side: M. hypoglossus anterior, capsule of Bursa synovialis paraglossalis, and connective tissue covering the ventral side of the lingual tip partially removed. For reconstruction of three-dimensional relationships of the structures, see Figs. 43–45). B: Modified from Fig. 30A (left side: ventral fibers of Fascia paraglossalis dorsalis and apical part of Proc. anterior paraglossalis removed to expose parts of the structures lying on the dorsal side of the paraglossale, e.g., part of the Fascia paraglossalis dorsalis shortly before it passes ventrally through the Foramen interparaglossale, and the aponeurosis of the M. hypoglossus anterior with the outline of the internal surface of the outer wall of the attached Bursa synovialis paraglossalis; right side: M. ceratoglossus, M. hypoglossus anterior, connective tissue, tendinous fibers of M. supraglossus, and fibers of the Fascia paraglossalis dorsalis attaching on the ventral surface of the paraglossale removed).

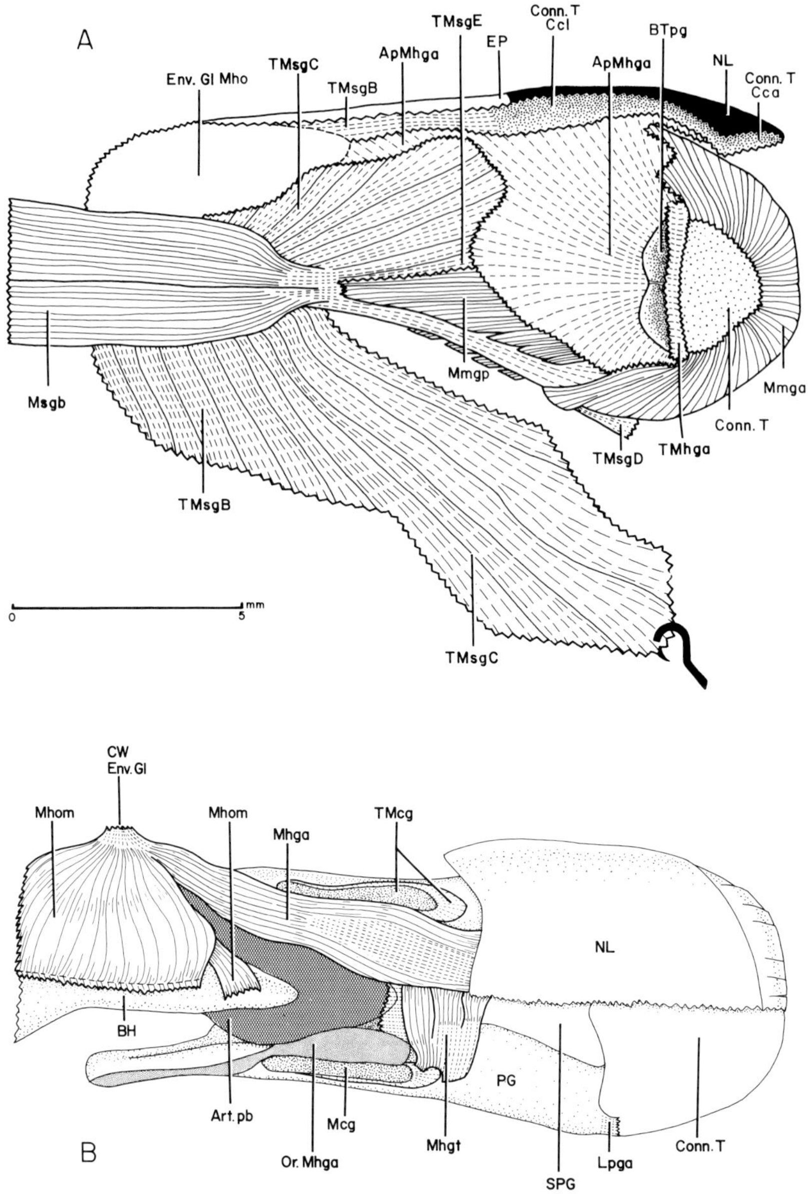

FIG. 31. A: Ventral view of the M. supraglossus-M. mesoglossus system of the lingual tip of *Psittacus erithacus,* Z3; modified from Fig. 30B (paraglossalia, basihyale, and Lig. paraglossale apicale removed. Left side: tendinous fibers of portion E of the M. supraglossus removed and caudal part of the aponeurosis of the M. hypoglossus anterior cut at the origin of the M. mesoglossus posterior to expose the ventral surface of the M. mesoglossus posterior. The tendinous fibers of portion C of the M. supraglossus are separated from portions D and E and are spread laterally with portion B. Right side: Window cut into the dorso-laterally running tendinous fibers originating from the aponeurosis of the M. hypoglossus anterior to expose the tendinous fibers of the M. supraglossus. For reconstruction of the spatial arrangement, see Figs. 45–47). B: Ventral view of the tongue's tip of *Psittacus erithacus,* Z1 (M. ceratoglossus, left M. hypoglossus obliquus medialis, left M. hypoglossus anterior, and connective tissue removed to show M. hypoglossus transversus lying directly on the ventral surface of the paraglossale and covering the Foramen interparaglossale).

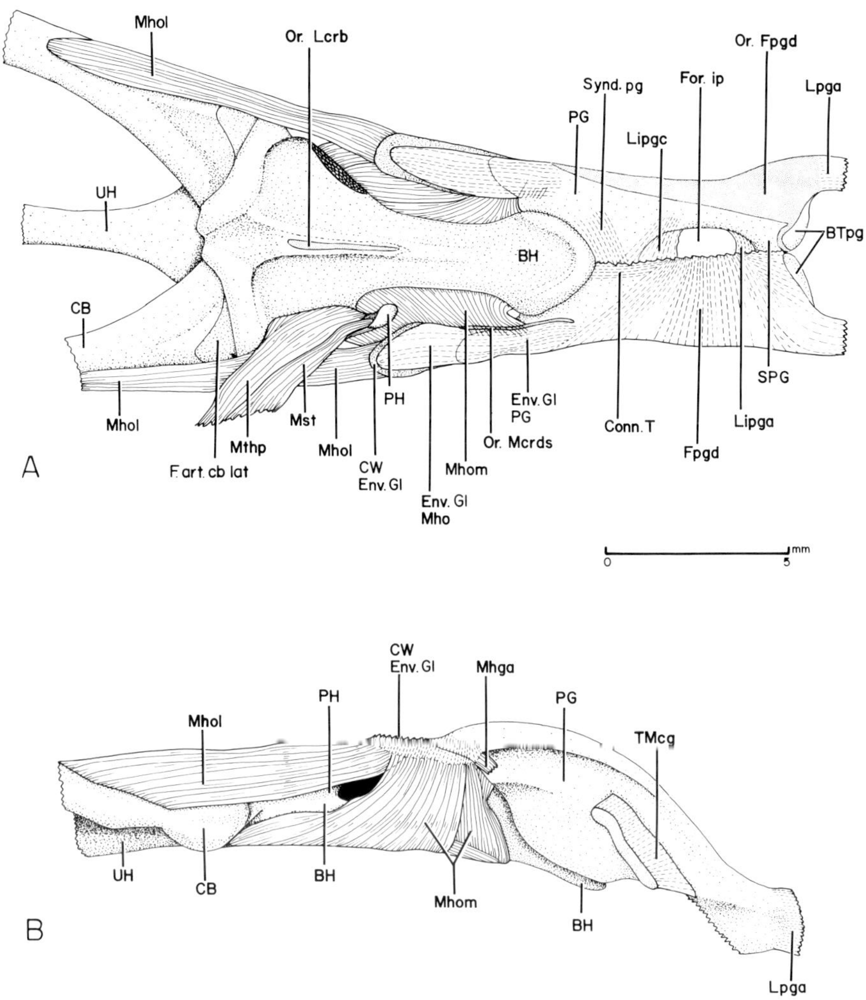

FIG. 32. Tip of the tongue of *Psittacus erithacus,* Z3. A: Dorsal view; modified from Fig. 25A (M. supraglossus, M. mesoglossus, and connective tissue removed; left side: Proc. parahyalis and Fascia paraglossalis dorsalis removed; right side: M. ceratoglossus removed). B: Lateral view; modified from Fig. 29B (M. supraglossus, M. mesoglossus, M. hypoglossus anterior, and connective tissue removed).

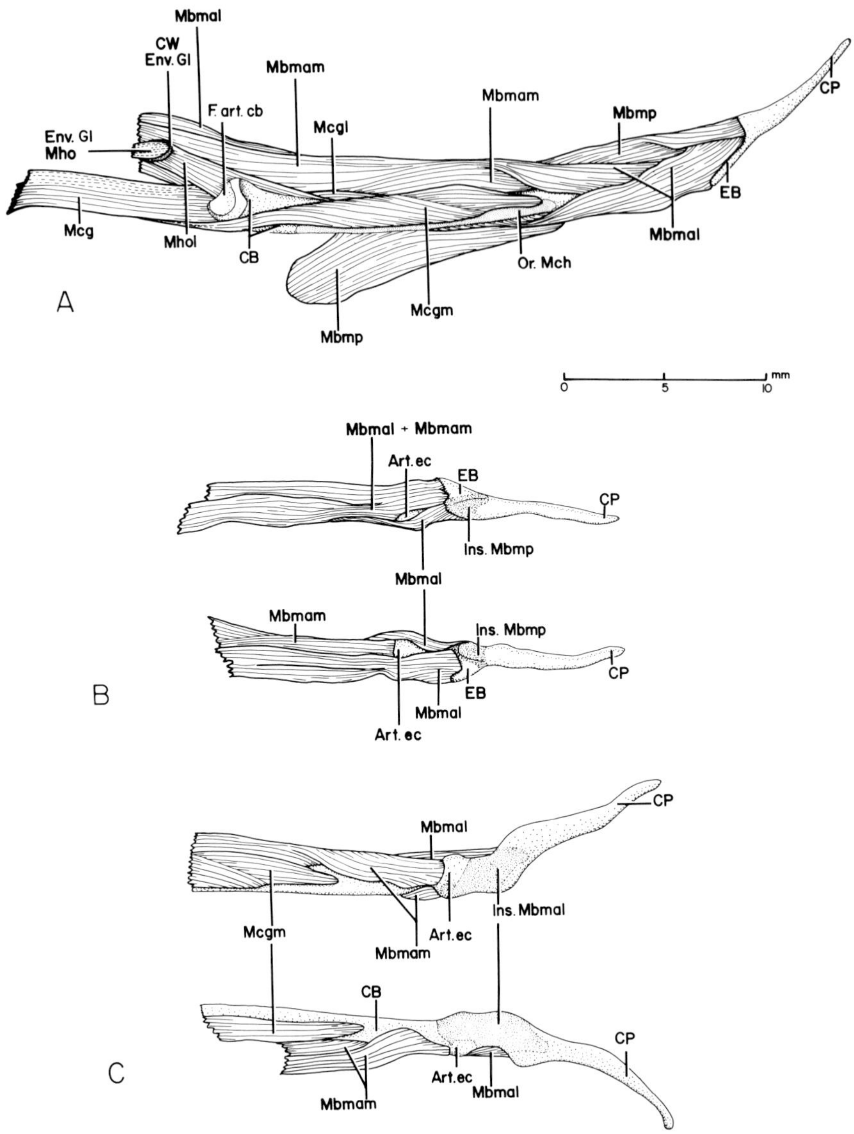

FIG. 33. A: Medial view of the right ceratobranchiale and epibranchiale with attached muscles of *Psittacus erithacus,* Z3; M. ceratohyoideus and M. supraglossus removed. B: Dorsal view of the left and right epibranchialia and caudal ends of the ceratobranchialia with attached muscles of *Psittacus erithacus,* Z3, modified from Fig. 33A (M. branchiomandibularis posterior removed). C: Medial view of the left and right epibranchialia and caudal ends of the ceratobranchialia with attached muscles of *Psittacus erithacus,* Z3; modified from Fig. 33B (M. branchiomandibularis anterior lateralis removed).

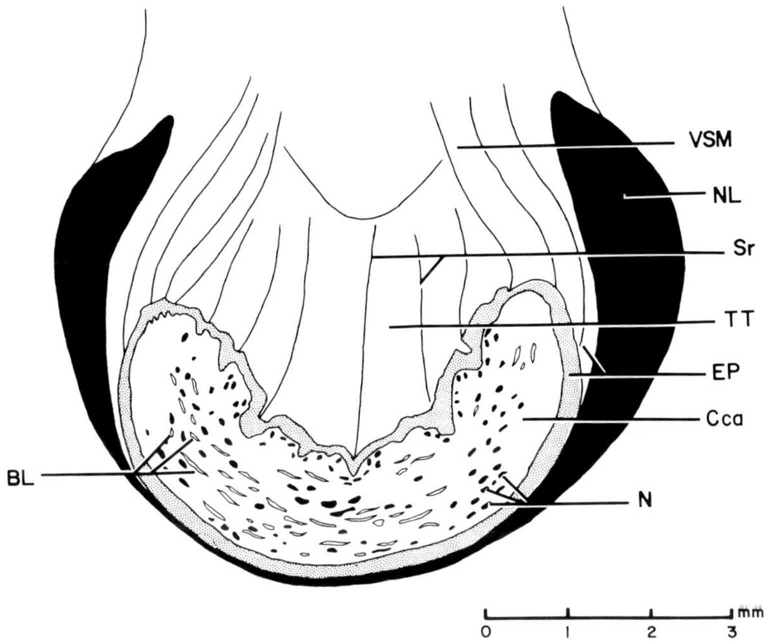

FIG. 34. Apical view of a cross-section through the lingual tip of *Psittacus erithacus,* Z4, at the level of the apical border of the lingual nail (cf. Fig. 43).

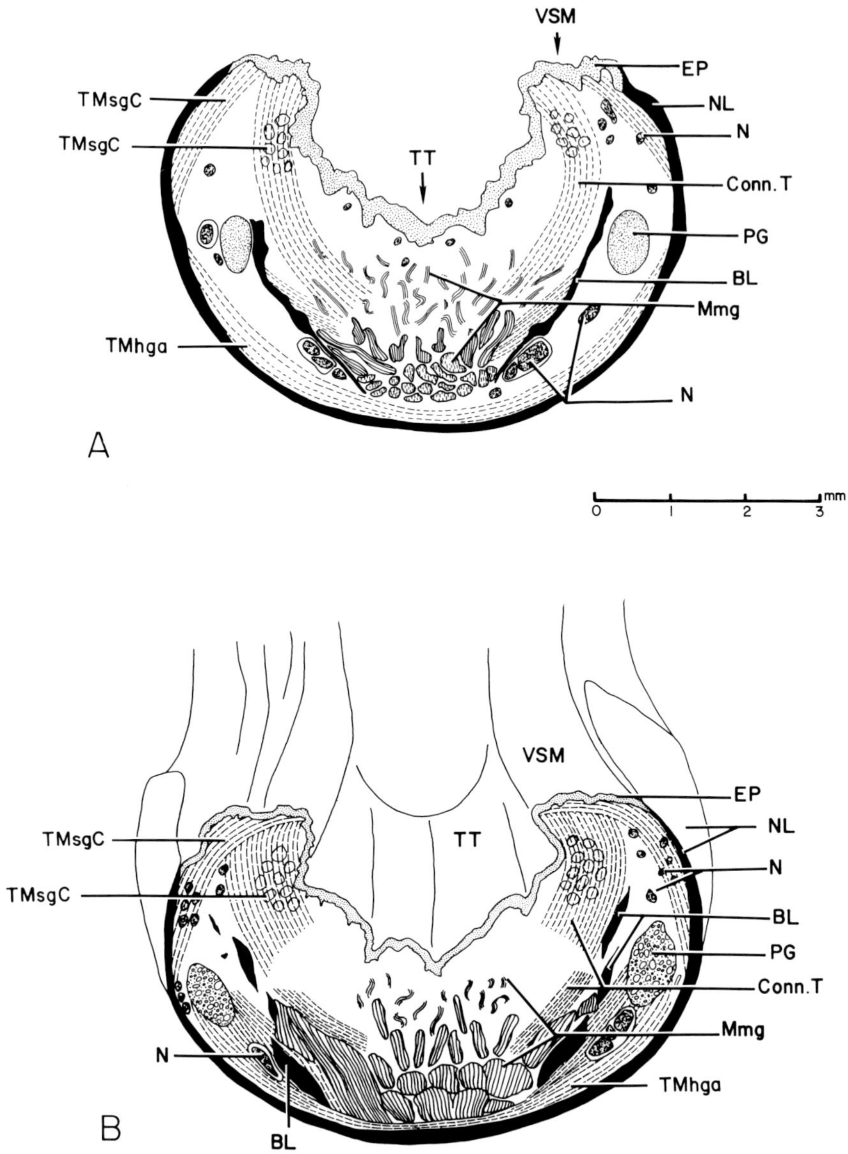

FIG. 35. Cross-sections through the lingual tip of *Psittacus erithacus,* Z4. A: At the level between the apical border of the lingual nail and apical tip of the paraglossale (caudal view). B: As in Fig. 35A, but somewhat more caudally (apical view); (cf. Fig. 43).

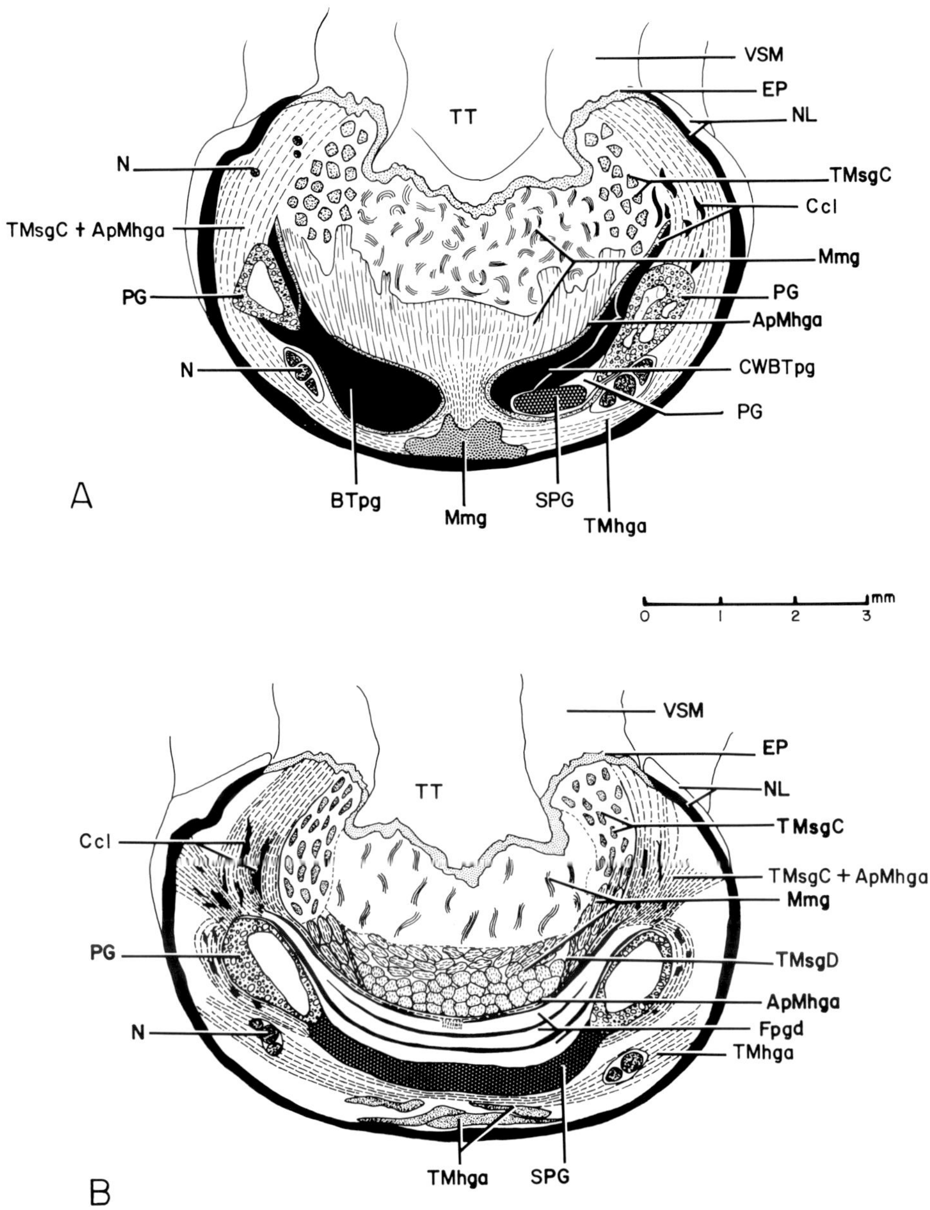

FIG. 36. Apical views of cross-sections through the lingual tip of *Psittacus erithacus,* Z4. A: At the level of the Bursae synoviales paraglossales (right side: BTpg not opened, view on the external surface of its outer wall through a window in the aponeurosis of the M. hypoglossus anterior; left side: BTpg opened, cross-section through underlying anterior margin of the Synchondrosis paraglossalis). B: At the level of the anterior portions of the Procc. anteriores paraglossales and the Synchondrosis paraglossalis (cf. Fig. 43).

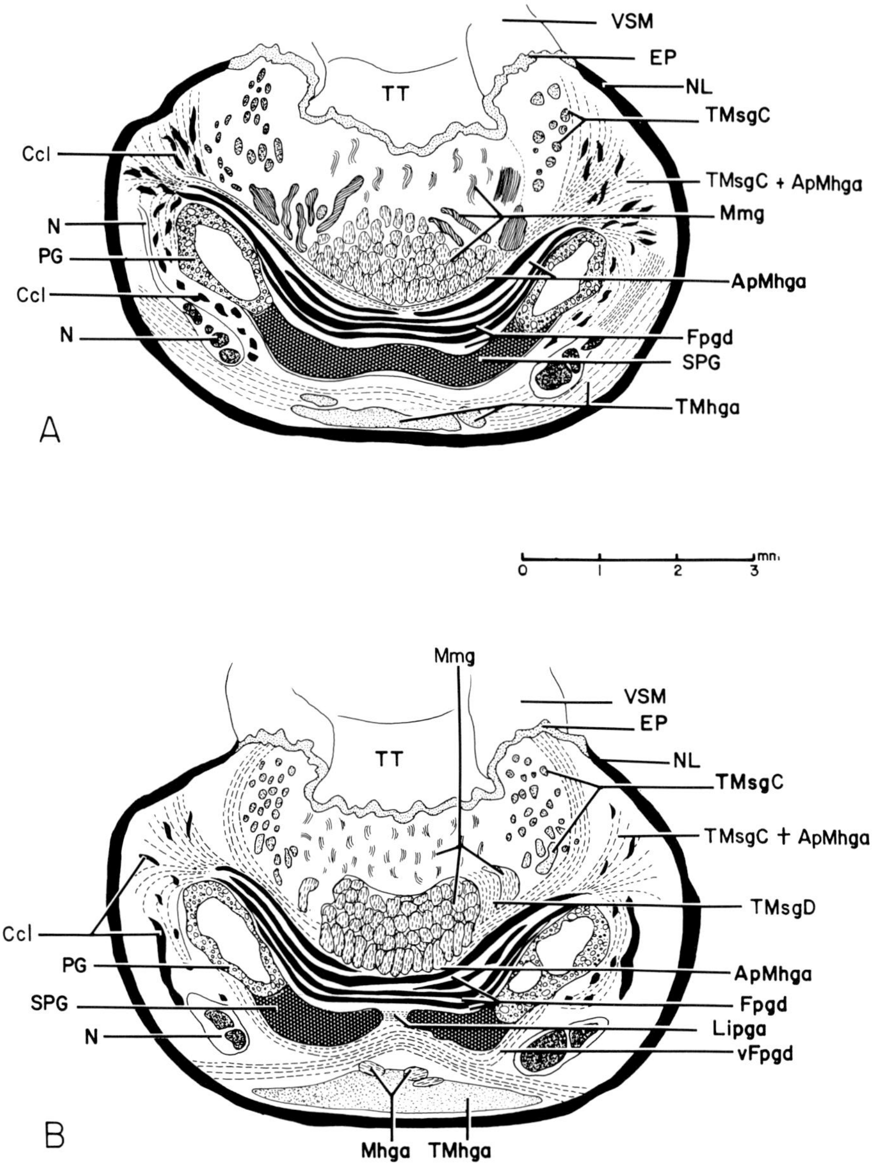

FIG. 37. Apical views of cross-sections through the lingual tip of *Psittacus erithacus,* Z4. A: As in Fig. 36B, but farther caudally. B: At the level of the caudal border of the Synchondrosis paraglossalis (cf. Fig. 43).

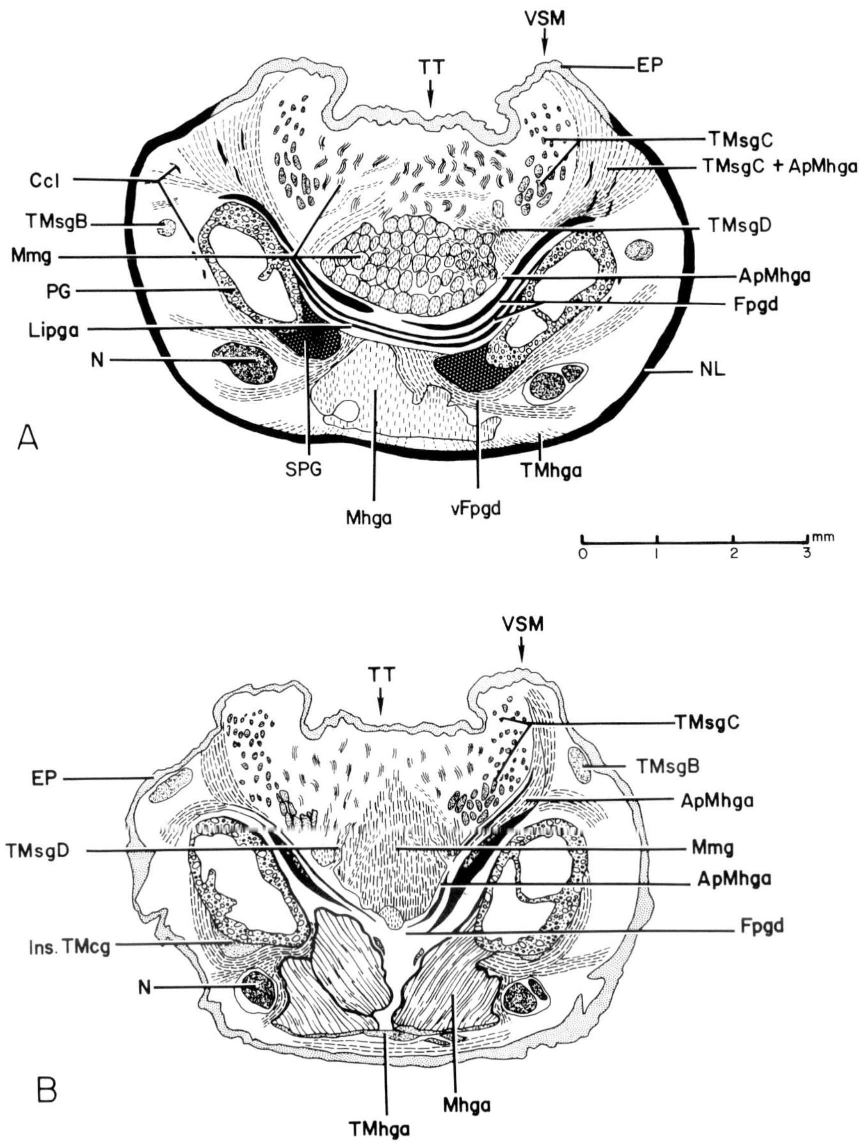

FIG. 38. Apical views of cross-sections through the lingual tip of *Psittacus erithacus*, Z4. A: At the level of the Foramen interparaglossale, just apically from the caudal border of the lingual nail. B: At the level of the center of the Foramen interparaglossale (cf. Fig. 43).

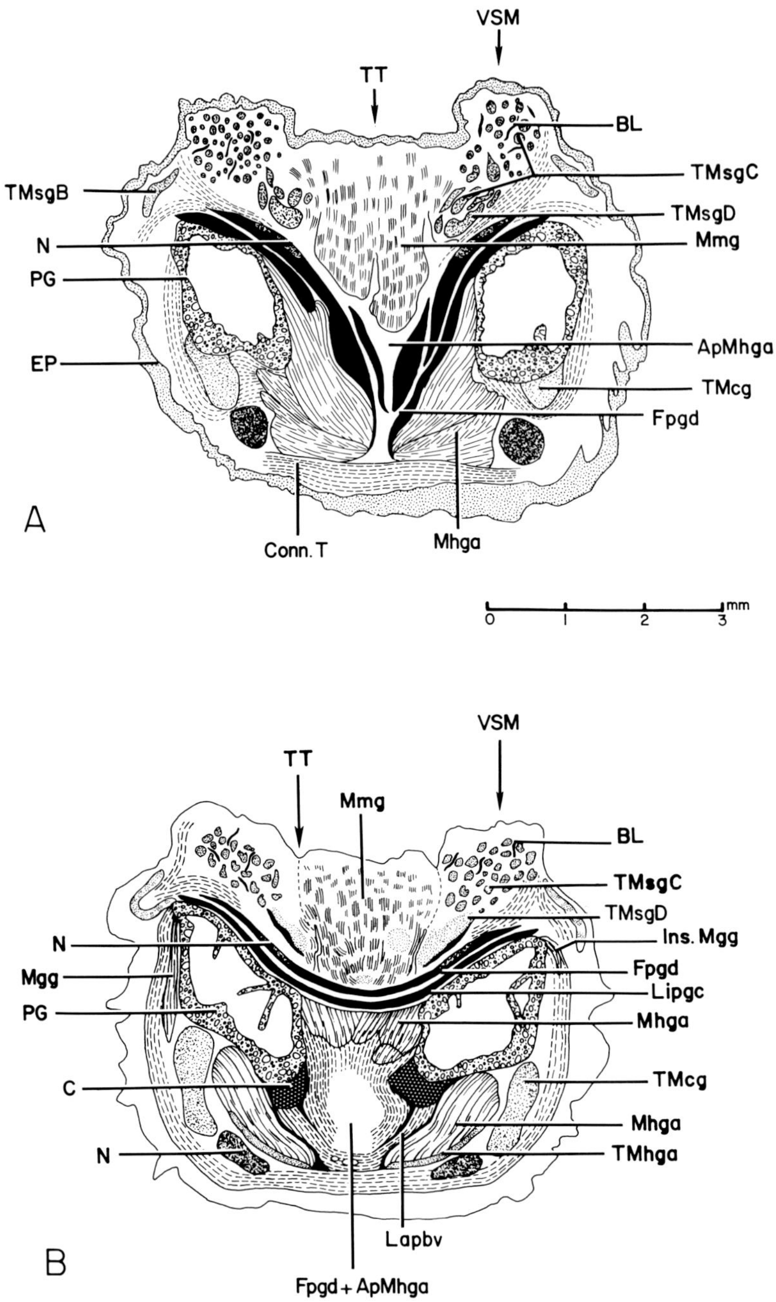

FIG. 39. Apical views of cross-sections through the lingual tip of *Psittacus erithacus,* Z4. A: As in Fig. 38B, but somewhat more caudally. B: As in Fig. 39A, but somewhat more caudally, i.e., at the level of the apical end of the insertion of the M. genioglossus.

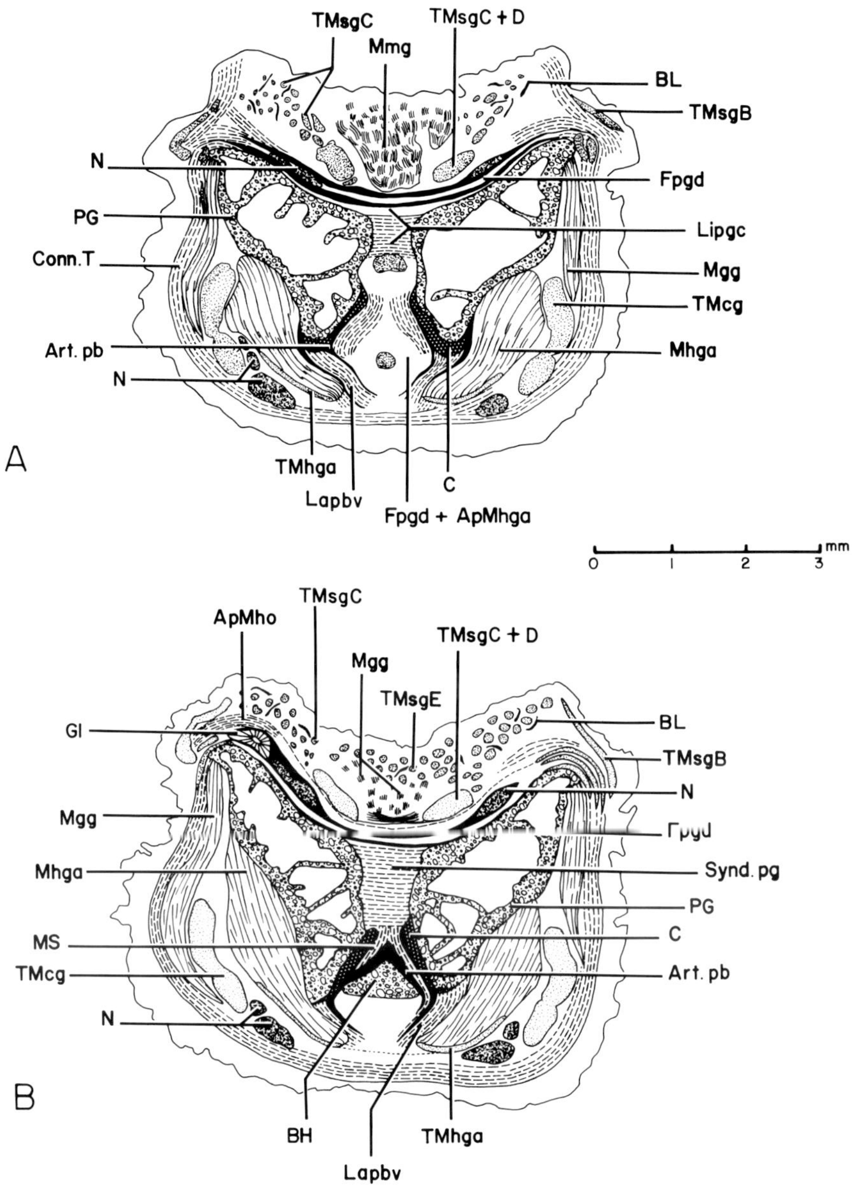

FIG. 40. Apical views of cross-sections through the lingual tip of *Psittacus erithacus,* Z4. A: At the level of the apical end of the Lig. interparaglossale caudale. B: At the level of the Syndesmosis paraglossalis and of the ventro-apical tip of the basihyale.

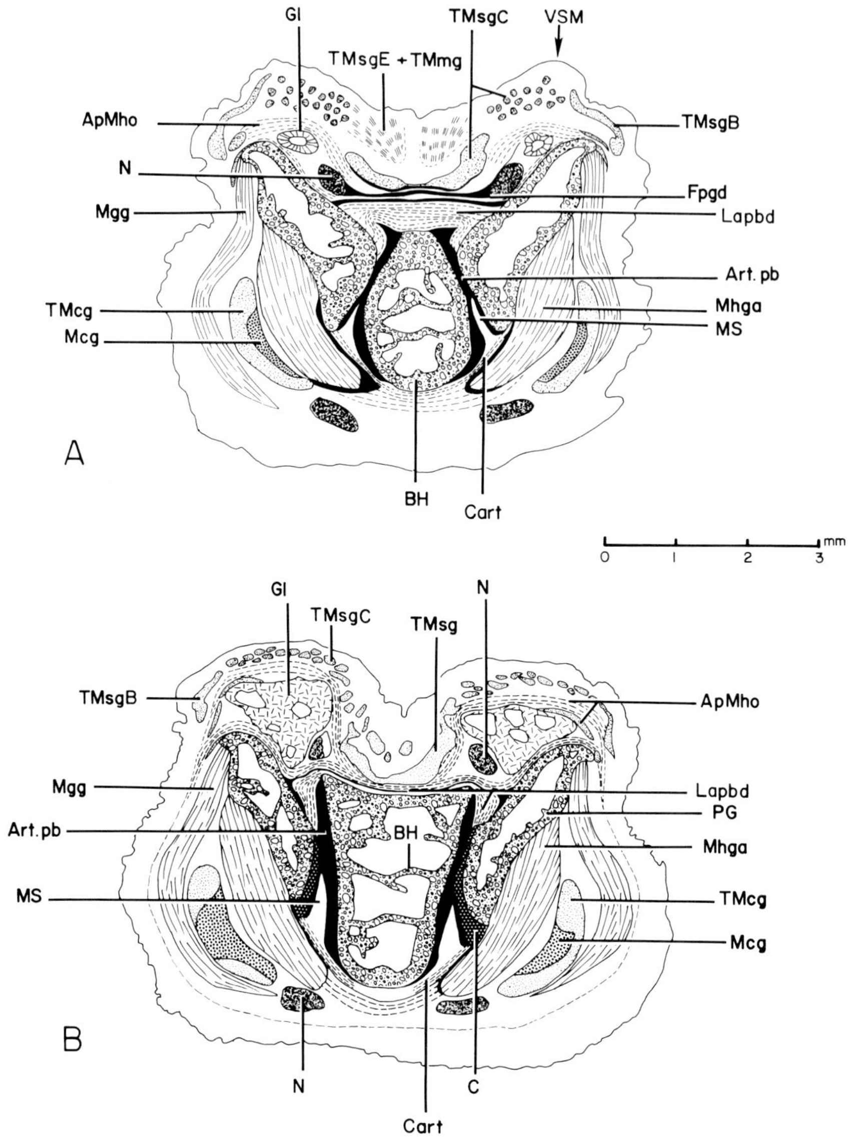

FIG. 41. Apical views of cross-sections through the lingual tip of *Psittacus erithacus*, Z4. A: At the level of the apical end of the Gl. lingualis. B: At the level of the caudal part of the Art. paraglosso-basihyalis, just apically from the apical end of the muscular part of M. supraglossus (cf. Fig. 43).

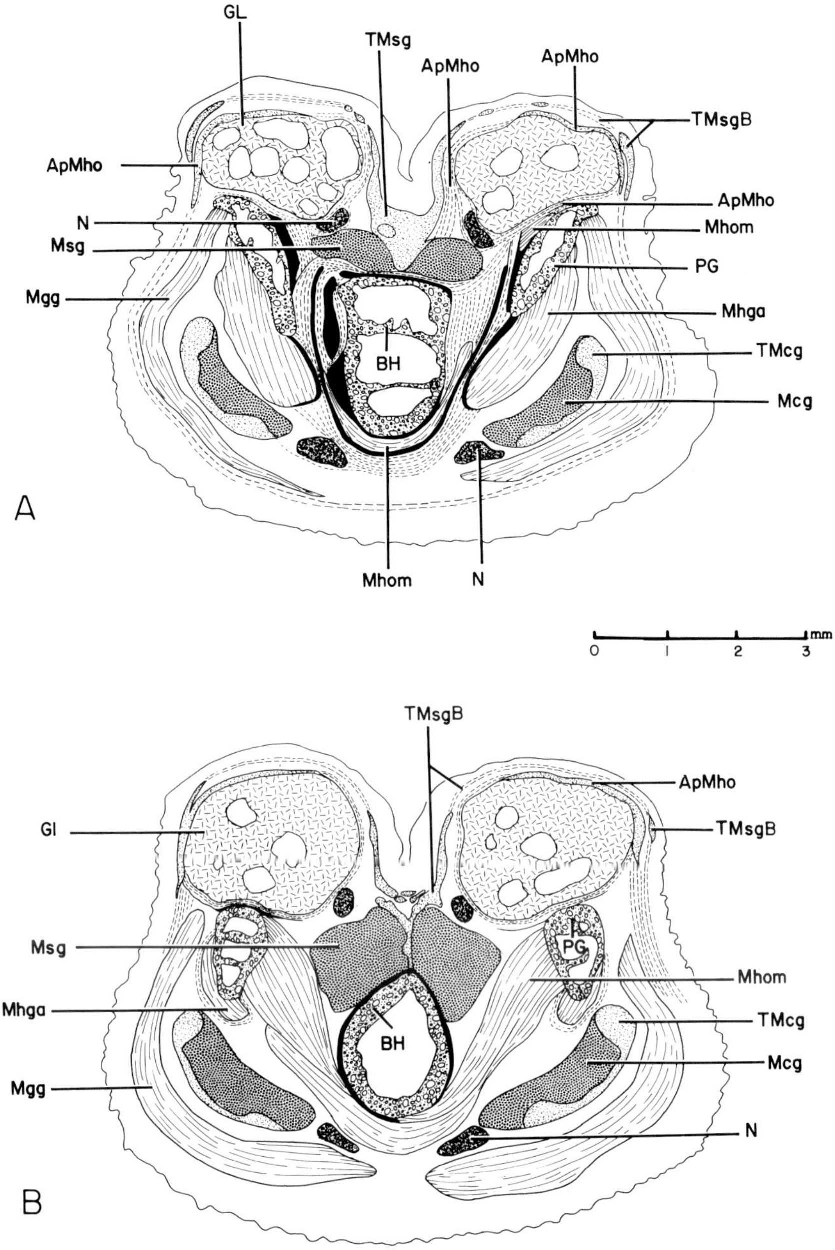

FIG. 42. Apical views of cross-sections through the lingual tip of *Psittacus erithacus,* Z4. A: At the level just caudally from the Art. paraglosso-basihyalis. B: At the level near the caudal end of Proc. posterior paraglossalis.

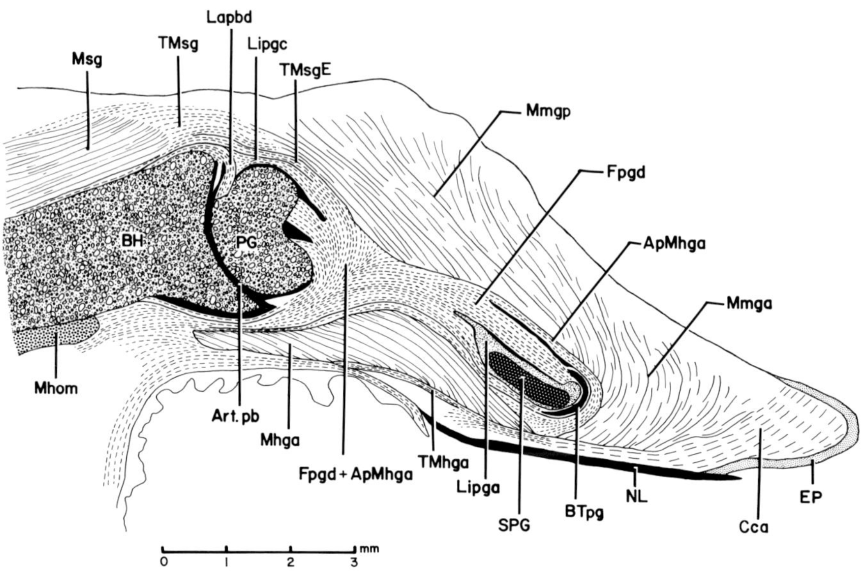

FIG. 43. Longitudinal, slightly parasagittal section through the tongue's tip of *Psittacus erithacus*, Z5.

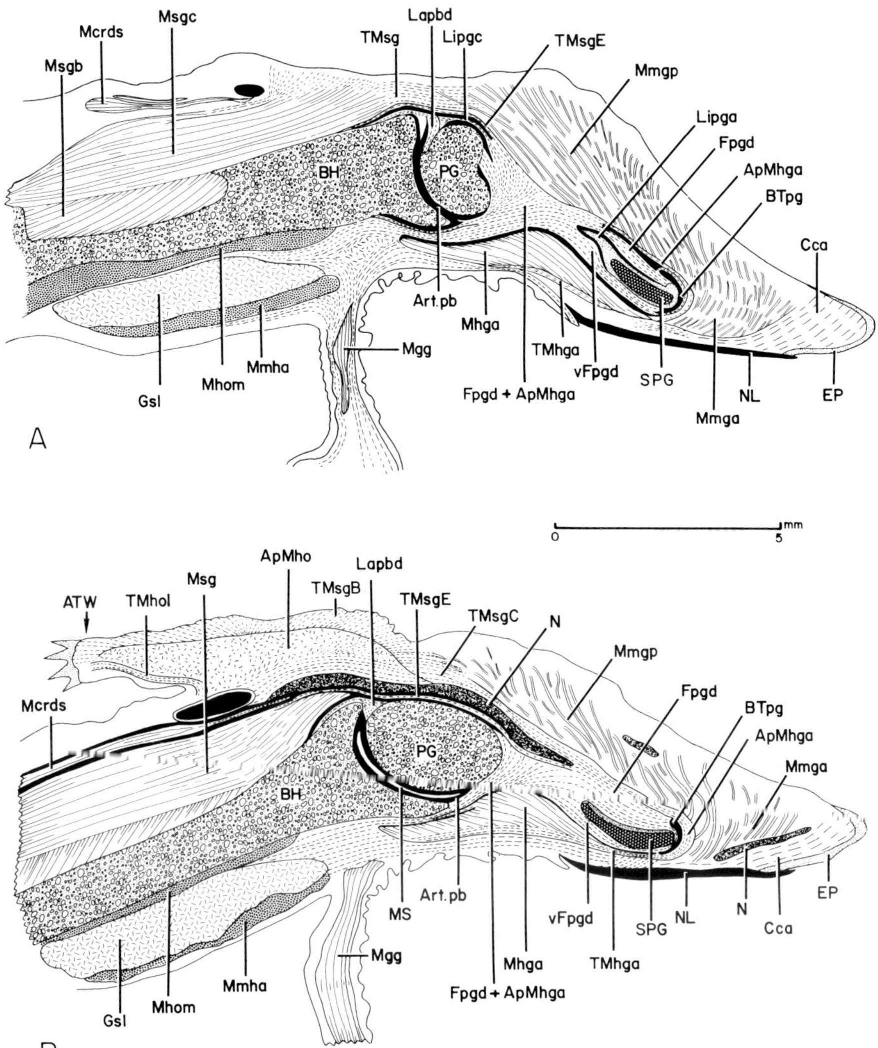

FIG. 44. Longitudinal sections through the anterior portion of the lingual apparatus of *Psittacus erithacus*, Z5. A: Slightly parasagittal (as in Fig. 43). B: Slightly more lateral than Fig. 44A, but still within the Foramen interparaglossale (cf. Fig. 38B).

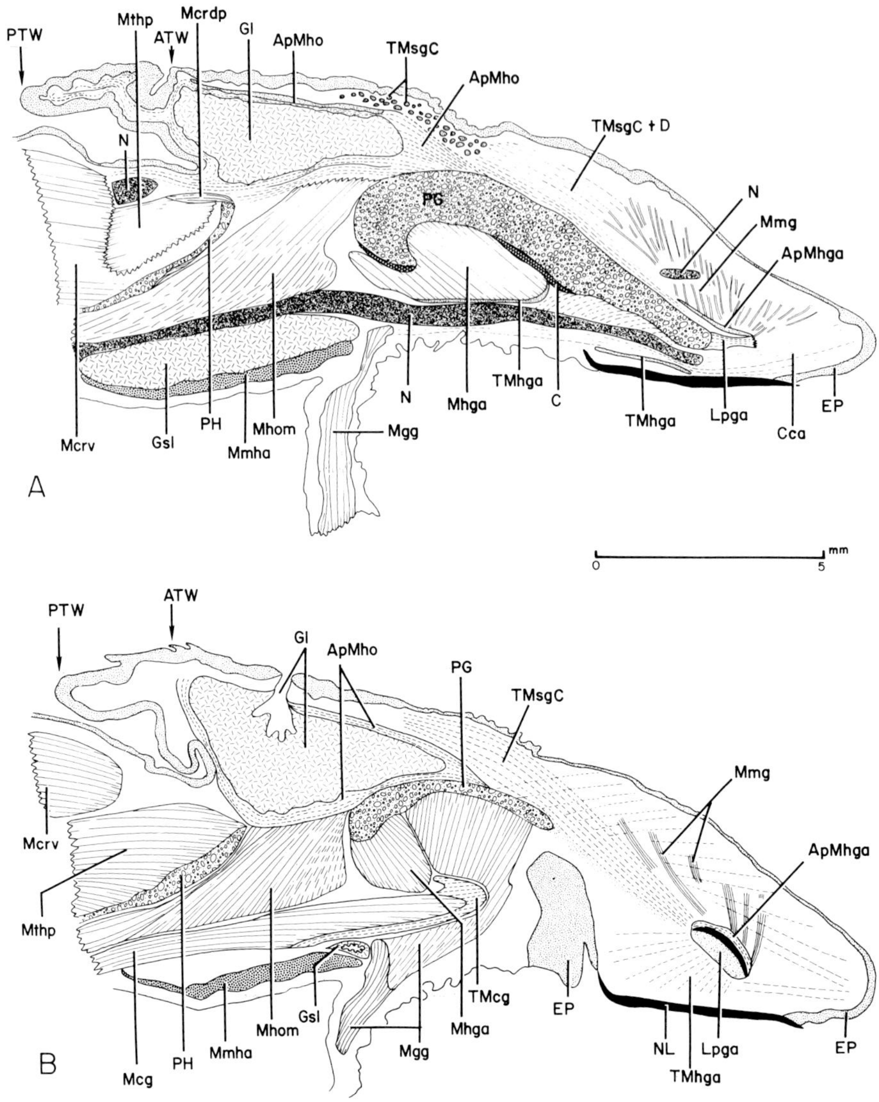

FIG. 45. Longitudinal sections through the anterior portion of the lingual apparatus of *Psittacus erithacus,* Z5. A: More lateral than Fig. 44B. B: More lateral than Fig. 45A.

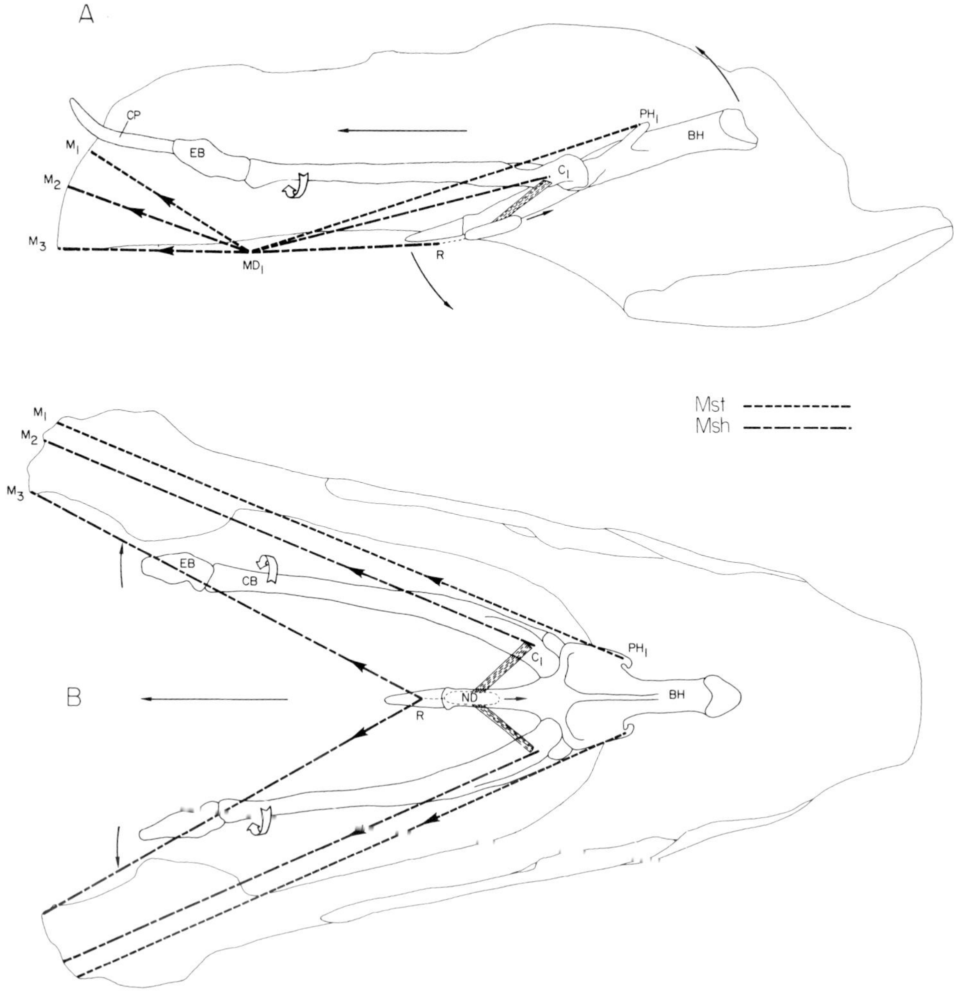

FIG. 46. Retraction of the protracted hyoid in *Psittacus erithacus.* A: Lateral view. B: Dorsal view. The contracting muscles are represented by force lines with arrows pointing in the direction in which the movable skeletal element is pulled towards the stationary element. The separate, free-standing arrows indicate the motions of the skeletal elements, or, in a few cases, external forces acting on the skeletal elements. C_1 insertion of M. serpihyoideus on ceratobranchiale, M_1–M_3 origins of extrinsic lingual muscles on mandible (M_1 dorsal end of origin of M. stylohyoideus, M_2 ventral end of origin of Mst and dorsal end of origin of Msh, M_3 ventral end of origin of Msh), MD_1 ventral edge of mandibular ramus with jaw muscles over which the Msh and Mst pass (shown only in 46A), PH_1 apical end of insertion of Mst on Proc. parahyalis, R caudal end of insertion of Msh on midsagittal raphe.

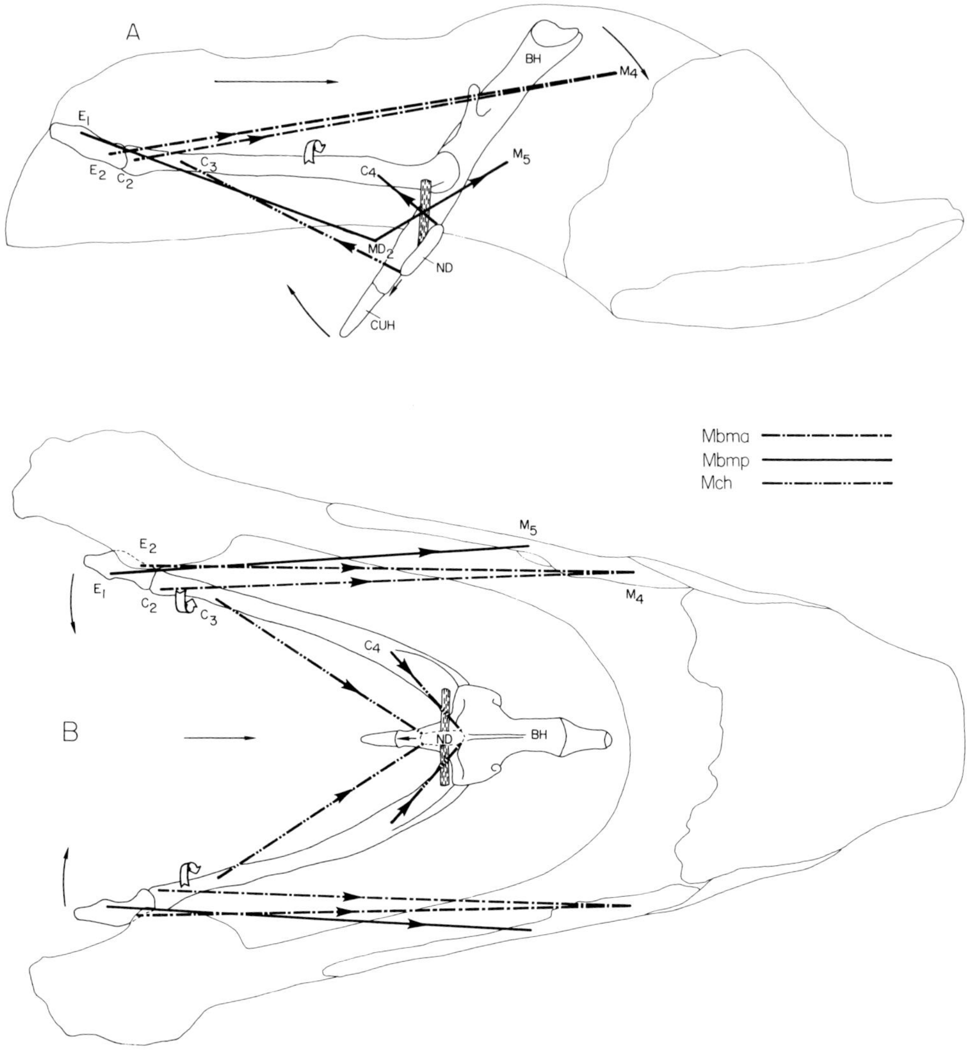

FIG. 47. Protraction of the retracted hyoid in *Psittacus erithacus*. A: Lateral view. B: Dorsal view. (For meaning of arrows see Fig. 46.) C_2–C_4 muscle attachments on ceratobranchiale (C_2 center of insertion of M. branchiomandibularis ant. med., C_3 caudal end of origin of M. ceratohyoideus, C_4 apical end of origin of Mch), E_1–E_2 insertions of Mbm on epibranchiale (E_1 center of insertion of M. branchiomandibularis post., E_2 center of insertion of Mbma lat.), M_4–M_5 origins of extrinsic lingual muscles on mandible (M_4 center of origin of Mbma, M_5 dorsal end of origin of Mbmp), MD_2 ventral edge of mandibular ramus with jaw muscles over which the Mbmp passes, ND Nodulus and connective tissue as insertion site for Mch.

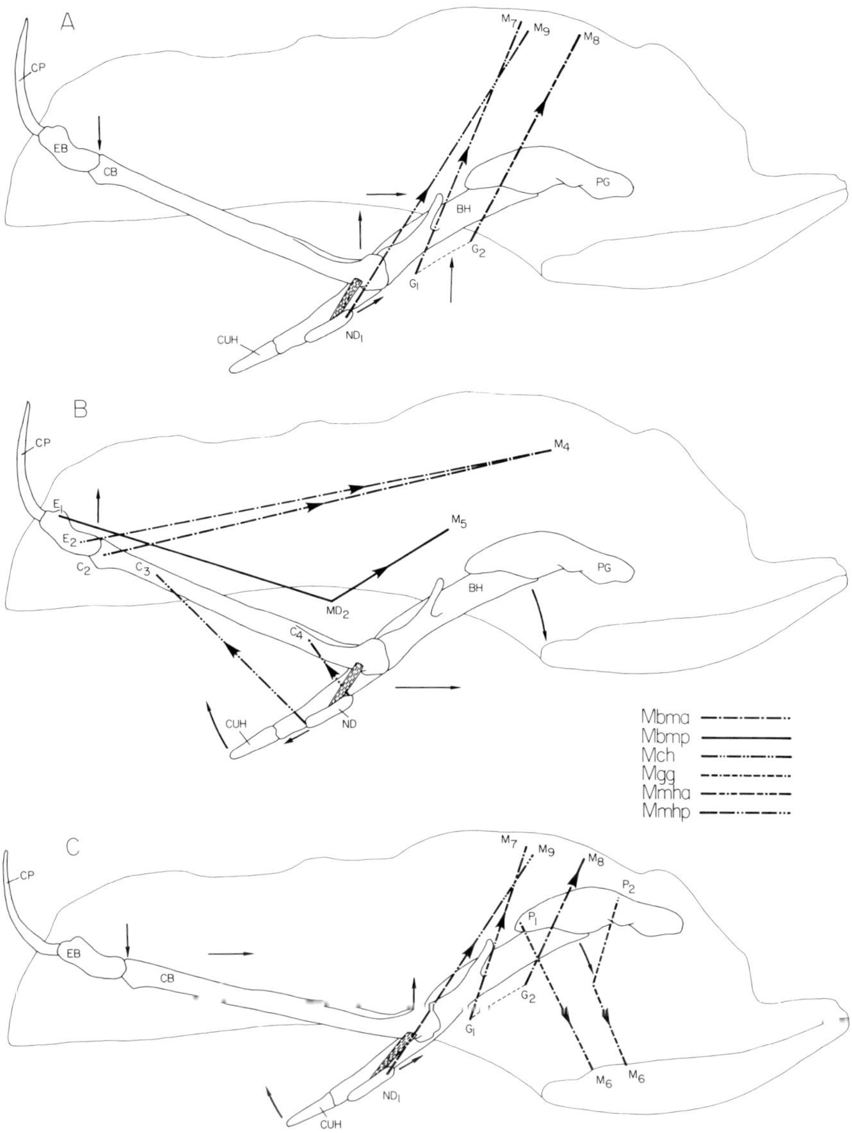

FIG. 48. Lateral views of the up-and-down movement of the hyoid with immobilized paraglossale in *Psittacus erithacus.* A–B: Raising and protraction of the retracted and lowered hyoid (A: Actions of the M. mylohyoideus; B: Actions of the M. branchiomandibularis and M. ceratohyoideus). C: Protraction of the hyoid without simultaneous raising; hyoid at intermediate position between protracted-raised and retracted-lowered position (see Fig. 48B for the simultaneous actions of the Mbm and Mch). (For meaning of arrows see Fig. 46.) C_2–C_4 muscle attachments on ceratobranchiale (C_2 center of insertion of Mbma med., C_3 caudal end of origin of Mch, C_4 apical end of origin of Mch), E_1–E_2 insertions of Mbm on epibranchiale (E_1 center of insertion of Mbmp, E_2 center of insertion of Mbma lat.), G_1–G_2 insertions of Mmha on Gl. sublingualis (G_1 caudal end of insertion, G_2 apical ends of insertion), M_4–M_9 origins of extrinsic lingual muscles on mandible (M_4 center of origin of Mbma, M_5 dorsal end of origin of Mbmp, M_6 caudal and apical ends of origin of M. genioglossus, M_7 caudal end of origin of Mmha, M_8 apical end of origin of Mmha, M_9 center of origin of Mmhp), MD_2 ventral edge of mandibular ramus with jaw muscles over which the Mbmp passes, ND–ND_1 insertion sites of lingual muscles on Nodulus (ND Nodulus and connective tissue as insertion site for Mch, ND_1 center of insertion of Mmhp), P_1–P_2 insertions of Mgg on paraglossale (P_1 caudal end of insertion, P_2 apical end of insertion).

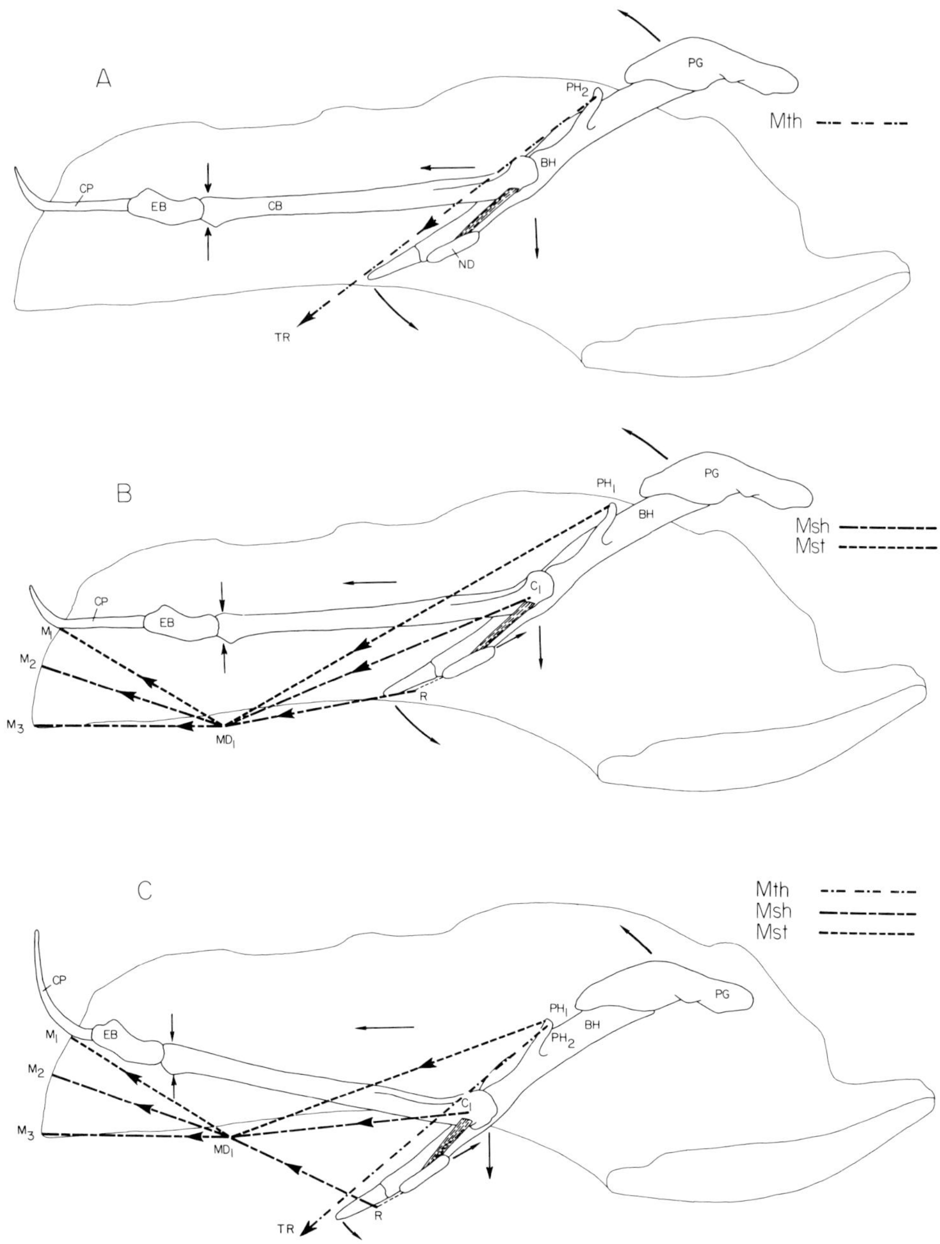

FIG. 49. Lateral views of the up-and-down movement of the hyoid in *Psittacus erithacus.* A–B: Lowering and retraction of the protracted and raised hyoid (A: Action of the M. tracheohyoideus; B: Actions of the M. serpihyoideus and M. stylohyoideus). C: Lowering and retraction of the hyoid at an intermediate position. (For meaning of arrows see Fig. 46.) C_1 center of insertion on ceratobranchiale of Msh, M_1–M_3 origins of extrinsic lingual muscles on mandible (M_1 dorsal end of origin of Mst, M_2 ventral end of origin of Mst and dorsal end of origin of Msh, M_3 ventral end of origin of Msh), MD_1 ventral edge of Ramus mandibularis with jaw muscles over which the Mst and Msh pass, PH_1–PH_2 insertions of extrinsic lingual muscles on Proc. parahyalis (PH_1 apical end of insertion of Mst, PH_2 apical end of insertion of M. tracheohyoideus parahyalis), R caudal end of insertion of Msh on midsagittal raphe, TR trachea.

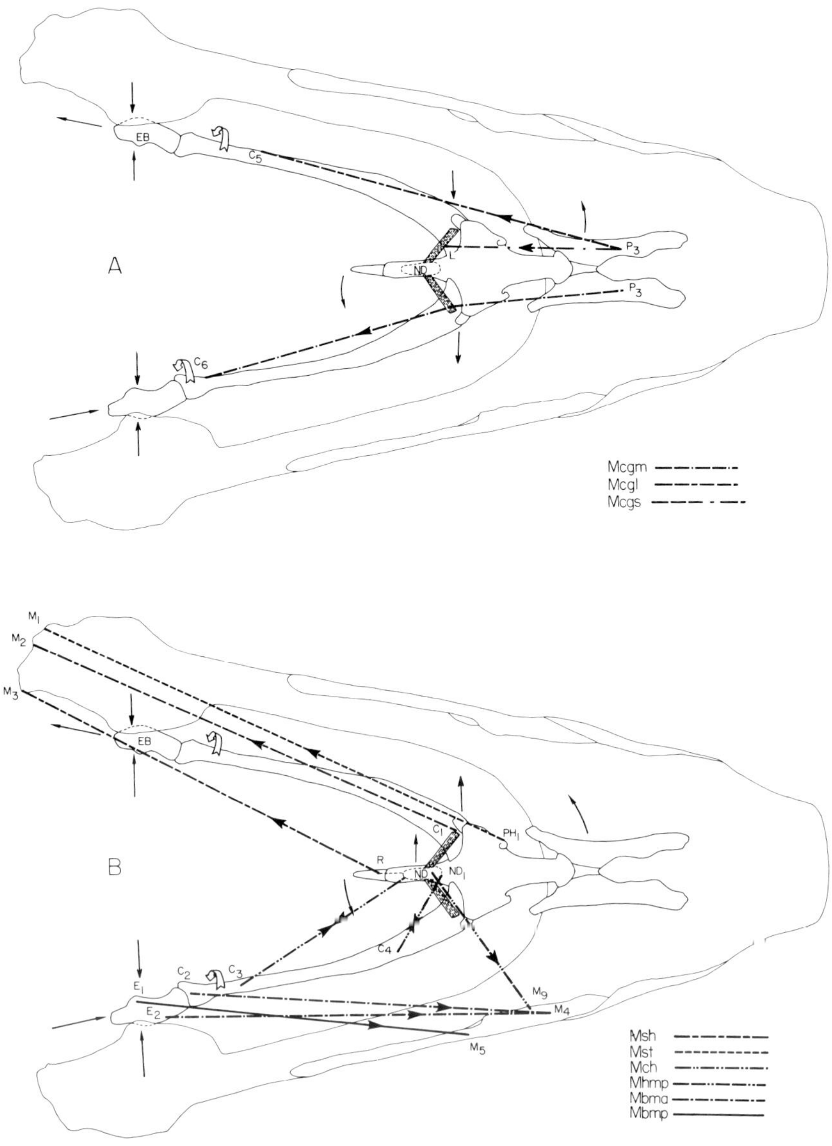

FIG. 50. Dorsal views of the sinistral movement of the hyoid with immobilized paraglossale in *Psittacus erithacus.* Hyoid at intermediate position between protracted and retracted position. A: Actions of M. ceratoglossus lateralis, medialis et superficialis. B: Actions of M. serpihyoideus, M. stylohyoideus, M. branchiomandibularis, M. ceratohyoideus, and M. mylohyoideus posterior. (For meaning of arrows see Fig. 46.) C_1–C_6 muscle attachments on the ceratobranchiale (C_1 center of insertion of Msh, C_2 center of insertion of Mbma med., C_3 caudal end of origin of Mch, C_4 apical end of origin of Mch, C_5 caudal end of origin of Mcgl, C_6 caudal end of origin of Mcgm), E_1–E_2 insertions of Mbm on epibranchiale (E_1 center of insertion of Mbmp, E_2 center of insertion of Mbma lat.), L center of origin of Mcgs on Lig. nodulo-ceratobranchiale, M_1–M_5 and M_9 origins of extrinsic lingual muscles on mandible (M_1 dorsal end of origin of Mst, M_2 ventral end of origin of Mst and dorsal end of origin of Msh, M_3 ventral end of origin of Msh, M_4 center of origin of Mbma, M_5 dorsal end of origin of Mbmp, M_9 center of origin of Mmhp), ND–ND_1 insertions of lingual muscles on Nodulus (ND insertions of Mch, ND_1 center of insertion of Mmhp), P_3 center of insertion of Mcg on Tuberositas paraglossalis, PH_1 apical end of insertion of Mst on Proc. parahyalis, R caudal end of insertion of Msh on midsagittal raphe.

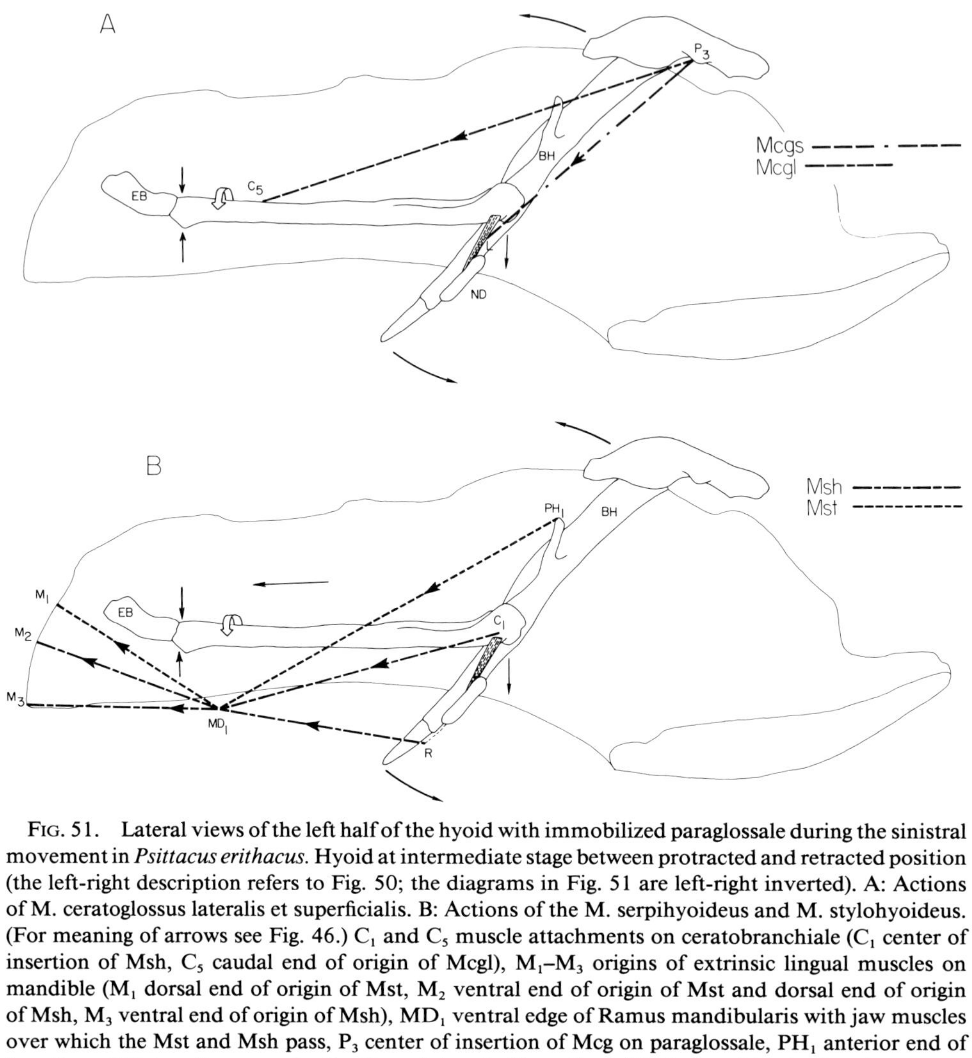

FIG. 51. Lateral views of the left half of the hyoid with immobilized paraglossale during the sinistral movement in *Psittacus erithacus.* Hyoid at intermediate stage between protracted and retracted position (the left-right description refers to Fig. 50; the diagrams in Fig. 51 are left-right inverted). A: Actions of M. ceratoglossus lateralis et superficialis. B: Actions of the M. serpihyoideus and M. stylohyoideus. (For meaning of arrows see Fig. 46.) C_1 and C_5 muscle attachments on ceratobranchiale (C_1 center of insertion of Msh, C_5 caudal end of origin of Mcgl), M_1–M_3 origins of extrinsic lingual muscles on mandible (M_1 dorsal end of origin of Mst, M_2 ventral end of origin of Mst and dorsal end of origin of Msh, M_3 ventral end of origin of Msh), MD_1 ventral edge of Ramus mandibularis with jaw muscles over which the Mst and Msh pass, P_3 center of insertion of Mcg on paraglossale, PH_1 anterior end of insertion of Mst on Proc. parahyalis, R caudal end of insertion of Msh on midsagittal raphe.

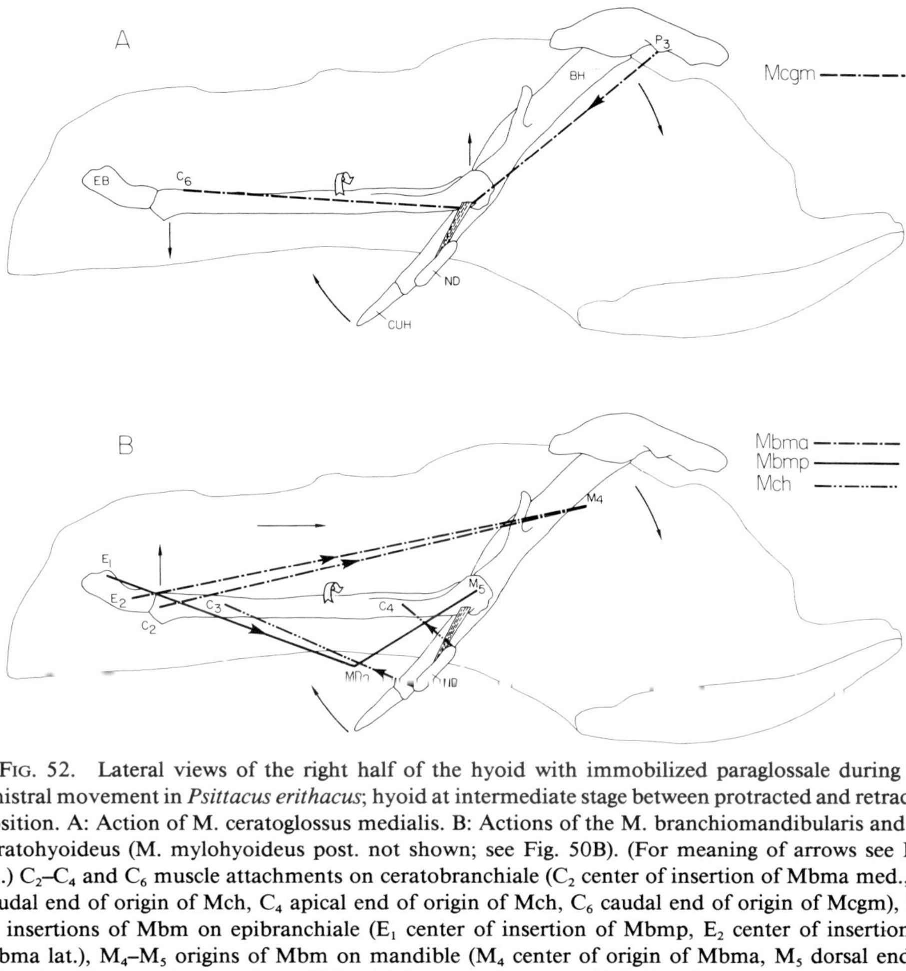

FIG. 52. Lateral views of the right half of the hyoid with immobilized paraglossale during the sinistral movement in *Psittacus erithacus*; hyoid at intermediate stage between protracted and retracted position. A: Action of M. ceratoglossus medialis. B: Actions of the M. branchiomandibularis and M. ceratohyoideus (M. mylohyoideus post. not shown; see Fig. 50B). (For meaning of arrows see Fig. 46.) C_2–C_4 and C_6 muscle attachments on ceratobranchiale (C_2 center of insertion of Mbma med., C_3 caudal end of origin of Mch, C_4 apical end of origin of Mch, C_6 caudal end of origin of Mcgm), E_1–E_2 insertions of Mbm on epibranchiale (E_1 center of insertion of Mbmp, E_2 center of insertion of Mbma lat.), M_4–M_5 origins of Mbm on mandible (M_4 center of origin of Mbma, M_5 dorsal end of Mbmp), MD_2 ventral edge of mandible with jaw muscles over which the Mbmp passes, ND Nodulus and connective tissue as insertion site for Mch, P_3 center of insertion of Mcgm on paraglossale.

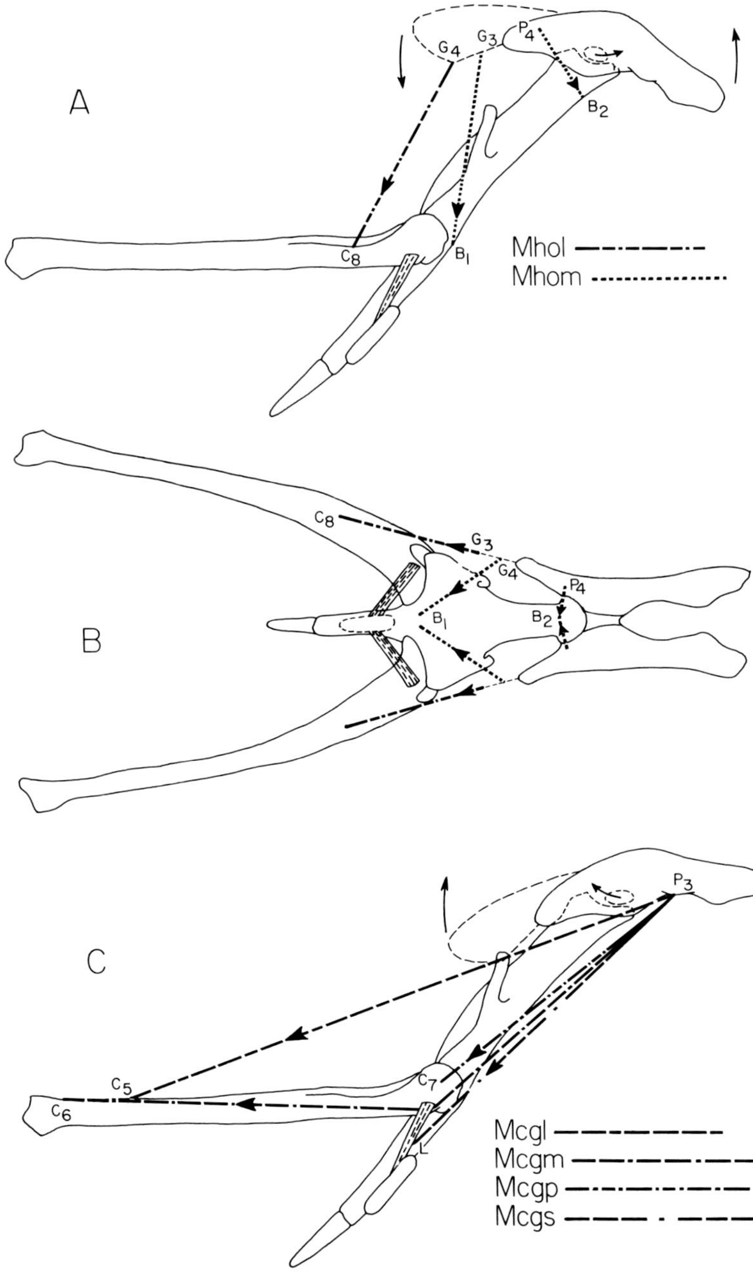

FIG. 53. Movements of the paraglossale with hyoid immobilized in intermediate position between protracted and retracted position in *Psittacus erithacus*. A–B: Extension and protraction of the flexed paraglossale (A: Lateral view; B: Dorsal view). C: Lateral view of flexion and retraction of the extended paraglossale. (For meaning of arrows see Fig. 46.) B_1–B_2 origins of M. hypoglossus obliquus med. on basihyale (B_1 caudo-medial end of origin, B_2 apical end of origin), C_5–C_8 origins of intrinsic lingual muscles on ceratobranchiale (C_5 caudal end of origin of M. ceratoglossus lat., C_6 caudal end of origin of M. ceratoglossus med., C_7 center of origin of M. ceratoglossus prof., C_8 caudal end of origin of M. hypoglossus obliquus lat.), G_3–G_4 insertions of Mho on tendinous envelope of Gl. lingualis (G_3 caudal end of insertion of Mhom, G_4 caudal end of insertion of Mhol), L center of origin of M. ceratoglossus supf. on Lig. nodulo-ceratobranchiale, P_3–P_4 insertions of intrinsic lingual muscles on paraglossale (P_3 center of insertion of Mcg, P_4 apical end of insertion of Mhom).

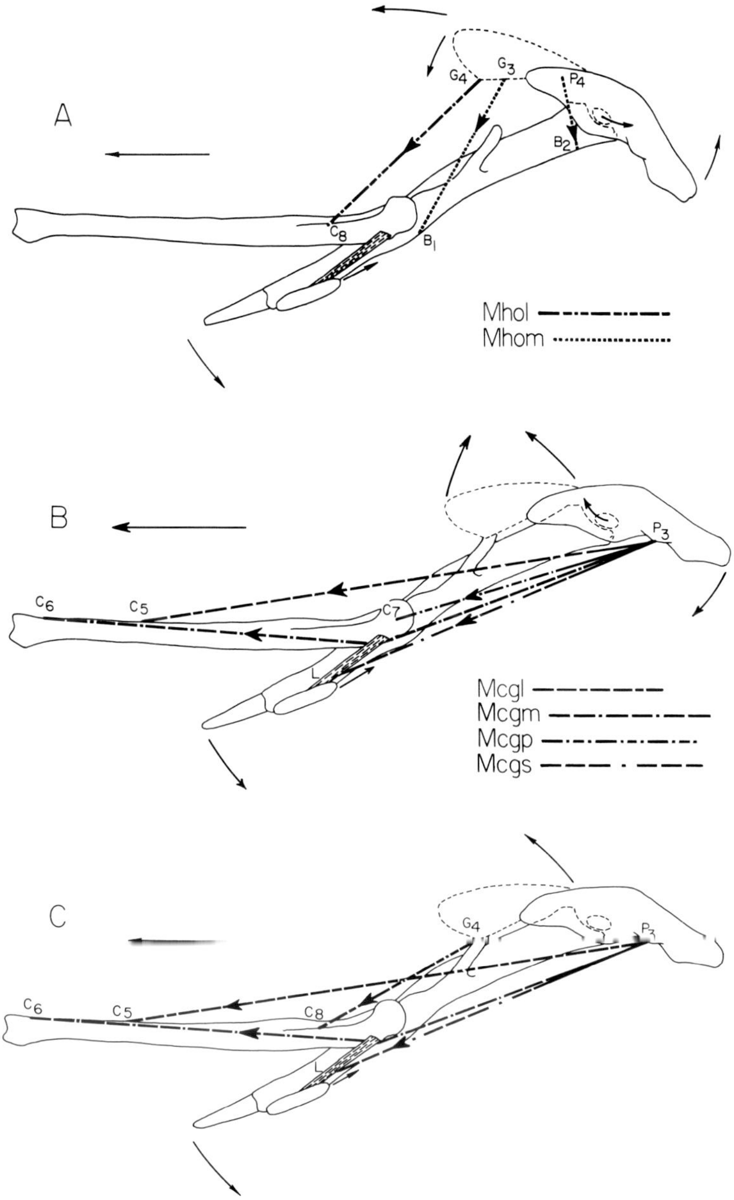

FIG. 54. Lateral views of movements of the paraglossale during retraction of the protracted hyoid in *Psittacus erithacus.* A: Extension and protraction of the flexed paraglossale. B: Flexion and retraction of the extended paraglossale. C: Immobilization of the extended paraglossale through simultaneous contractions of the M. ceratoglossus and M. hypoglossus obliquus (Mhom not shown). (For meaning of arrows see Fig. 46.) B_1–B_2 origins of Mhom on basihyale (B_1 caudo-medial end of origin, B_2 apical end of origin), C_5–C_8 origins of intrinsic lingual muscles on ceratobranchiale (C_5 caudal end of origin of Mcgl, C_6 caudal end of origin of Mcgm, C_7 center of origin of Mcgp, C_8 caudal end of origin of Mhol), G_3–G_4 insertions of Mho on tendinous envelope of Gl. lingualis (G_3 caudal end of insertion of Mhom, G_4 caudal end of insertion of Mhol), L center of origin of Mcgs on Lig. nodulo-cerato-branchiale, P_3–P_4 insertions of intrinsic lingual muscles on paraglossale (P_3 center of insertion of Mcg, P_4 apical end of insertion of Mhom).

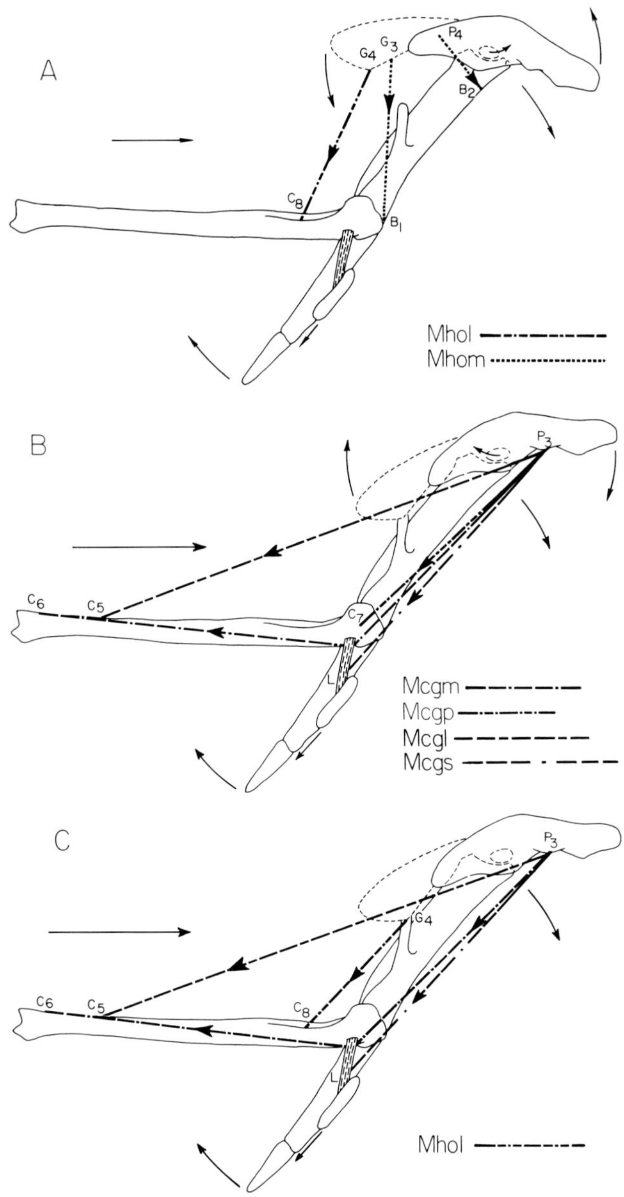

FIG. 55. Lateral views of movements of the paraglossale during protraction of the retracted hyoid in *Psittacus erithacus.* A: Extension and protraction of the flexed paraglossale. B: Flexion and retraction of the extended paraglossale. C: Immobilization of the paraglossale through simultaneous contractions of the M. hypoglossus obliquus (Mhom not shown) and M. ceratoglossus. (For meaning of arrows see Fig. 46.) B_1–B_2 origins of Mhom on basihyale (B_1 caudo-medial end of origin, B_2 apical end of origin), C_5–C_8 origins of intrinsic lingual muscles on ceratobranchiale (C_5 caudal end of origin of Mcgl, C_6 caudal end of origin of Mcgm, C_7 center of origin of Mcgp, C_8 caudal end of origin of Mhol), G_3–G_4 insertions of Mho on tendinous envelope of Gl. lingualis (G_3 caudal end of insertion of Mhom, G_4 caudal end of insertion of Mhol), L center of origin of Mcgs on Lig. nodulo-ceratobranchiale, P_3–P_4 insertions of intrinsic lingual muscles on paraglossale (P_3 center of insertion of Mcg, P_4 apical end of insertion of Mhom).

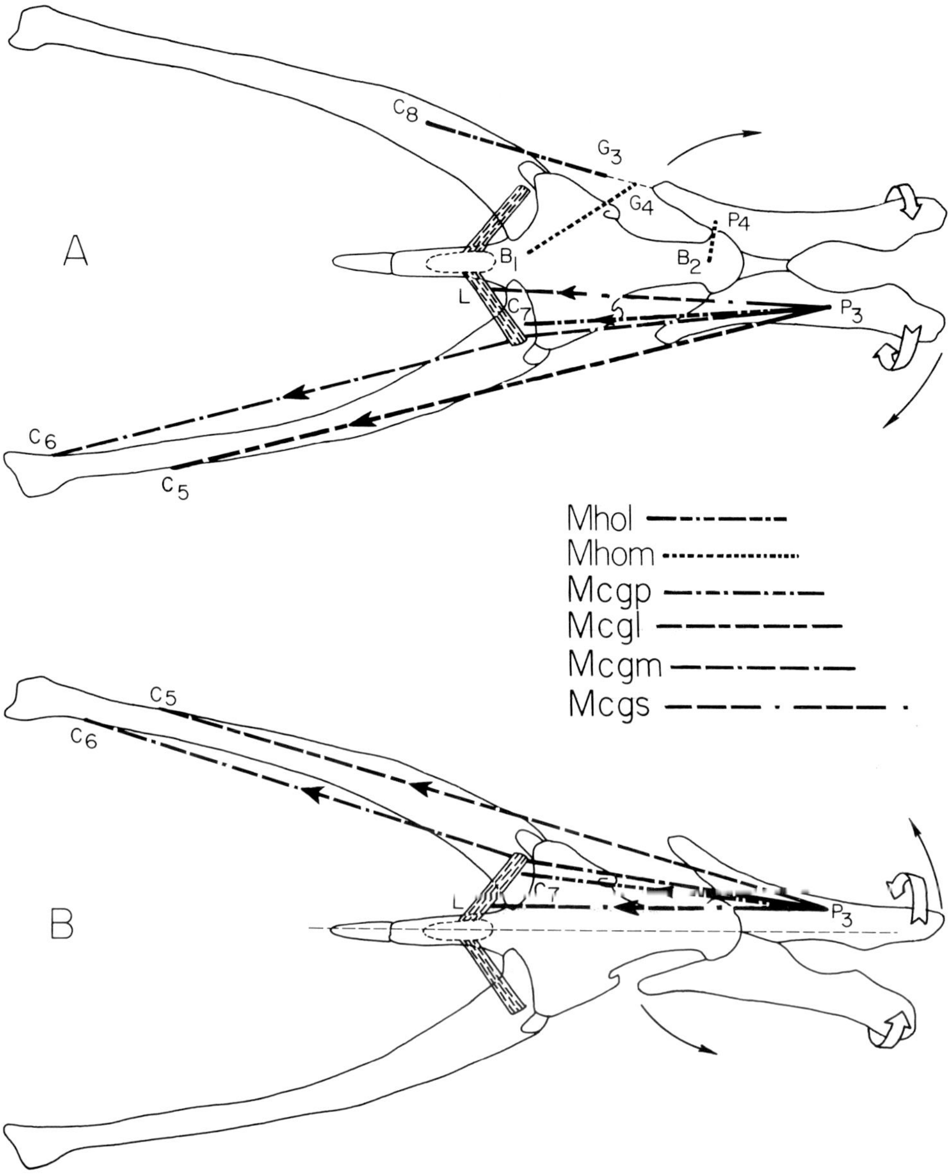

FIG. 56. Dorsal views of lateral movements of the paraglossale with the hyoid immobilized in intermediate position in *Psittacus erithacus.* A: Turning of the straight paraglossale to the right. B: Straightening of the laterally turned paraglossale. (For meaning of arrows see Fig. 46.) B_1–B_2 origins of Mhom on basihyale (B_1 caudo-medial end of origin, B_2 apical end of origin), C_5–C_8 origins of intrinsic lingual muscles on ceratobranchiale (C_5 caudal end of origin of Mcgl, C_6 caudal end of origin of Mcgm, C_7 center of origin of Mcgp, C_8 caudal end of origin of Mhol), G_3–G_4 insertions of Mho on tendinous envelope of Gl. lingualis (G_3 caudal end of insertion of Mhom, G_4 caudal end of insertion of Mhol), L center of origin of Mcgs on Lig. nodulo-ceratobranchiale, P_3–P_4 insertions of intrinsic lingual muscles on paraglossale (P_3 center of insertion of Mcg, P_4 apical end of insertion of Mhom).

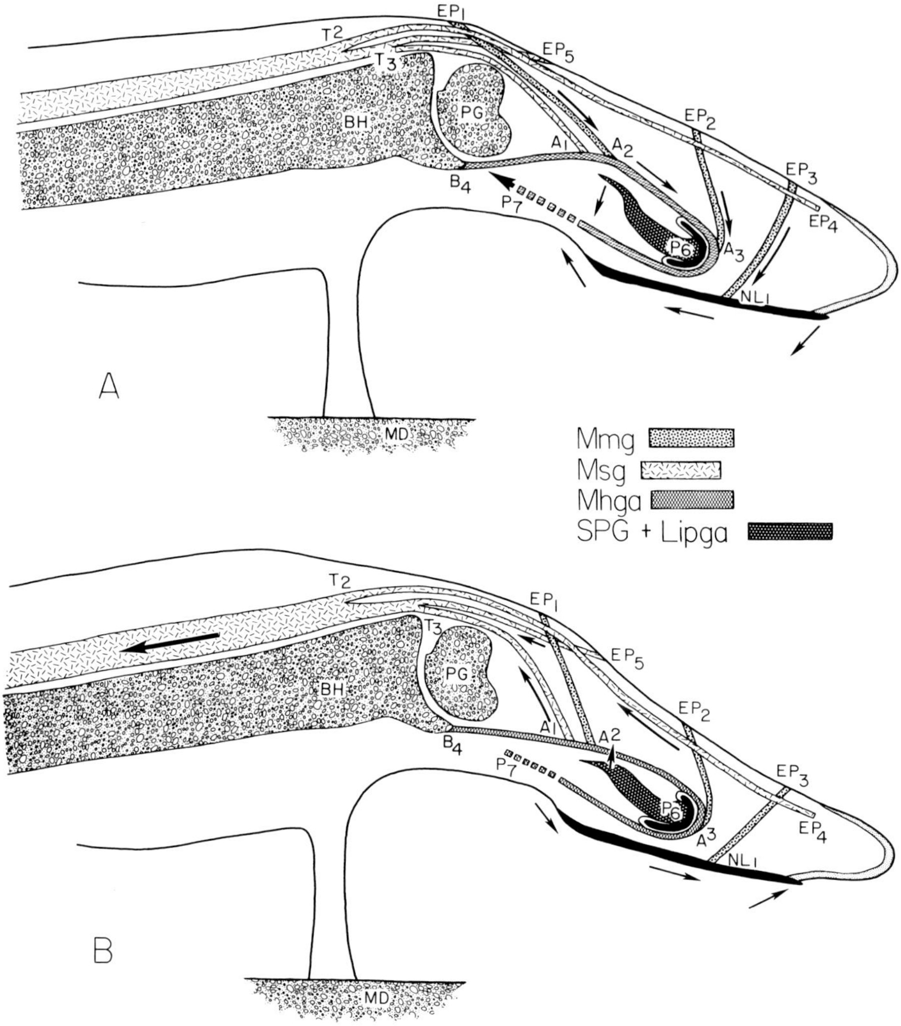

FIG. 57. Midsagittal views of movements of the epithelial surface of the tongue's tip in *Psittacus erithacus.* A: Depression of central area of the dorsal epithelial surface through contraction of M. hypoglossus anterior and M. mesoglossus. B: Flattening of the dorsal epithelial surface through contraction of M. supraglossus. (For meaning of arrows see Fig. 46.) A_1–A_3 attachments of intrinsic lingual muscles on aponeurosis of Mhga (A_1 insertion of tendinous portion E of Msg, A_2 caudal end of origin of Mmgp, A_3 apical end of origin of Mmgp), B_4 origin of Mhga on basihyale, EP_1–EP_5 insertions of intrinsic lingual muscles on epithelial surface of tongue's tip (EP_1 caudal end of insertion of Mmgp, EP_2 apical end of insertion of Mmgp, EP_3 central insertion of Mmga, EP_4 apical end of insertion of tendinous portion C of Msg, EP_5 insertion of tendinous portion E of Msg), NL_1 central origin of Mmga on lingual nail, P_6 apical edge of Synchondrosis paraglossalis over which the tendon of Mhga passes, P_7 direction of caudo-dorsal end of insertion of Mhga on paraglossale, T_2–T_3 origins of tendinous portions of Msg from central tendon (T_2 origin of tendinous portion C, T_3 origin of tendinous portion E).

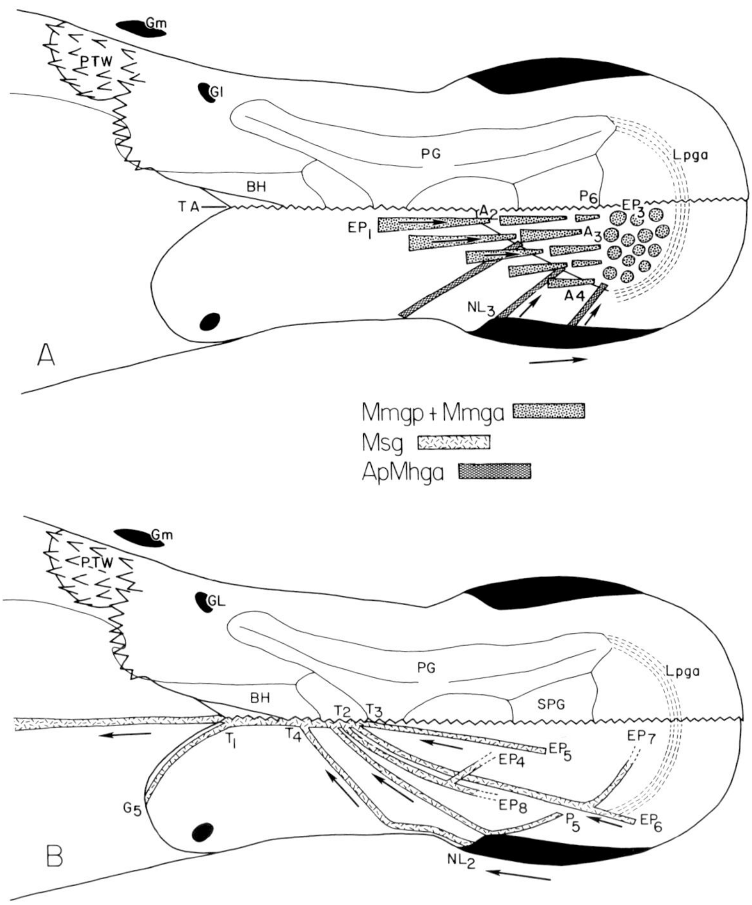

FIG. 58. Dorsal views of movements of the epithelial surface of the tongue's tip in *Psittacus erithacus*. A: Depression of central area of the dorsal epithelial surface through contraction of M. hypoglossus ant. and M. mesoglossus (left side: superposition of skeletal elements and epithelial surface structures of the tongue's tip; right side: Mmg, aponeurosis of Mhga, tendinous fibers originating caudo-laterally from the aponeurosis of the Mhga to insert on dorsal border of lingual nail, and caudal contour of Gl. lingualis). B: Flattening of the dorsal epithelial surface through contraction of M. supraglossus (left side: superposition of skeletal elements and epithelial surface structures of the tongue's tip; right side: tendinous portions of Msg and caudal contour of Gl. lingualis). (For meaning of arrows see Fig. 46.) A_2–A_4 origins of muscles and tendinous fibers on aponeurosis of Mhga (A_2 caudal end of origin of Mmgp, A_3 apical end of origin of Mmgp, A_4 origin of tendinous fibers originating caudo-laterally from the aponeurosis of Mhga to attach on dorsal border of lingual nail), EP_1 and EP_3–EP_8 insertions of muscles and tendinous fibers on epithelial surface of the tongue's tip (EP_1 caudal end of insertion of Mmgp, EP_3 central insertion of Mmga, EP_4 insertion of tendinous portion C of Msg on central area of the tongue's tip, EP_5 insertion of tendinous portion E of Msg on central area of the tongue's tip, EP_6 insertion of main tendon of portion D of Msg, EP_7 insertion of tendinous portion D of Msg on central area of the tongue's tip, EP_8 direction of insertion of tendinous portion C of Msg on V-shaped mound), G_5 insertion of tendinous portion A of Msg on Gl. lingualis, NL_2–NL_3 insertions of tendinous fibers of intrinsic muscles on lingual nail and lateral part of V-shaped mound (NL_2 insertion of tendinous portion B of Msg, NL_3 insertion of tendinous fibers originating caudo-laterally from aponeurosis of Mhga), P_5 insertion of tendinous portion C of Msg on paraglossale, P_6 apical edge of Synchondrosis paraglossalis over which the tendon of Mhga passes, T_1–T_4 origins of tendinous portions of Msg from central tendon (T_1 origin of tendinous portion A, T_2 origin of tendinous portions C and D, T_3 origin of tendinous portion E, T_4 origin of tendinous portion B).

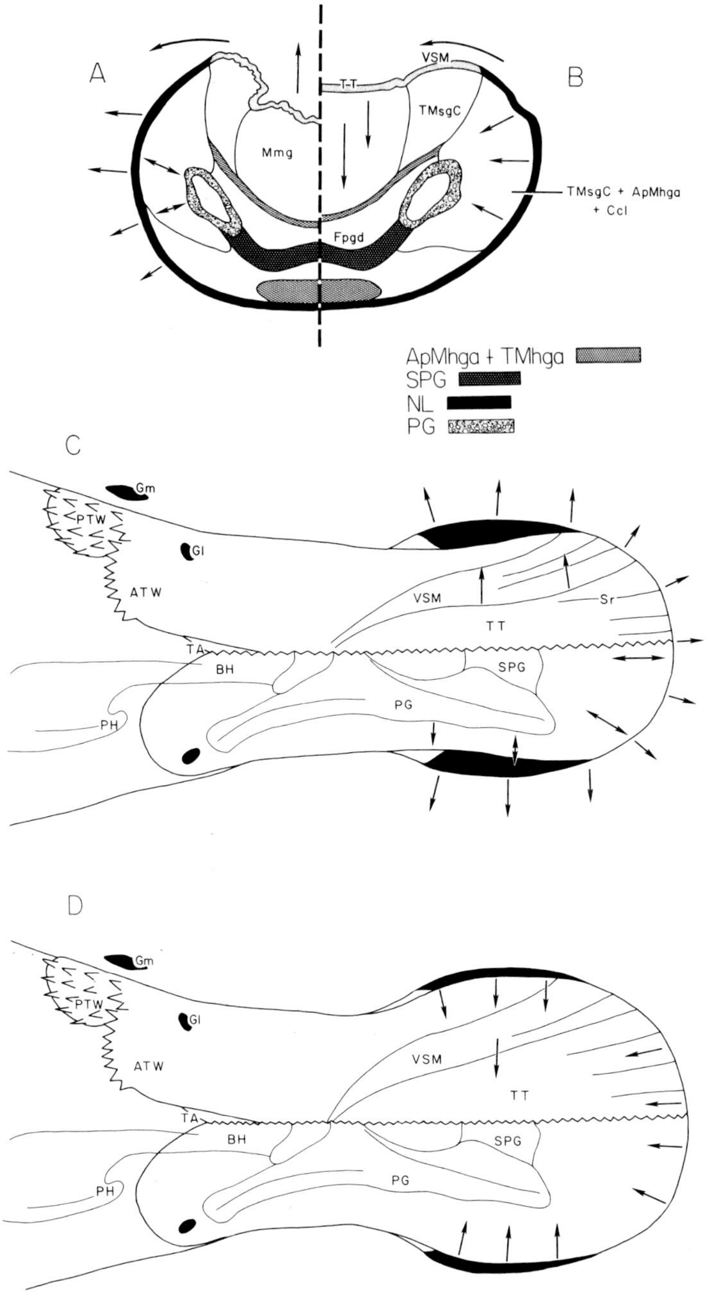

FIG. 59. Movements of the tongue's tip brought about by the Corpora cavernosa in *Psittacus erithacus*. A–B: Cross-sections through the tongue's tip, adapted from Fig. 37A (A: Broadening and flattening of deflated tip of tongue through inflation of the Corpus cavernosum lat.; B: Narrowing of inflated tip of tongue and relaxation of central area through deflation of C. c. lat.). C–D: Dorsal views of the tongue's tip. Left sides: epithelial surface structures; right sides: superposition of skeletal elements, caudal outline of Gl. lingualis and outline of anterior part of the tongue's tip. (C: Inflation of the Corpora cavernosa lateralia et apicale; D: Deflation of Cc. cc. latt. et ap.). (For meaning of arrows see Fig. 46.)

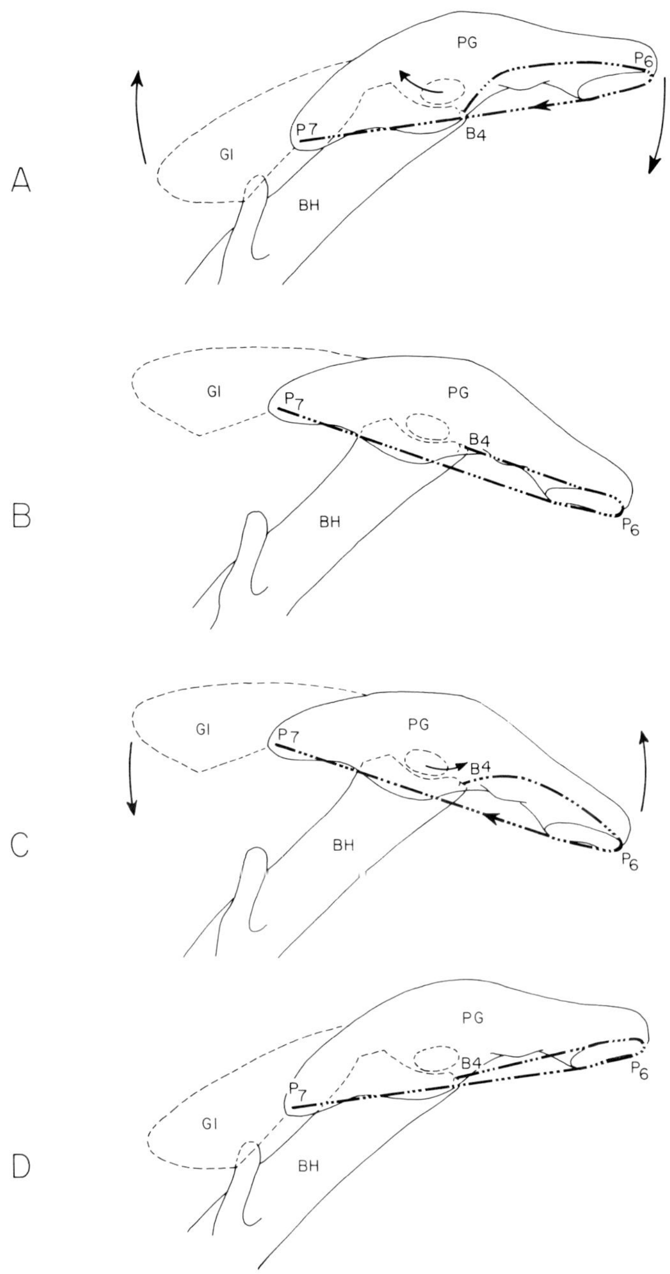

FIG. 60. Lateral views of the functions of the M. hypoglossus anterior during movements of the paraglossale in *Psittacus erithacus*. A–B: Contraction of Mhga during flexion of the paraglossale (A: Initial stage with paraglossale extended and aponeurosis of Mhga relaxed; B: Final stage with paraglossale flexed and aponeurosis of Mhga taut). C–D: Contraction of Mhga during extension of paraglossale (C: Initial stage with paraglossale flexed and aponeurosis of Mhga relaxed; D: Final stage with paraglossale extended and aponeurosis of Mhga taut). (For meaning of arrows see Fig. 46.) B_4 origin of Mhga on basihyale, P_6 apical edge of Synchondrosis paraglossalis over which the tendon of Mhga passes, P_7 caudo-dorsal end of insertion of Mhga on paraglossale.

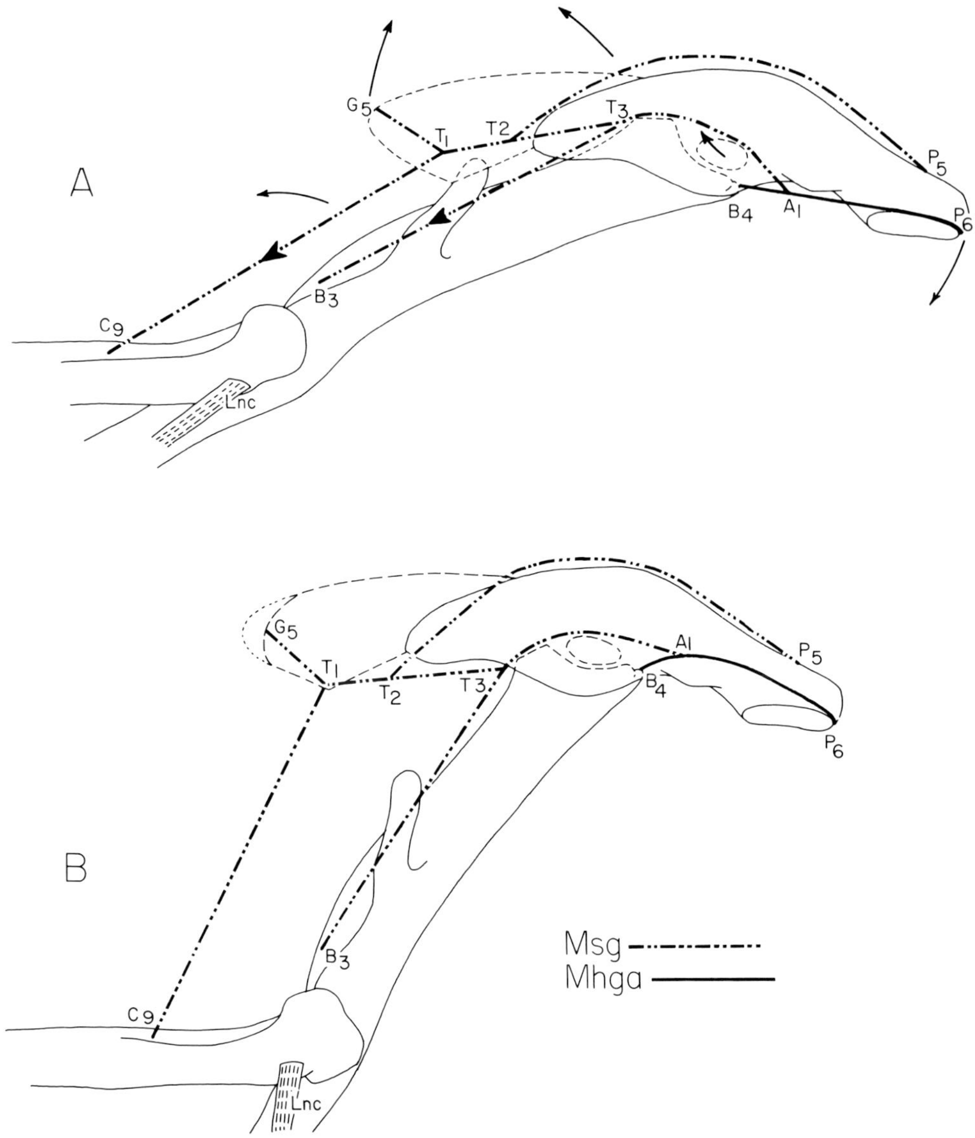

FIG. 61. Lateral views of the contraction of the M. supraglossus during retraction of the hyoid and flexion of the paraglossale in *Psittacus erithacus*. A: Initial stage with protracted hyoid, extended paraglossale, taut aponeurosis of M. hypoglossus ant., and uncompressed Gl. lingualis. B: Final stage with retracted hyoid, flexed paraglossale, relaxed aponeurosis of Mhga, and compressed Gl. lingualis. (For meaning of arrows see Fig. 46.) A_1 insertion of tendinous portion E of Msg on aponeurosis of Mhga, B_3–B_4 origins of intrinsic lingual muscles on basihyale (B_3 caudal end of origin of Msg bas., B_4 origin of Mhga), C_9 caudal end of origin of Msgc on ceratobranchiale, G_5 insertion of tendinous portion A of Msg on Gl. lingualis, P_5 insertion of tendinous portion C of Msg on paraglossale, P_6 apical edge of Synchondrosis paraglossalis over which the tendon of Mhga passes, T_1–T_3 origins of tendinous portions of Msg from central tendon (T_1 origin of tendinous portion A, T_2 origin of tendinous portion C, T_3 origin of tendinous portion E).

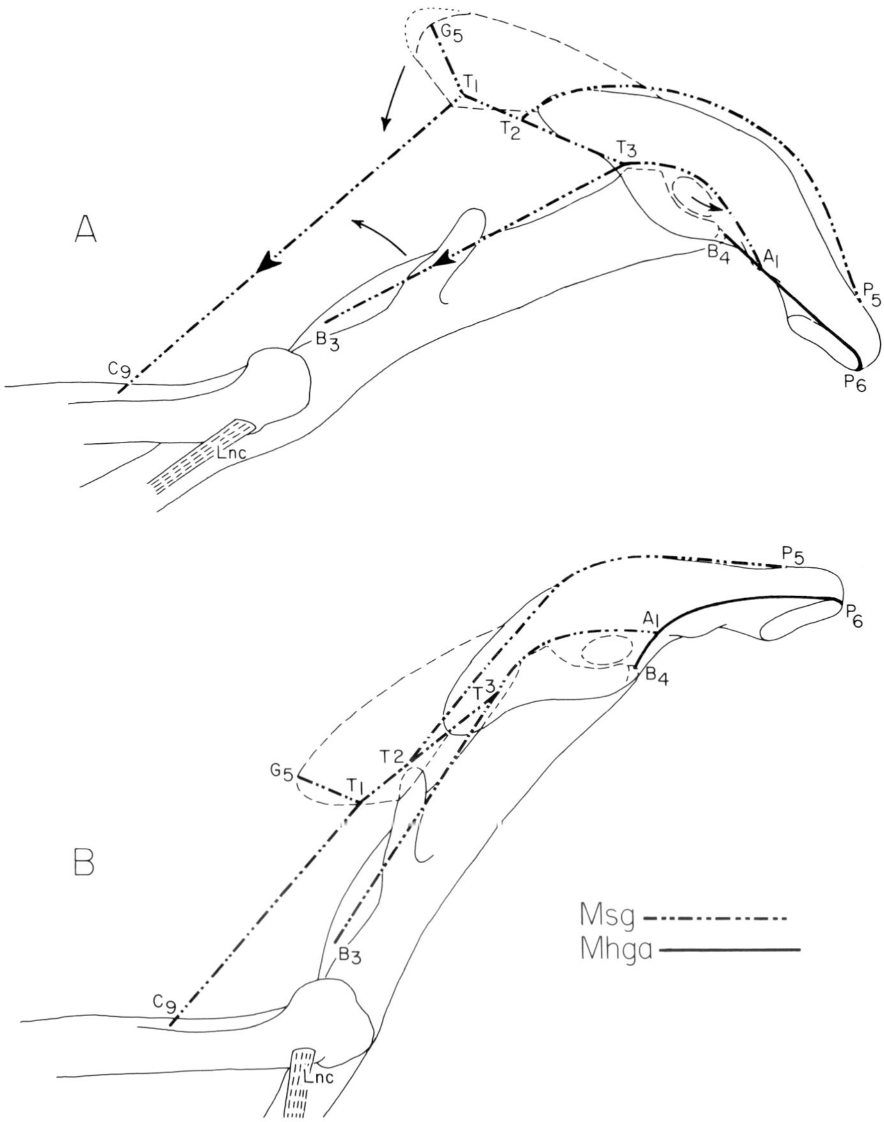

FIG. 62. Lateral views of the contraction of the M. supraglossus during retraction of the hyoid and extension of the paraglossale in *Psittacus erithacus*. A: Initial stage with protracted hyoid, flexed paraglossale, taut aponeurosis of M. hypoglossus ant., and compressed Gl. lingualis. B: Final stage with retracted hyoid, extended paraglossale, relaxed aponeurosis of Mhga, and uncompressed Gl. lingualis. (For meaning of arrows see Fig. 46.) A_1 insertion of tendinous portion E of Msg on aponeurosis of Mhga, B_3–B_4 origins of intrinsic lingual muscles on basihyale (B_3 caudal end of origin of Msgb, B_4 origin of Mhga), C_9 caudal end of origin of Msgc on ceratobranchiale, G_5 insertion of tendinous portion A of Msg on Gl. lingualis, P_5 insertion of tendinous portion C of Msg on paraglossale, P_6 apical edge of Synchondrosis paraglossalis over which the tendon of Mhga passes, T_1–T_3 origins of tendinous portions of Msg from central tendon (T_1 origin of tendinous portion A, T_2 origin of tendinous portion C, T_3 origin of tendinous portion E).

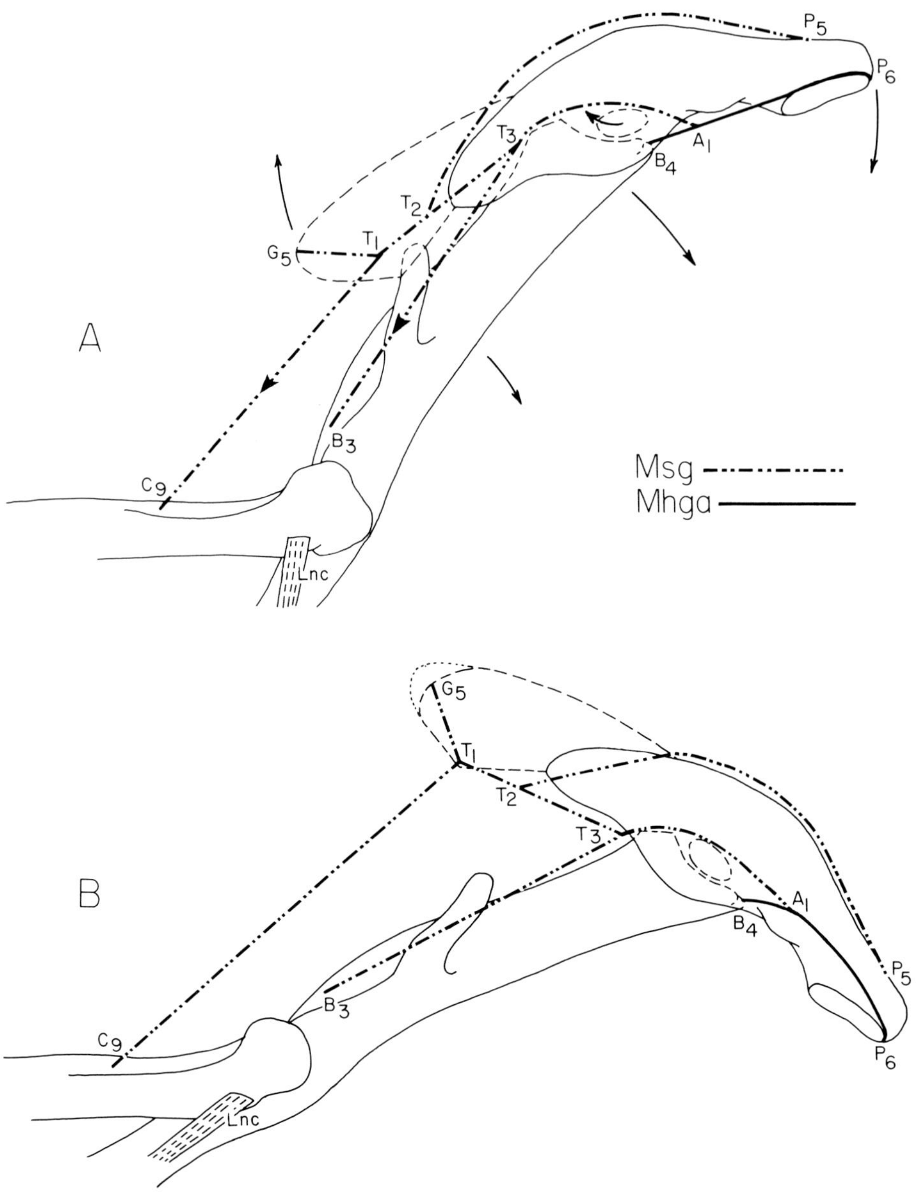

FIG. 63. Lateral views of the contraction of the M. supraglossus during protraction of the hyoid and flexion of the paraglossale in *Psittacus erithacus*. A: Initial stage with retracted hyoid, extended paraglossale, taut aponeurosis of M. hypoglossus ant., and uncompressed Gl. lingualis. B: Final stage with protracted hyoid, flexed paraglossale, relaxed aponeurosis of Mhga, and compressed Gl. lingualis. (For meaning of arrows see Fig. 46.) A_1 insertion of tendinous portion E of Msg on aponeurosis of Mhga, B_3–B_4 origins of intrinsic lingual muscles on basihyale (B_3 caudal end of origin of Msgb, B_4 origin of Mhga), C_9 caudal end of origin of Msgc on ceratobranchiale, G_5 insertion of tendinous portion A of Msg on Gl. lingualis, P_5 insertion of tendinous portion C of Msg on paraglossale, P_6 apical edge of Synchondrosis paraglossalis over which the tendon of Mhga passes, T_1–T_3 origins of tendinous portions of Msg from central tendon (T_1 origin of tendinous portion A, T_2 origin of tendinous portion C, T_3 origin of tendinous portion E).

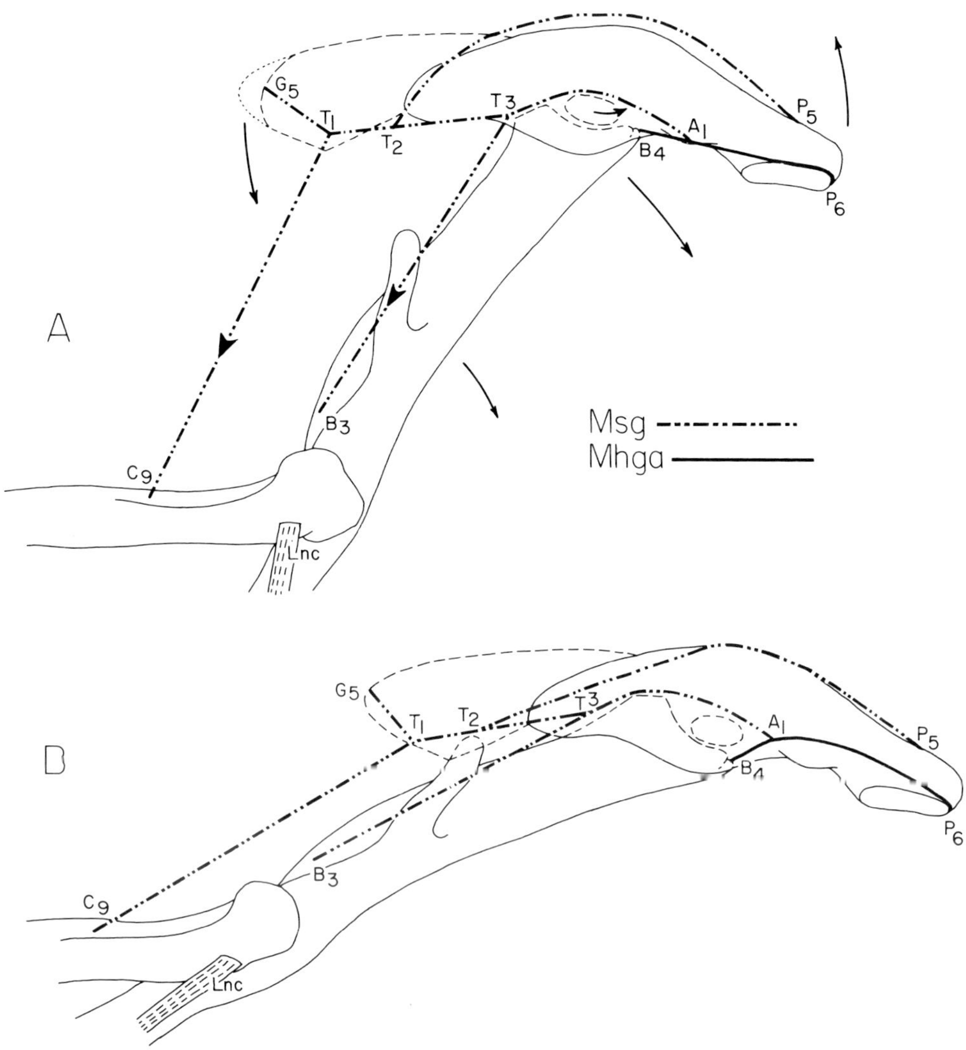

FIG. 64. Lateral views of the contraction of the M. supraglossus during protraction of the hyoid and extension of the paraglossale in *Psittacus erithacus*. A: Initial stage with retracted hyoid, flexed paraglossale, taut aponeurosis of M. hypoglossus ant., and compressed Gl. lingualis. B: Final stage with protracted hyoid, extended paraglossale, relaxed aponeurosis of Mhga, and uncompressed Gl. lingualis. (For meaning of arrows see Fig. 46.) A_1 insertion of tendinous portion E of Msg on aponeurosis of Mhga, B_3–B_4 origins of intrinsic lingual muscles on basihyale (B_3 caudal end of origin of Msgb, B_4 origin of Mhga), C_9 caudal end of origin of Msgc on ceratobranchiale, G_5 insertion of tendinous portion A of Msg on paraglossale, P_6 apical edge of Synchondrosis paraglossalis over which the tendon of Mhga passes, T_1–T_3 origins of tendinous portions of Msg from central tendon (T_1 origin of tendinous portion A, T_2 origin of tendinous portion C, T_3 origin of tendinous portion E).